RECIPROCAL LOGIC II

How Order Is Conserved as Mass Becomes Increasingly Complex

The Primary Subtractive Colors

By

Frederick S. Johnston, Jr.

The Primary Additive Colors

Library of Congress Control Number: 2008904942
ISBN: Softcover 978-1-4363-4719-8
Hardcover 978-1-4363-4720-4

This book was printed in the United States of America.

To order additional copies of this book, contact:
Xlibris Corporation
1-888-795-4274
www.Xlibris.com
Orders@Xlibris.com

2-D MEGA-LOGIC

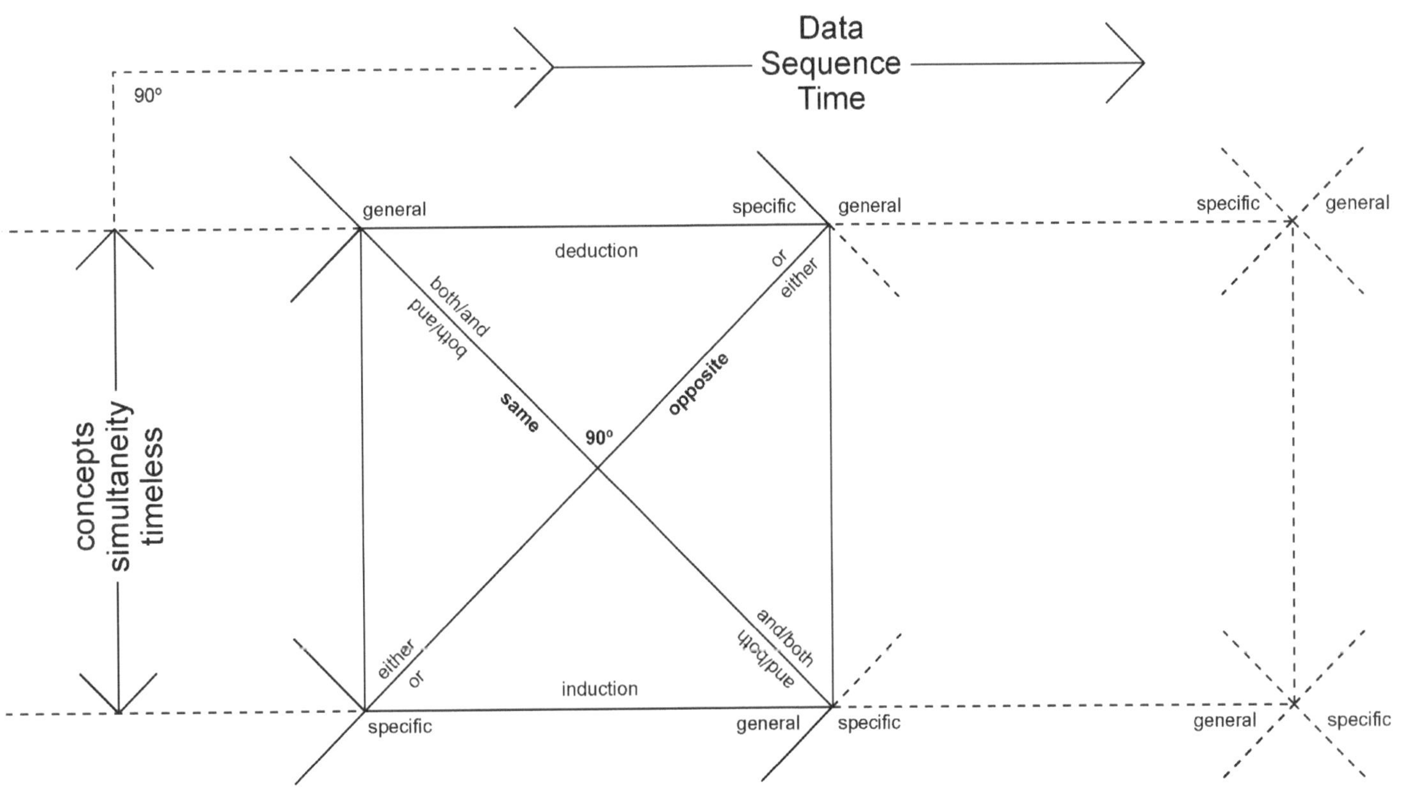

2-D LOGIC OF HUMAN BRAIN-MIND CONSCIOUSNESS

2-D MEGA LOGIC

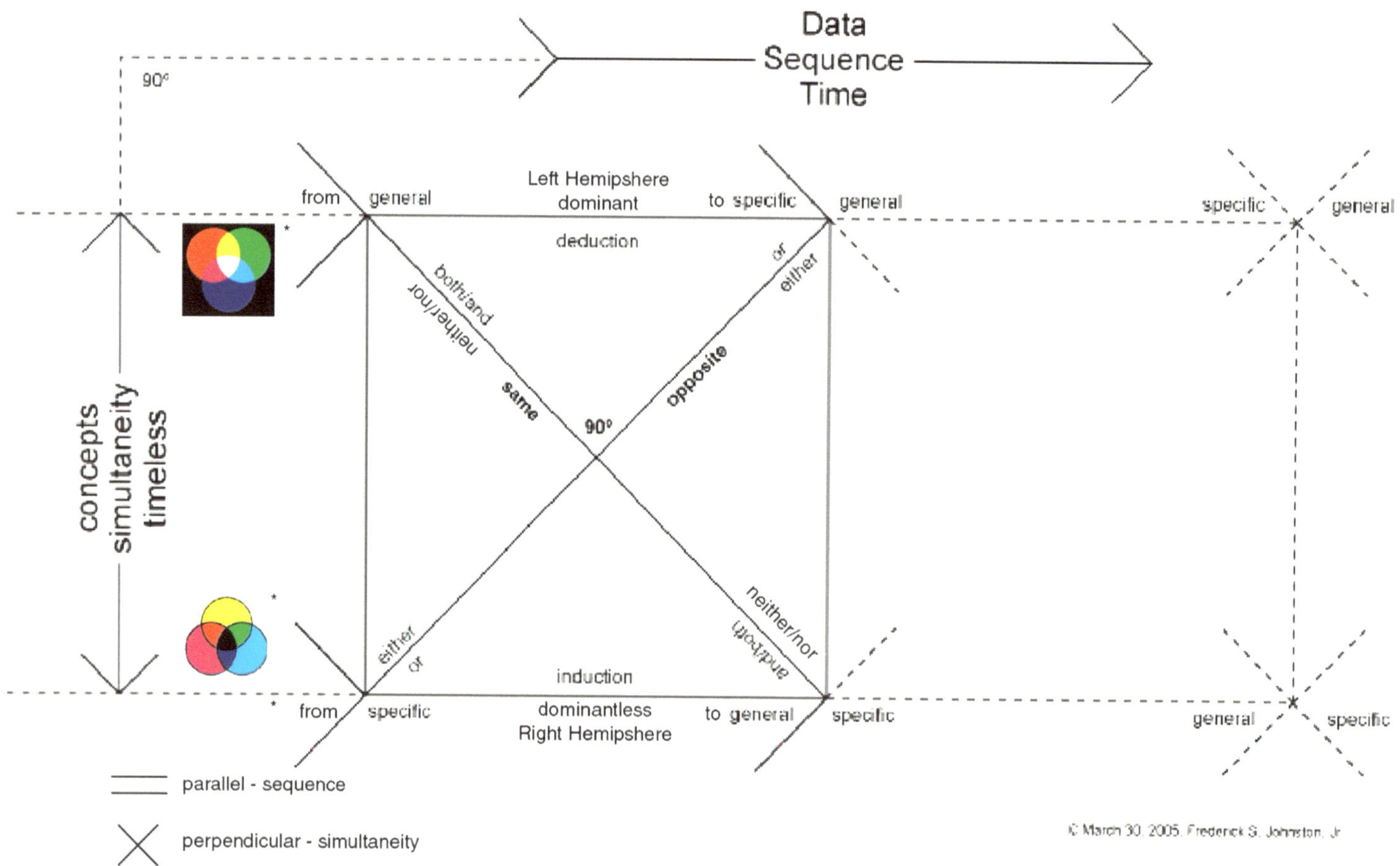

* See animation on the Reciprocal Logic Web site: www.reciprocallogic.com

3-D Megalogic

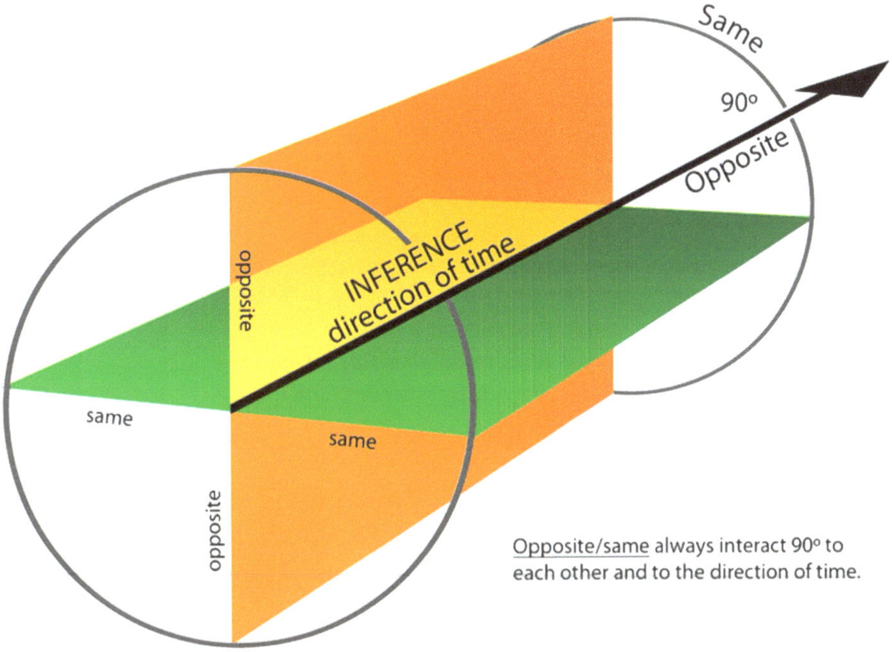

Same
90°
Opposite
INFERENCE
direction of time
opposite
same
same
opposite

Opposite/same always interact 90° to each other and to the direction of time.

In his hand,
He took the golden Compasses, prepar'd
In God's Eternal store, to circumscribe
This Universe, and all created things:
One foot he center'd, and the other turn'd
Round through the vast profunditie obscure,
And said, thus farr extend, thus farr thy bounds,
This be thy just circumference, O World.

—Milton, *Paradise Lost*, Book VII

CONTENTS

Introduction

Trying to understand the logical relationship between pairs of opposing emotions, opposing predicates, opposing perceptions, and opposing concepts in the Western world began with Zeno and more in depth with Heraclites of Ephesus. Later, the Greek philosopher Aristotle tried to put an end to these futile efforts with his three laws of thought by simply requiring one to make a decisive choice of either *one* of the *two* opposing aspect. Such a decision seemed necessary to Aristotle in order to comply with his discovery of the logic of deduction that brought data from a generalization to a specific conclusion without any obvious opposition. Aristotle, however, was still haunted by the generalization having to be limited to a small collection of relatable data since no single human mind could ever encompass everything at once. This meant that the general had to be somewhat specific, again revealing the conceptual conflict between the specific and the general.

This dilemma was not taken seriously until much later because deductive conclusions were always consistent with and conformed to sequence of any specific mass. The processing of data by deduction to a specific conclusion jump-started science because it reflected the logic of nature. The acceptance of deduction despite its conceptual paradox has given the human mind insight into and control of the laws and processes of mindless nature.

Much later, Sir Francis Bacon convinced logicians and scientists that deduction was only *part* of the logic of nature, the other half being induction, the reverse processes of deduction. Induction, therefore, was the processing of information from the specific to a general conclusion, showing again that the concepts of specific and general were **opposing** concepts of the **same** but a more complex form of logic called inference.

Despite its two conceptually opposing conclusions, inference became the exclusive logic of science because it greatly accelerated and enhanced the discovery of highly complex processes of nature that deduction alone was unable to do. Why fret over this puzzlement of the conceptual paradox in inference when it was so amazingly successful in solving the problems of nature? So the *inconclusiveness* of the *conclusions* of inference was simply swept under the rug. Out of sight, out of mind.

Despite the obvious and consistent success of inference, some philosophers still wonder why inference requires both a specific conclusion and a general conclusion thereby being self-opposing. Inference seems to contradict its own ability to be conclusive.

A recent book that tries to overcome the logical paradox in inference, *The Rejected Cornerstone*, by Jane Weir, iUniverse Inc., 2005, attempts to demonstrate that conceptual paradox is not a contradiction but instead a natural relationship between conflicting concepts. The author, however, does not reveal how or why this paradoxical "natural relationship" resides within and seems to be required for the logic of inference. Why would the *inconclusiveness* of paradox be a prerequisite for logical *conclusions*?

In this book, I will demonstrate that all *concepts*, such as **specific** and **general**, *interact* to create their own form of logic. This *inconclusive* logic of *interactions* itself *interacts* with the *conclusive* form of inference to create an even more powerful and *complexly* enhanced megalogic that conserves the *simple* interaction of conclusive/inconclusive.

Without the logic of *simultaneous* interactions there could be no logic of *sequence* by which inference processes information because sequence/simultaneity *interact*. More profoundly, we will see that the *simplicity* of this megalogic will enable scientists to delve deeper into the more complex conceptual interactions that form nature's Universal laws and forces.

My previous work on reciprocal logic and its seven addenda was primarily a collection of scientific evidence of how nature's conceptual interactions are logically formed. Here, I have summarized much of these random works of scientific evidence along with new scientific evidence and organized them into simpler format that makes it easier for my reader to climb the learning curve of how successive increases in complex systems remain logically ordered. These scientific references are to be read as part of the text because they are evidence of the logic of conceptual *interactions* that are always paradoxical and thereby inconclusive. Addendum V, for some long-forgotten reason, I mistakenly used the letters of the alphabet instead of numbers to identify its Figures, so they don't fit neatly in with the others. You will find lettered Figures between pages 281 and 290. They are important so please make a note of it for your ready reference.

You will find that *successive* increases in the complexity of *mass* is best explained by *simultaneous* interactions of concepts even though these *timeless* relationships run against the grain of inference that *sequences* data in *time*. I therefore have introduced an inordinate number of schemas in the text that reveal what the sequence of words alone cannot adequately convey. These FIGURES, along with Figures at the end of the book taken from previous addenda, should be pondered even though they will slow down your reading. But since "a pictures *is* worth a thousand words," marrying pictures and words is the easiest and most effective way to learn. With **both** the *simultaneously* interacting pictures **and** the *sequential* reading of words in the text the conceptual paradox in inference will no longer be so *illogical*. Instead, the *interaction* of the opposing logics—<u>sequence/simultaneity</u>—will become a more complex *megalogic* that has heretofore escaped the human mind.

Sir Francis Bacon said, "Some books should be tasted, some should be swallowed and some should be chewed and digested." This book needs to be ruminated. With diligence and patience, you will be able to master the *logic* between *data sequence* and *conceptual simultaneity* that will lift your mind to a higher level of ordered complexity. You will discover a world far richer, more complex, and mentally enhancing than inference alone can ever offer.

Because this work is more *circular* and referential than *linear* and inferential, some ideas and their interactions will seem to be introduced too soon and some too late. Therefore, you may need to read this work again before everything logically falls into place *simultaneously*. Above all, think *concepts* instead of perceptual *objects* of mass.

Do not expect to read an emotionally stirring story or an engrossing theme that leads to a conclusion. You can, however, expect to reach a lofty height of *mind* that transcends your *brain*, a *conceptual* state more beautiful and enlightening than any story can ever tell.

THE GEOMETRIC CONCEPTS OF MIND AND MINDLESS NATURE

The point of philosophy is to start out with something so simple as to seem not worth stating, and end up with something so paradoxical that no one will believe it.

—Bertrand Russell

Philosophy gets on my nerves. If we analyze the ultimate ground of everything, then everything finally falls into nothingness. But I have decided to resume my lectures again and look the Hydra of doubt straight in the eye, and it can be quite ominous if one values one's life.

—Ludwig Boltzmann

To win an argument has never been the primary aim of the logicians approach to human thinking . . . for proof always takes the form of argument.

—Alburey Castell, *A College Logic*

Concepts, or ideas, are strange beasts. They are slippery and paradoxical. Ideas arrive in the mind spontaneously. They can just pop up out of nowhere without any causal effect. *Ideas* are ephemeral and amorphous, and in contrast to *factual* knowledge, they can transform their definition into the very **opposite** idea. Any concept, therefore, cannot be precisely or conclusively defined because over time it will seemingly contradict itself.

This weird behavior of the *mind's* conceptual thought also applies to *mindless* nature. For example, can a celestial object be labeled a planet or a star, **either** one conclusively without **opposition**? Read "A Planet? Maybe It's a Star" by Kenneth Chang, the *New York Times*, August 4, 2006, in which Chang writes, "The line dividing planets and stars has become blurry."

Labeling a person as, say, "genius" tends to make everyone think that he or she is proficient in all areas of knowledge. "Idiot savants" are lopsidedly brilliant in a specific subject. And even polymaths are not geniuses in all subjects. When we label a person with a *concept*, it is never precise and complete, but instead a mere *interpretation* that is subject to valid **opposition**. The *idea* can be *dominant* but never absolute.

One would think that the scientific concept of *mass* could be defined without **opposition**. After all, *scientific* or deductive reasoning is mathematically precise, conclusive, and predictable if the position, time, and mass of the object are known. Read *The Intelligibility of Nature: How Science Makes Sense of the World* by Peter Dear, University of Chicago Press, 2006.

Any object—mass that is perceivable, such as a chair or a table—are **specific** parts of a more complex and **general** idea of *furniture*, which **specificity** in turn is *part* of a more **generalized** set of *objects* such as houses, apartments, and condominiums, which **specificities** in turn are but *parts* of the more complex and **general** idea of a city, which in turn is part of . . . Reciprocally, a chair in **general** is also more complex than a **specific** material from which it is made, and the **general** material is more complex than the **specific** molecules from which it is made, and the molecules in **general** are more complex than the **specific** atoms from which they are made. Regardless of the level of complexity from where you start, there is always something either more **specific** or more **general**. The concepts of **specific** and **general** therefore seem to be tailor-made for the logic of complexity. At any level of complexity, **neither** the **specific** of the **general** idea is absolute to the exclusion of the other. **Either** one is only an *interpretation* of the observer—i.e., a *perceptual* and *conceptual* point of view.

Because **specific/general** interact, reciprocally **either** concept can be transformed into the other without making any **difference** to the laws of nature. Again, a **general** interpretation at any given level of complexity or scale becomes **specific** at a further distance, from which a more complex **generality** surrounds it; and at a still further distance, this **generality** becomes a **specific** point of view and so on. These *increasingly more* complex conceptual interpretations can be **reversed** to *increasingly less* complexity. **Either opposing** transformation makes **no difference** to *mindless* nature that *timelessly* interacts **both opposing** interpretations *perpetually* and universally the **same**. This allows the mind to choose **either opposing** transformation at any *time* and from any **specific** point of view.

Correspondingly, our body's **specific** *mass* is part of the more complex earth in **general**; but this **generality** of earth is *simultaneously* a **specific** planet that is part of a **general** set of planets that forms a more complex **specific** solar system that is *simultaneously* a mere part of the more complex **general** set of solar systems that form our **specific** galaxy, we call the Milky Way, that is *simultaneously* **specific** to a more complex **general** universe of galaxies. Again, it makes **no difference** to nature's law and forces whether the idea of *mass* is viewed as being **either specific or general** at any level of complexity.

An informative read on the reciprocal interactions of **specific/general** at various levels and scales of complexity is, "Now You See It, Now You Don't," by Neil D. Theise, *Nature*, June 30, 2005.

Even the *scientific* idea of *mass* must be *interpreted* as **either** *acceleration* **or** *inertia*, **either** concept of which is a valid part of the more complex idea of *mass*. *Mass* can also be a *part* of the more complex *conceptual* interaction such as that seen in the formula $E = mc^2$ where mass/energy perpetually interact. *Mass* can also interact with *massless*, such as a *star* in *space*. Without *space*, the star would have no place to be seen; and without the *star*, space would be nothing to see. The **specific** *star* and the **generality** of *space* **both** need each other *simultaneously* for **either** one to be known. Star/space as **specific/general** interact with the **specific** star being dominant.

The concept of *entropy* is simply nature's way of "balancing her books." Entropy constantly tries to *balance heat* energy from mass with the *coldness* cosmic space. As *heat* expands from mass, it becomes increasingly *heatless*, giving one the impression that everything in the universe will eventually become absolutely *cold*. However, this will not happen, as we shall later see. Just as mass/energy perpetually *interact*, so too do heat/cold. Entropy constantly tries to *balance*—i.e., make the **same**, these **opposing** concepts. There is no such thing as absolute *heat* or absolute *coldness* just as there is no such thing as absolute *mass* or absolute *energy*. Refer to http://www.entropysite.com and see Figure 9.

The concept of *inertia* in *mass* is another *balancing* act of nature. An *inertial* mass is one where all the forces acting on it are the **same**. A **specific** mass in an "inertial frame of reference" "feels" no force, so it *interprets* itself to be **either** *motionless* **or** *moving* in at uniform speed in a straight line without any resistance. **Either** *interpretation* is valid but *inconclusive* because **both opposing** interpretations, motion/motionless, are the **same** more complex idea of *inertial* mass. Figure 76 shows how the *concepts* of motion/motionless are each valid interpretations of *inertia*. An astronaut in a stable orbit around the earth—called "free fall"—feels no force acting on him even though the *force* of gravity acts on him. In free fall, an astronaut is *balanced* between the **opposing** ideas of **both** *acceleration* **and** *inertia* as if they were the **same** more complex and enhanced idea of *force*. Read about these so-called inertial sweet spots in "Navigating Celestial Currents," by Erica Klarreich, *Science News*, April 16, 2005.

When two masses collide in space, their collective *momentum,* **both** *mass* **and** *velocity*, is *conserved*—i.e., remains the **same**. In other words, the post-collision idea of *momentum* is **no different** than their precollision *momentum*. The *conceptual interaction* of mass/velocity is *conserved*—remains the **same**—after as before their *physical collision*, which as we shall see is **no different** than *conceptual conflict*. Figure 112 shows the conservation of momentum regardless of any **similarities/differences** between any *two* colliding weightless objects.

An example of the interaction between *acceleration* and *inertia* is seen in the detonation of a cartridge *constrained* within the barrel of a rifle. The *heavy weight* of the rifle in proportion to the *light weight* of the bullet makes the bullet, on detonation of the cartridge, *accelerate forward* faster than the *heaver* rifle *accelerates backward* (recoils) so that the two **opposing** forces are *balanced*, precisely the **same**, even though the rifle and the bullet are **different** kinds of mass and of **different** weights. The *dominance* of *inertia* in the heavy *mass* rifle and the *dominance* of *acceleration* in the light *mass* of the bullet make the *force* in **both opposing** concepts of *mass* the **same**. This also reveals Newton's third law of motion that with every *action* of mass, there is an *equal* and **opposite** *reaction*. Substitute the concept of **sameness** for *equal*, and you have two **opposing** *parts* (acceleration/inertia) of the **same** more complex idea of force. How the **opposing** concepts of an *interaction* can become the **same** is the primary focus of this book and will be explained in detail as we proceed.

There is an old philosophical conundrum that states, "What would happen if an *irresistible* force met an *immovable* object?" Physicists, however, now know that in our *macro*-world, *each* one of these two concepts is *relative* instead of *absolute*, that **both** of these concepts *interact* reciprocally. **Either opposing** *concept* is only an *interpretation*, not a precisely definable and **unopposable** *fact*. If **both** *movable* **and** *immovable* were *absolute*, their collision would be a *contradiction*, which is meaningless.

Because *concepts* interact reciprocally **either** one can substitute for the other *alternately* at any time without making any **difference** to the laws of nature. We often speak of ideas as being "overthrown" as if the new idea excluded the old idea, or solved the problem of the previous idea, such as when Einstein's *relativity* "overthrew" Newton's *absolute* interpretations of *space* and *time*. Einstein made each concept *relative* by interacting the **opposing** concepts of time/space as the **same** more complex metric.

Newton's *absolute* interpretation of nature's concepts is the simplest, the most intuitive, and the most useful way to calculate *local* (**specific**) motions of mass. It posits nature from the *inertial* point of view. Einstein's *relative* interpretation is more complex and *universal* (**general**) and mathematically accurate, particularly for solving celestial and cosmic *accelerations*.

Trying to solve a conceptual conflict by eliminating **either** idea from its *interaction* at any level of complexity is futile. No matter how forcefully an idea is squelched, it always remains seething in the background, waiting for an opportune moment for its **opposing** *interpretation of* how the world ought to be.

Our paradigm of inference that solves problems, answers question, and processes data to *conclusions* are **opposed** to *inconclusive* concepts and their interactions. Because we recognize inference as our only form of logic, we think that *ideas* are like *data* that can infer a conclusion. Misinterpreting *ideas* as being *data*, we think our *conceptual choice* is *decisive*. When two or more people discuss news, or *information*, their conversation is usually *amicable*. But when one expresses an *opinion* (idea) that interprets the news. the **opposing** idea comes to the fore, making the conversation divisively *hostile*. Each *conceptual* **opponent** tries to argue his or her interpretation conclusively, but even if one is persuaded by an argument, there will always be someone else who is **opposed** to it. The *conceptual interaction* itself is perpetual, preventing **either** concept from escaping its part of the interaction. Read the editorial, "Americans Expect Congress to Find Solutions to Pressing Problems," the *Tampa Tribune*, January 4, 2007.

Trying to solve the problem of an ideological **opponent** has always produced disastrous consequences, *chaos* instead of logical *order*. Conceptual interactions are not a problem of *mindless* nature, but our *minds* have unwittingly and illogically made them a "problem" by trying to find a solution to their conflict, thereby placing **both** conceptual **opponents** in an *insoluble* predicament. With only the logic of *data* processing to guide us—our paradigm of inference—the conceptual *solution* becomes a *problem* that is *insoluble*. We are stuck with conceptual paradox even though we detest its *inconclusiveness*.

Real problems that arise from data are subject to orderly *conclusions*. This is particularly so with *deduction* that sequences *data* **from** the **general to** a **specific** *conclusion*. *Induction,* on the other hand, reverses this process by sequencing *data* **from** the **specific to** a **general** *conclusion*. The human *mind* must dominantly choose **either** logical process *alternately* in *time*, because if **both** processes interacted *simultaneously* as seen on the inside front cover, the *conclusions* of **specific/general** would interact at 90° *inconclusively*.

Mindless nature, however, places both reciprocally **opposing** processes—**both** deduction **and** induction—in *simultaneous double reversal* to *create* the more complex and enhancing form of inference. This *timeless* framework remains the **same** so that the mind can *sequence* **similar-different** data **either** deductively **or** inductively to a *conclusion* at any *time*. The two reciprocally **opposing** processes of inference do not go in the 180° contradictory directions as *time*-*antitime*. Instead, **both** sequences run *parallel* in *time* in **opposition** to their *timeless* interaction. Inference is formed by doubling deduction then placing both logical processes in simultaneous reversal. This creates an antinomy – a logical paradox – that the mind overcomes by choosing **either** deduction **or** induction alternately in time to arrive at a conclusion. As long as **both opposing** conclusions tell the **same** story **either** conclusion is valid.

Despite the *antinomy in* inference its logic empowers the human mind to discover the universal laws and forces that govern all mass **specificities** the **same** at any time or place. Although we can never interact our *minds* concepts as complex as the **sameness** of the gravitational force, we nevertheless can always reciprocally choose between **either opposing** interpretation to get *valid* part of of the **same** force.

Reciprocal logic teaches us that our minds must reciprocally *interact* **opposing** *ideas*. Our persistent attempts, indeed our deductive and inductive imperative, is to bring *ideas* along with data to an **unopposable** conclusion. But since *ideas* always *interact*, they never conclude as facts do. The history of *ideas*—whether of religion, philosophy, politics, or science—is sufficient evidence that *ideas* cannot be inferred and that *ideological* conflict is *perpetual* and *insoluble*. The most cogent and lambent arguments have never excluded the **opposing** idea. Because *concepts interact perpetually*, **either** choice that tries to solve the conflict by excluding the **opponent's** *idea* always exacerbates instead of mitigates the conceptual conflict.

It is futile therefore to kill a *conceptual* **opponent**. The death of the *object* mass is not the death of the *concept* of mass. The dead person's **opposing** idea is *conserved* as part of an *interaction.*

To show that even the most learned and logically oriented minds still do not have a clear understanding of conceptual interactions, read "Devoid of Content," by Stanley Fish, the *New York Times*, May 31, 2005. Professor Fish rightly said that *ideas* should not be considered the **same** as information, but he failed to explain that ideas make the *form* and *not* the *content* so that the reader can make his own *interpretation* of the information. Read also "Toiling at the Information Factory," by Charley Reese, the *Tampa Tribune*, August 31, 2005.

The neurological scientist, Christof Koch, said in "The Quest for Consciousness: A neurological Approach," *Cerebrum*, Summer 2004, "Historically, philosophy does not have an impressive track record of *answering* questions about the natural world in a decisive manner, whether it's the origin and evolution of the cosmos, the origin of life, the nature of the mind, or the nature versus-nurture debate. This failure is rarely talked about in polite, academic company. Philosophers, however, excel in *asking* conceptual question from a point of view that scientists don't usually consider."

The fact that no one has ever solved conceptual conflict by excluding the *idea* he or she **opposes** makes one wonder why we persistently keep trying. If trying to solve conceptual **opposition either** *within* **or** *between* human minds has never worked, maybe it can't, and we should simply accept the paradox of interactions and the antinomy in the logic of inference. But the brain's imperative for inferential *conclusions* has prevented the mind from acquiescing to the *inconclusiveness* of conceptual paradox.

Aristotle was right with his three laws of thought that said we must choose **either** one of **both opposing** *concepts*. In other words, for practical purposes, we should not be ambivalent and indecisive. Unfortunately, he failed to add the word *dominantly* so that **either** choice would be *relative* instead of *absolute* to the exclusion of its **opposite**.

Aristotle can be forgiven of his *conceptual* error of being *conclusive* with *concepts*; otherwise his thoughts would have run counter his discovery of the logic of deduction where *data* can be **specifically** *concluded*. By not making a *logical* distinction between *data* and *concepts*, Aristotle left concepts *conclusive*. Although *science* with its *conclusive facts* was able to progress in ordered complexity, *philosophy* remained primitive and chaotic without any logic for its *inconclusive concepts*.

The logic of inference is *straightforward* **from** data **to** conclusion so that from the view of inference, the *circular* reasoning of conceptual interactions is fallacious. From the reciprocal point of view, *circular* reasoning is *valid* while *straightforward* reasoning is *invalid*, or fallacious. So we have two **opposing** forms of the **same** more complex idea of *logic*.

Because inference cannot process *concepts* conclusively, something is either wrong with inference or something is missing. Since inference has been so successful in bringing *data* to logical conclusions, it can't be wrong, so inference must be an *incomplete* part of a more complex logical interaction. What is missing is the very antinomy—the *logical* paradox—in inference with its own *logical* rules for *concepts* that we have simply overlooked. This logic of conceptual *interactions* has been under our nose all along; but the very success of inference and our animal need to **specify**, to distinguish one thing from another, has *hidden* reciprocal logic. The very **specificity** of our unique brain/minds and the *conclusions* of inference *logically* **oppose** the *inconclusive* square of antinomy's **sameness** of concept. As to geometry, the *parallel* form of inference **opposes** the *diagonaled* square of the antinomy. Parallel/perpendicularity interact at 90° to form the more complex and enhancing megalogic seen on the inside front cover.

Just as inference alone is divided into **either** deduction **or** induction with deduction being *dominant*, so too the even more complex megalogic is divided into **either** inference **or** reciprocal logic with inference *dominant*. *Mindless* nature however *simultaneously* interacts **both** inference **and** reciprocal logic *timelessly* so that our *minds* can chose **either** logic at any *time* to validly *interpret* the world.

Despite our having already divided inference into its reciprocally **opposing** processes of **either** deduction **or** induction so that we can come to a *conclusion*, dividing the more complex megalogic into **either** inference **or** reciprocal logic is going to be even less popular. Since inference is *dominant*, making this leap to a higher order of logical complexity will be greatly resisted. If we have been *dominantly* **opposed** to **sameness** in inference, we will be even more **opposed** to the idea that *mindless* nature makes the megalogic of inference/reciprocity universally the **same**. How nature does this is by *doubling* this interaction and placing **both** in *simultaneous reversal*, just as we created inference by placing both processes in *simultaneous double reversal*.

The megalogical form on the inside front cover helps us understand *why mind* and *mindless* nature interact *concepts* by the **same** logic, *why* the laws of nature are the **same** everywhere *simultaneously,* and *why* the universal forces of nature treat all mass **specificities** the **same**. But this logic rules out a *personal* God. An *impersonal* God leaves us in a *meaningless* universe while our *concepts* demand that we get *meaning* from information. Even a deductive conclusion begs for meaning. Our minds are caught in the paradox of meaning/meaningless as dominant/dominantless, as well as the more profound insight that our *minds* need for *meaning* **opposes** the *meaningless* universal laws and forces of *mindless* nature.

Locally, we are also **opposed** to **sameness** by the need to choose **either** of **both opposing** emotions, **either** of **both opposing** predicates, as well as **either** of **both opposing** concepts. But if we can understand that the concept of **oppositeness** is *dominant* instead of absolute, then **sameness** can interact *dominantlessly* and then the acceptance of conceptual paradox will not be so formidable. The acceptance of paradox will be a boon to human thought. More complexly, the logical paradox, the antinomy in inference, will bring *order* out of conceptual *chaos* and give us a more complex and enhancing megalogic that overarches (dare I say *unites*) the two *division* of knowledge—*data* and *concepts*. Instead of believing only what we want to believe to preserve our *interpretation* of the world, or choosing only the concept that makes us emotionally comfortable, our choice of *concept* will be *dominant* instead of *absolute*, leaving room for rational dialogue and negotiation. There will be no solution to the conflict, but it will no longer be a problem incapable of being solved.

Now let's see if my conjecture that concepts reciprocally interact, that our paradigm of inference is but an incomplete *part* of a more complex megalogic, and more profoundly, that *mind* and *mindless* nature \interact concepts by the **same** logic.

THE CONCEPTUAL INTERACTIONS OF NATURE'S FORCES

The eternal mystery of the Universe is its understandability.
—Albert Einstein

The more the universe seems comprehensible, the more it seems pointless.
—Steven Weinberg

The universe we observe has precisely the properties we should expect if there is, at bottom, no design, no purpose, no evil and no good, nothing but blind pitiless indifference.
—Richard Dawkins

A mans said to the Universe
"Sir," I exist."
"However, replied the Universe,
"This fact has not created in me a *sense of obligation."*

—Stephen Crane

The Universe is not hostile, nor yet is it friendly, It is simply indifferent.

—John Haynes Holmes,

A Sensible Man's View of Religion

One can distinguish any *object* of mass from all other **similar-different** *objects*, but the *concept* of mass has no **similarities-differences** to **specify**. As previously stated, try to define the *idea* of *mass* conclusively, and its very definition will seemingly contradict itself. Give the *idea* of *force* a precise definition of **either** *acceleration* **or** *inertia*, and it too will eventually reveal the other part of its interaction.

To understand *concepts* and their interactions that form nature's *forces*, let's begin with the simplest forces of *mass* seen in the *micro*-world. The two quantum particles of *mass* and their respective forces can be identified only as *concepts*. Theses to fundamental particles—known as fermions—have no **similarities-differences** to distinguish them as *objects*, only the *concepts* of **either** electron **oppositeness or** quarks **sameness**, **either** choice of which excludes any possibility that the other part can ever exist. **Either** idea is *absolute* to the exclusion of the other.

For example, the electron is conceived by the mind as **either** *up* **or** *down*, with *nothing*, no **sameness** of concept, in between. Our *idea* of the isolated electron, therefore, is *absolute* **oppositeness**. Its force is concentrated at a *point* that *infinitely* expands *inversely* proportional to the distance, the **same** as the more complex force of gravity.

Quarks, the other fermion, are the *absolute* geometric concept of **sameness**. There is no way that its *concept* can be divided into **either opposing** part. There is no force at the center *point* of the quarks *circle*. The force is **generalized** and *pointless*. Contrary to the electron's force, the quarks' force accelerates *directly* proportional to the square of the distance, and its force of expansion is *constrained* within a *finite* circle. The quarks' force is *absolute* to the exclusion of the electron's force. Also, the electron is easily *quantified* while quarks are subject only to *qualification*. Because the isolated electron and the isolated quarks are each conceptually *absolute* to the exclusion of the other they cannot *relate* by interacting reciprocally. More to logic, the electron cannot *interact* in **opposition** to the **sameness** of the quarks because they are only *parts* of reciprocal logic, the logic of conceptual interactions.

The *micro*-electrons seemingly *absolute* concept is always subject to *random change* between its absolutely **opposing** interpretations, while the *micro*-quarks idea of absolute **sameness** is *constant*, never subject to *change*. This is the **opposite** of our *relative* force of gravity that is always subject to *constant change* in *time*. *Absolute* concepts of *micro*-mass does not interact <u>change/constant</u> as **<u>opposite/same</u>** respectively, so there is no complex force of gravity at this *simple* level of *complexity*. *Time* is excluded from the *absolute parts* of **either** electron **oppositeness** or quark **sameness**. For example, when the isolated electron changes between **either** of its two absolute **opposing** interpretations it does so instantaneously, in no *time*, while the absolute **sameness** of quarks is forever conserved *timelessly*. Before an electron is measured by some *macro*-device it can be conceived as being in two places at once, a superposition where *simultaneity* instead of *sequence*, or continuum, reigns. More on this later, but for now suffice to say that in the quantum world the two *parts* of mass are **either** *singular* **or** *dual*, **either** *point* **or** *pointless*, **either** *particle* **or** *wave*. In the more complex *macro*-world instead of *excluding* each other the fermions parts *include* each other by interacting. They *relate* as singular/dual, point/pointless, mass/energy, as **specific/general**, each functioning as **either/both**, and interacting *simultaneously* in *time*, as seen on the inside front cover. Perhaps this is why the number *two,* as **both**, is the only prime number that is *even*, all the rest being *odd*, although *odd* and *even* sequence hence forth *alternately* in whole natural numbers infinitely. The *concept* of **specificity** as **opposition** is also a singularity that is *conserved*, or *timeless,* as well as the **sameness** of quarks *concept*, even though the **specificities** of *object* mass *change* appearances and complexity.

The question arises, how does *mindless* nature transform the *absolute concept* of quarks **sameness** to the *relative* **sameness** of the proton as an *object* of mass? How does the *constraining* circle of **sameness**, called the "proton," lose its *absolute* concept of **sameness** in order to *interact* with electron **oppositeness**. The *absolute* **sameness** of the proton may be taken away by the neutrinos *contradictory* concepts. In one neutrino, **both** *left* **and** *right* hands go absolutely *right* and in the other neutrino **both** *right* **and** *left* hands go absolutely *left*. **Either** neutrino—"little neutral one"—gets its name from being **neither** *left*—**nor** right-*handed*. These self-*contradictory* neutrino "hands" take away the quarks *absolute* concept leaving the proton with the *relative* concept of **sameness** for **either** electric charge to **oppose**.

Either *absolute* neutrino makes it difficult for physicists to determine whether it is **either** *mass* **or** *massless* because it keeps randomly contradicting itself. As would be expected, the *absolute* **sameness** of a neutrino is reluctant to interact with *relative* concepts. Most of the neutrinos from the sun go all the way through the mass of the earth without colliding with a single atom. And why not since its *absolute* hand has been excluded from the *relative* mass of the proton? The fact that neutrinos are clustered closely around but outside galaxies suggests that they have been emitted from the galaxies mass that are dominated by protons and neutrons. Read, "Tiny, Plentiful and Really Hard to Catch," by Kenneth Chang, the *New York Times*, April 26, 2005. With the *concept* of *absoluteness* having departed from the proton via the neutrinos, its *relative* concept of *positive* charge (+) is eager to be **opposed** by the *negative* (-) electric charge.

Back to the *micro*-electron, its seemingly *absolute* concept of **oppositeness** is surrounded by so-called virtual particles, little specks of energy popping in and out of the vacuum of space randomly so quickly that they cannot be thought of as being subject to *time* or having any existence except mathematically. This *spatial* **sameness**, a *nothingness*, is conceptually ready and willing to help the electron interact in **opposition** to quarks **sameness** at the first opportunity.

By the **same** token, *quarks* also are "surrounded by quark-antiquark pairs that zip in and out of existence . . . . . . These fleeting 'sea quarks' . . . contribute to the proton's magnetism and charge." This quotation was taken from "Ups and Downs," in the Research section of *Nature*, September 8, 2005. These "sea quarks" suggest that quarks **sameness** of *concept* is also willing and eager to be **opposed** by *by the* electron at a higher level of *mass* complexity so that **both** the electron **and** the proton can conceptually interact as **opposite/same** functioning as **either/neither**, dominant/dominantless respectively.

The level of complexity required for the emergence of *relative*, or *macro-mass* interactions cannot occur without the help of the *massless* photon's **both/and** function. As we shall see later in more detail, the **sameness** of the quarks **neither/nor** function, and the **sameness** of the photons **both/and** function form a more complex and enhanced **sameness** of geometric concept that the more complex electron can **oppose**.

The *simplest* conceptual interaction in our *complex macro*-mass is between the **opposing** (-)/(+) electric charges that *attract* each other, *interacting* in a "mutual embrace." Two electric charges that are the **same**, whether (-/-) or (+/+), will not interact each other, instead, they *repulse* each other. The force of *attraction* between the **opposing** electric charges and the **sameness** of the magnetic field forms the more complex electromagnetic force. See the animation of this at the Web site: http://www.reciprocallogic.com.

Relative macro-mass *dominantly* **opposes sameness** just as the *micro*-electron *absolutely* **opposes** the idea of quarks **sameness**. FIGURE 345 taken from *Chemistry*, November 1977, confirms that in the *micro*-mass realm, the *collision* between two protons spinning in the **same** direction *absolutely* **oppose** each other, while two protons spinning in **opposite** directions pass through each other as if they were the **same**.

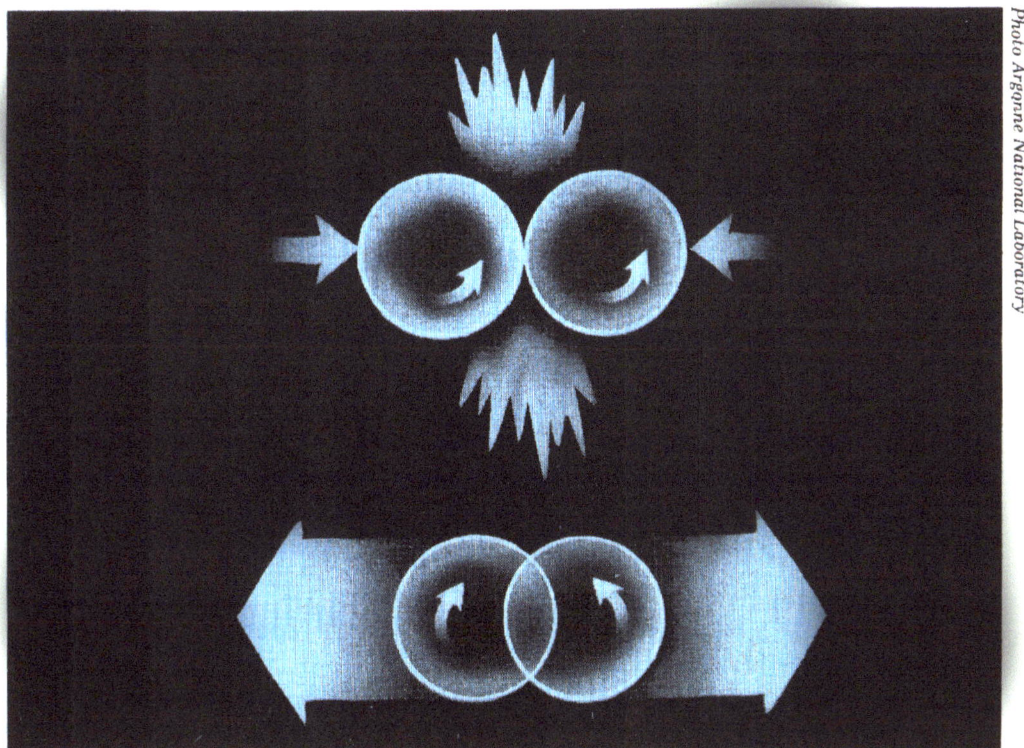

The upper diagram shows two protons, both spinning in the same direction, colliding violently at high energy. The lower drawing, however, shows the same high energy protons spinning in opposite directions: they appear to pass through each other

FIGURE 345

The natural order of mass is that is that the concept of **sameness** *repulsive* while the concept of **oppositeness** is *attractive*. In other words, conceptual conflict is the normal regardless of its complexity. See Figures 294a and b. Again refer to http://www.reciprocallogic.com.

The two electric charges are *separable*, *alternating* in *time* in **opposition** to the concept of **sameness**, which concept functions as **both** *North* **and South** poles creating the *pointless* magnetic field. The **sameness** of **both** poles is *inseparable and timeless*. Since the poles are *simultaneously* formed, no matter how many times you divide or multiply them, they remain the **same** N/S magnetic field, an inseparable **generality** with n**either opposing** electric charge. See Figure D in Addendum V.

Because the *separable* electric *charges* are **opposed** to the *inseparable magnetic* poles, they *interact* at 90° to form the aptly named electromagnetic force. See Figure E in Addendum V where point/pointless, or point/circle, interact as more than the sum of its parts, so complexly enhanced that the two **opposing** *hands* of the electromagnetic force are created, as seen in Figure F in Addendum V and Figure 20.

To show that Figure E in Addendum V is but part of a more complex and enhanced interaction, see Figure G where the electron *point* becomes a *circle* and the magnetic *circle* becomes a *point*, the exact **reverse** of Figure D! Figures E and G are reciprocally ordered **both** *within* **and** *between* each other.

In our *macro*-realm of mass, the *interaction* of **opposite/same** (-/+) as electric charge, if *doubled* and placed in *simultaneous reversal*, would form the more complex idea of **sameness** (+). But this does not occur at our level of complexity. Our *macro*-mass realm is limited to the reciprocally **opposing** hands of the *complex* electromagnetic force that should be *logically* depicted as the **electro/magnetic** force that conserves the *simple* conceptual interaction of **opposite/same** respectively.

The isolated electric *charge* and the isolated *chargeless* magnet are each *static* so **neither** one can *move*. But when **both** *charge* **and** *magnetic* field *interact* their respective **opposite/same** geometric *concepts*, they create the electromagnetic force that is more complex and *dynamic* than the sum of its *static* parts. The electromagnetic force orders all the interactions of our *macro*-mass realm. The *simple* interaction of **opposite/same** is conserved at any level of mass complexity. This is the key to understanding how logical order is conserved as mass becomes increasingly complex. Darwinian evolution is only the *macro*-mass *part* of the logical hierarchy of complexity. Please keep this in mind as we move along.

The *simultaneous* interaction of the electromagnetic force is *dominated* by the *sequence* of the electric current. Without the *electric* current, all life as well as the complexity of our modern world would come to a screeching halt. The highly complex interactions of molecular biology and of life's neurological systems are almost totally dependent on the *sequencing* the **opposing** electric *charges* pulsating in *time* with the *dominantless* concept of **sameness** *timelessly* tagging alone in support. Just as the constrained **sameness** of the inferential *concepts* – its antinomy – empowers the data to be logically processed, so too the *constrained* **sameness** of the complex molecular and neurological systems empowers the **opposing** information to be logically processed. You will see more than ample evidence of this in due time.

The *dominance* of **oppositeness** is seen in the highly complex *interactions* of molecular biology where *acids* are divided into **either** *ion* **or** *anion*, while *bases* are neutral as **neither** *ion* **nor** *anion*, and where acid/base is analogous to the *interaction* of **oppose/same**. The famous chemist P. W. Atkins said of this interaction, "All the phenomena of the world are merely the reaction of one kind of *acid* with one kind of *base*." Nobel Prize–winning biochemist Albert Szent-Györgyi said, "Life is nothing but an electron looking for a place to rest." Acid/base *interact* simply as **opposite/same** regardless of elemental or molecular complexity, which is why pH—the *ratio* of *acid* to *base*—is so important to any chemical process.

From the *simplest* fermions to the most *complex* geometric concepts of life, all *mass* is *dominantly* **opposed** to **sameness**. For example, there could have been no evolution of complex animal life without the **same** idea of *sex* being divided into the reciprocally **opposing** concepts of **either** *male* **or** *female*. The **specificity** of *mass* **opposes sameness**, which is why evolution has favored the emergence of **either** *male* (-) **or** ***female*** (+) sex. Without the *dominance* of **opposing** sexes, complex animal life could not have evolved beyond mere replication. Besides, what could be perceived if all mass was the **same**? **Sameness** in *mass* has to be *dominantly* **opposed** because all *mass* **specificities** are **similar-different**, no two ever being the **same**.

Either opposite or same can validly substitute for the other one. Therefore two males or two female can be sexually *attracted*. But this **sameness** of sex is not in the best interest of evolution, which is why reciprocally **opposing** concepts have has been conserved *dominantly* in nature *vertically* at each successive level of complexity. Read "But Is It natural?" under News, *Nature*, October 26, 2006.

Evidence that our *mind's concepts* and the isolated electron's *mindless concepts* are *vertically* the **same** from their **opposing** (relative/absolute) points of view is that any attempt to measure the isolated electron makes it *spin* **either** *clockwise* **or** *counterclockwise*. Read, "Electrons Are Not Ambidextrous," by Andrzej Czarnecki and William J. Marciano, *Nature*, May 26, 2005.

Before this macro/micro-*interaction* occurs, the electron spins **neither** *right* **nor** *left*, superimposed as if **both** were the **same** idea of, say, *spinless*. But when the mind's *relative concepts* collide with the electron's *absolute concepts* the electron divides into **either** *right* **or** *left* spin. Also, when the *macro*-observer shines light's **both/and** function on the electron's **neither/nor** function, this **super-sameness** of concept empowers the electron to reveal a more complex and enhanced concept of **opposition**. The metaphor for this among scientists for the simpler interaction between absolute/relative is that if the electron were a cat, it would be **neither** alive **nor** dead until our intruding *macro-minds* opened the *micro*-"box." Only then is the cat found to be **either** *alive* **or** *dead*. When we "open the box" on the isolated electron, it reveals its *random* self-contradictory function of **either-or**, thwarting our mind's *deliberate* choice of **either/or**.

Read "Quantum Trickery: Testing Einstein's Strangest Theory," by Dennis Overbye, in the Science section of the *New York Times*, December 27, 2005. Overbye explains:

> To a scientist, a "cat state" is the condition of being two diametrically opposed conditions at once, like black and white, up and down, or dead and alive . . . [such states] were each spinning clockwise and counterclockwise at the same time. Moreover . . . they were all doing whatever it was they were doing together in perfect synchrony. Should one of them . . . decide to spin only one way, the rest would instantly fall in line whether they were across a test tube are across the galaxy . . . . The idea that measuring the properties of one particle could instantaneously change the properties of another one (or a whole bunch) far away is strand tot say the least—almost as strange as the notion of particles spinning in two directions at once . . . ."The more success the quantum theory has, the sillier it seems," Einstein once wrote to a friend. The full extent of its silliness came in the 1020's when quantum

> theory became quantum mechanics. In this new view of the world, as encapsulated in a famous equation by the Austrian Erwin Schrödinger, objects are represented by waves that extend throughout space, containing all possible outcomes of an observation—here, there, up or down, dead or alive. The amplitude of this wave is a measure of the probability that the object will actually be found to be in one state or another, a suggestion that led Einstein to grumble famously that God doesn't throw dice. Worst of all from Einstein's point of view was the uncertainty principle, enunciated by Werner Heisenberg in 1927. Certain types of knowledge, of a particle's position and velocity, for example, are incompatible: the more precisely you measure one property, the blurrier and more uncertain the other becomes . . . John S. Bell a particle physicist . . . wound up proving that no deeper theory could reproduce the predictions of quantum mechanics . . . [Bell's suggested experiment] agreed with quantum mechanics and not reality as Einstein had always presumed it should be. Apparently a particle in one place could be affected by what you do somewhere else. "That's really weird," Dr. Albert said, calling it "a profoundly deep violation of an intuition that we've been walking with since caveman days." . . ."The discovery that individual events are irreducibly random is probably one of the most significant findings of the 20th Century," Dr. Zeilinger wrote.

The electron's *mindless concept* of **oppositeness** and the *mind's concept* of **oppositeness**—absolute/relative respectively—caused the physicist P. A. M. Dirac to exclaim, "What is it that the electron is mindful of man!" The physicist G. Wald exclaimed, "A physicist is an atom's way of knowing about atoms." And the physicist Carl Baker exclaimed, "What is a man that the electron is mindful of him?" In the same vein, the physicist James Clerk Maxwell understood that "the only laws of matter are those which our mind must fabricate, and the only laws of the mind are fabricated for it by matter." Simon Kochen, Andrew Wile's department chairman, Princeton University, said, "We evolved from this universe and our brains work in some kind of conformity with the universe. There must be some kind of harmony between the way we think and the way the universe works."

This "harmony" is that our *mind's* concepts interact as *part* of the **same** logic as *mindless* nature's conceptual interactions.

The simplest and easiest way to untangle "quantum entanglement" is to remove the concept of *time*. The word *event* should not be used because it implies that something occurred during the *sequence* of *time*. The *absolute concepts* of the quantum world are not only timeless but also quantized; i.e., *non-sequential* and *irreversible*; therefore they cannot be **opposed** by the *irreversible* concept of *time*. *Micro*-concepts *cannot* sequence in *time* **from** one place **to** another predictably and conclusively. The electron "jumps" from one energy level to another, as seen in Figure 73, **either** in *no time* **or** without traversing the *distance* between! In the quantum realm, the concepts of *time* and *distance* do not interact to create the more complex idea of *speed*, or *velocity* by which the idea of *mass* can continually *sequence*. The *interaction* between the **opposing** concepts of *relative* mass and *absolute* mass is also *timeless* on the *vertical* rise of complexity.

Physicists are dedicated to mathematics because deduction in inference is derived from the electron dominance with its capacity to *quantify*. *Numbers* give precise explanations to physical events and solutions to factual problems. Mathematical proof rules out any *interpretation* of mind, assuring cientists that their proofs are in harmony with—or the **same** as—*mindless* nature. As Dr. Steven Bardwell, director of Plasma Physics for the Fusion Energy Foundation said in "The New Math-destroying Cognitive Development," *Fusion* magazine, May 1981, "The subject of mathematics, as all great mathematicians have known, is not numbers and their manipulation; it is the human mind as the mirror of the universe." The mind's *dominance* of deduction however makes mathematics the "mirror."

Although deductive *proof* is *dominant*, it is but a *logical interpretation*, only the part of inference. Without induction, inference is *incomplete*. Physicists are also beginning to realize that their excessive bias toward the *sequencing* of data in inference does not adequately explain nature's *simultaneous* interactions. Something is missing, but scientists do not know what it is. The "what" of course is the *concept* of **sameness**, the *simultaneity* of which is **opposed** by the *sequencing* of *data* in inference.

Read the essay "The Mental Universe," by Richard Conn Henry, physicist, *Nature*, July 7, 2005. The above title might mislead you to believe that a *mental* universe is ordered by Intelligent Design, or by a personal God who could not treat all mass **specificities** the **same**. However a *mind*, no matter how great its *intelligence*, cannot be the **same** everywhere *simultaneously*.

Let us now examine *mass* from its *weightless* interpretation. The motions of *weightless* mass are more non-intuitive than those of *weighted* mass that we have been discussing because *weightlessness* is *dominated* by the *simultaneous interaction* of **opposite/same** instead of the *sequential dominance* of inference.

If a human in outer space, such as an astronaut, were of the **same** mass and elasticity as the earth, and if they gently collided, **both** would bounce away from each other at the **same** velocity along the **same** 180° line but in **opposite** directions, *ideally* demonstrating Newton's third law of motion. If the astronaut and the earth were **both** *spinning* at the **same** rate in the **same** direction at the **same** time of their *collision*, **both** would become *spinless*—i.e., *spinning* **neither** *right* **nor** *left*. If they spun at the **same** rate in **opposite** directions, they would continue *spinning*. See the animation of this with *weighted* mass at http://www.reciprocallogic.com.

Perhaps the best way to demonstrate that weight/weightless are two reciprocally **opposing** interpretations of the **same**, but more complex and enhanced **sameness** of gravitational mass is seen in Figure 78. At the top half of this Figure, a weighted observer is transformed from an accelerated observer into an inertial observer as the earth increases its rate of spin. The bottom half is the exact **opposite**. Here, an observer spinning in a weightless hollow tube is transformed from an inertial observer into an accelerated observer. Thus, weight and weightless are two reciprocally **opposing** interpretations of the **same** more complex idea of mass. Also, acceleration and inertia are two reciprocally **opposing** interpretations of the **same** more complex idea of force. **Either opposing** concept of any interaction can be transformed into the other without making any **difference** to the laws of nature. If we placed **both** the top **and** the bottom interpretations of this Figure in simultaneous double reversal they would become a more complex and enhanced **sameness no different** than the more complex **sameness** of Universal gravity. The human mind, however, must interpret **either** the top half **or** the bottom half alternately in time, always **opposed** to the more complex **sameness** of mindless nature.

Since all *concepts* are reciprocally ordered, we can do the **same** thing with *weighted* mass as we did with *weightless* mass in Figure 78 by dividing the relatively spherical earth into its two *Northern* and *Southern* hemispheres. From the traditional interpretation of *north* as *up* and *south* as *down* we can **reverse** our interpretation so that if one lived in, say Australia, their "down under" view would become "up above" view. If most people had lived in the Southern Hemisphere, that hemisphere would be dominantly *up* because of the force of gravity makes all of us look *up* from the center of the earth no matter where we stand of its surface. *Up* in **either** hemisphere is *down* in the **opposite** hemisphere. Place the *interaction* of up/down in simultaneous double reversal, and they become *simultaneously* the **same** for mindless nature.

Likewise, hurricanes, or cyclones, in the *Northern* Hemisphere rotate *counterclockwise* but rotate *clockwise* in the **Southern** Hemisphere, two **opposing** rotations of the **same** phenomenon. Interacting clock/anticlock hurricanes simultaneously would *diminish* these **opposing** *rotations* to a *rotationless* state. Again, our *minds* must *alternate* these reciprocally **opposing** views in *time* while *mindless* nature's *timeless* force of gravity makes no interpretation, it force interacts with **both** earth's hemispheres *simultaneously* the **same**.

As to *weightless* mass in outer space, no matter how **different** two closely orbiting bodies may be in size, shape, density, material, elasticity, or different combinations thereof, *each* feels the **same** force of gravity acting on it. The Universal and *timeless* force of gravity has no preference for **either** orbiting body, interacting **both** masses *simultaneously* the **same**. **Both** such masses are *logically* ordered to move in the **same** geometry of an ellipse.

Let me explain gravity's force of **sameness** between two closely interacting *weightless* bodies in greater detail. With any two such orbiting bodies *interact*, the *greater* mass *dominates* and actually gets the choice of **either** *point* of **both** *points* (foci) of the ellipse. It makes **no difference** to nature which one of the two foci of the *dominant* mass chooses since **both** foci are *simultaneously* required to form the **same** ellipse. To see how the ellipse is constructed, see FIGURE 346. The *dominant* mass circumscribes a *small* ellipse, while the *dominantless* mass circumscribes a *larger* ellipse. The reason for this is that the *greater* mass has the greater *inertia* so it *resists change*, thereby forming a *smaller* ellipse. Conversely, the *lesser* mass has the greater *acceleration* that *favors change* so it forms a *larger* ellipse. Each mass's orbit interacts acceleration/inertia, so b**oth** orbiting masses *simultaneously* interact acceleration/inertia in *double reversal*, creating a more complex and enhanced **sameness** of geometric concept with no **opposition** between them. See Figures 97 and 14. The **sameness** of gravity between the two orbiting bodies and the **sameness** of space harmoniously interact, **opposed** by the **similarities-differences** of elemental content between the two weightless masses**.** Because the simple *conceptual* interaction of **opposite/same** is conserved, it makes **no difference** to nature what kind of weightless *objects* orbit each other.

The Mechanical Construction of An Ellipse

An ellipse has the important property shown below. There are two points, symmetrically distributed on the long axis, known as foci and denoted by S and K. The sum of the distances S to P and K to P, where P is *any* point on the ellipse, is always the same. In fact, if one attaches a piece of string to two fixed points and moves a pencil in the manner of the second figure, the point of the pencil traces out an ellipse.

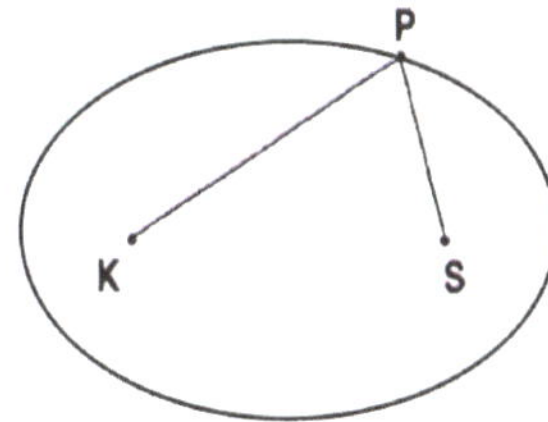

There are two points on the long axis of an ellipse known as the foci points S and K in this figure, with the property that the sum of the distances S to P, K to P, is the same whatever point on the ellipse may be chosen for P.

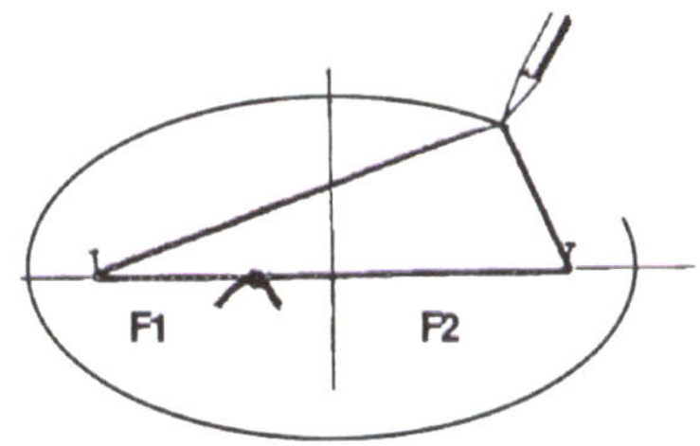

The fact the first figure illustrated above permits the mode of construvction of an ellipse shown here.

Source:* *ASTROLOGY/COSMOLOGY
Fred Hoyle
W. H. Freeman And Company

FIGURE 346

The ellipse has the remarkable ability to change sizes and shapes and yet keep its *constant* geometric form, conserving the *simple* interaction of change/constant.

Since no *concept*, no matter how complex, can stand alone, including the gravitational force of *contraction* to a point, there should be a force of Universal *expansion* with which the Universal *contraction* of gravity can *interact* to create an even more complex universal **Sameness** that **similar-different** Universal *time* frames can **oppose**. If there were no force of Universal *expansion* that is Universally the **same** everywhere *simultaneously*, then there would be no Universal *symmetry* for *asymmetry* to **oppose.** Without **both** Universal *contraction* **and** Universal *expansion* there would be no *logical* correlation with the inferential *contraction* to a point with the **specific** conclusion and *expansion* from a point with the **general** conclusion, and my conjecture of "how logical order is conserved as mass becomes increasingly complex" would be disproved.

So let us again go through the proposed proof that all successive increases in *complexity* interact the geometric concepts of **opposite/same**, *conserved* from the *electron point* of view and *quarks pointless* view respectively, and that complexity is *logically created* on the vertical by this interaction being *doubled* and placed in *simultaneous double reversal* at each successive step up the ladder.

As we have already seen, the *simplest* and must *obvious* interaction in our *macro*-world is between any two **opposing** electric charges (-/+) that "mutually embrace," or *attract*, while two electric charges that are the **same**, whether (-/-) or (+/+), *repulse* each other, as seen in the animation in http://www.reciprocallogic.com.

At the next higher level of complexity, the geometric concepts of *attraction* and *repulsion* again *interact* as **opposite/same** respectively. At this level of complexity, if things get too *close* to each other, a *repulsive* force steps in; and when things get too *far* apart, an *attractive* force steps in, **both** interacting to *conserve* the proportional *constraint* between the **opposing** forces of *mass*. The physicist Richard Feynman said that if only one sentence could be preserved from our vast knowledge of science, it should be, "All things are made of atoms—little particles that move around in perpetual motion, attracting each other when they are a little distance apart, but repelling upon being squeezed into one another." Translated into the words of reciprocal logic, the reciprocally **opposing** concepts of **either** *attraction* **or** *repulsion* are proportionally constrained and **neither** one can exclude the other from their *interaction*.

At the next higher level of mass complexity in our *macro*-world, two electric charges are *divided* so that they can alternate in *time* to interact in reciprocal **opposition** at 90° to the *indivisible* **sameness** of the N/S magnetic poles. This more complex electromagnetic force can be logically described as the interaction of **electro/magnetism**.

If the electromagnetic force's "mutual embrace" of **opposite/same** is more complexly enhanced than the less complex isolated **opposing** (-/+) electric charges and the **sameness** of the N/S magnetic poles, then the *less* complex parts and the more complex interaction and its two parts should look the **same** on the *vertical*. You will see this in Figure 294a. Keep in mind that **sameness** is conserved on the *vertical* rise of complexity **opposed** not only by the **similarities-differences** of *mass* but also the concepts of *change* in *time* on the *horizontal*.

The *magnetic* field can **reverse** its N/S poles to S/N and back again *alternately* in time the **same** just as the two *electric* charges, but the magnetic poles never *separate* during this reversion in order to *conserve* their timeless geometric concept of **sameness**. Evidence of the Earths magnetic field *reversing* its poles *alternately* without dividing is seen in the records of the magnetized rocks over deep geological time.

Just as the logical form of inference is *divided* into two reciprocally **opposing** *processes*, so too is the electromagnetic force *divided* into two reciprocally **opposing** *hands*. And just as with inference, our *macro*-mass force of electromagnetism *alternates* its hands in *time* to create a *current*, as seen in Figure 20.

If **both** hands of thc **electro/magnetic** force could be place in simultaneous double reversal, it would create the more \complex and enhanced force of gravitational **sameness**. Another way to envision this is to take the left/right *hands* of Figure 20 and place them in simultaneous *double reversal*. This *handless* interaction would automatically exclude all **opposition** except the *irreversible hand* of *time*. But the human mind cannot fathom this. We are stuck in the macro-mass realm of the **electro/magnetic** force where *time* dominates, as seen in the megalogic on the inside front cover.

Distinguishing the forces between the various levels of complexity, scientists, unfortunately, have changed their names as if they were **different** instead of the **same**. It takes no stretch of the imagination, however, to see that the forces of **either** *attraction* **or** *repulsion* respectively and **either** *pull* **or** *push* is **no different** than the *contracting* force of *mass* and the *expanding* force of *space* respectively. Understanding that these **different** terms have the **same** meaning allow us to see clearly through the confusion of **sameness** on the vertical.

Now we have again reached the complexity of gravity's *contracting* Universal force of **sameness** by successive increases in the complexity of **opposite/same,** let us see if there is any compelling scientific evidenced for a Universal force that *expands* all mass **specificities** Universal the **same**. In other words, two universal forces that are the **same** but interact in **opposite** – contracting/expanding – directions.

For **both opposing** universal forces of **sameness** to become a more complex Universal force of **Sameness**, the *expanding force* should *accelerate* at the **same** *rate* as gravity's universal force of *contraction* but in the **opposite** way. FIGURES 347 and 348 depict how gravity's **sameness** of *contraction* appears to an observer *from* any mass point of view. This *proportional* expansion however has been replaced by *accelerated* expansion.

The **opposite** of this would be *expansion toward* any mass point of view, and would be *pointless*, which is why it has been impossible to *perceive* and so difficult to *conceive*. This "anti-gravitational" force has been discovered with new more accurate and powerful instruments and more sophisticated computer programs. Even so, its discovery has come as a complete surprise to astronomers and cosmologists because **both** accelerated Universal forces are *conceptually* the **same**, *indistinguishable* from each other. We observe and feel only the gravitational force that logically orders not just our point of *mass* but all points of mass Universally the **same**. Read "By X-raying Galaxies, Researchers Offer New Evidence of Rapidly Expanding Universe," by Dennis Overbye, the *New York Times*, May 19, 2004, in which "little is known about what is propelling the cosmos outward."

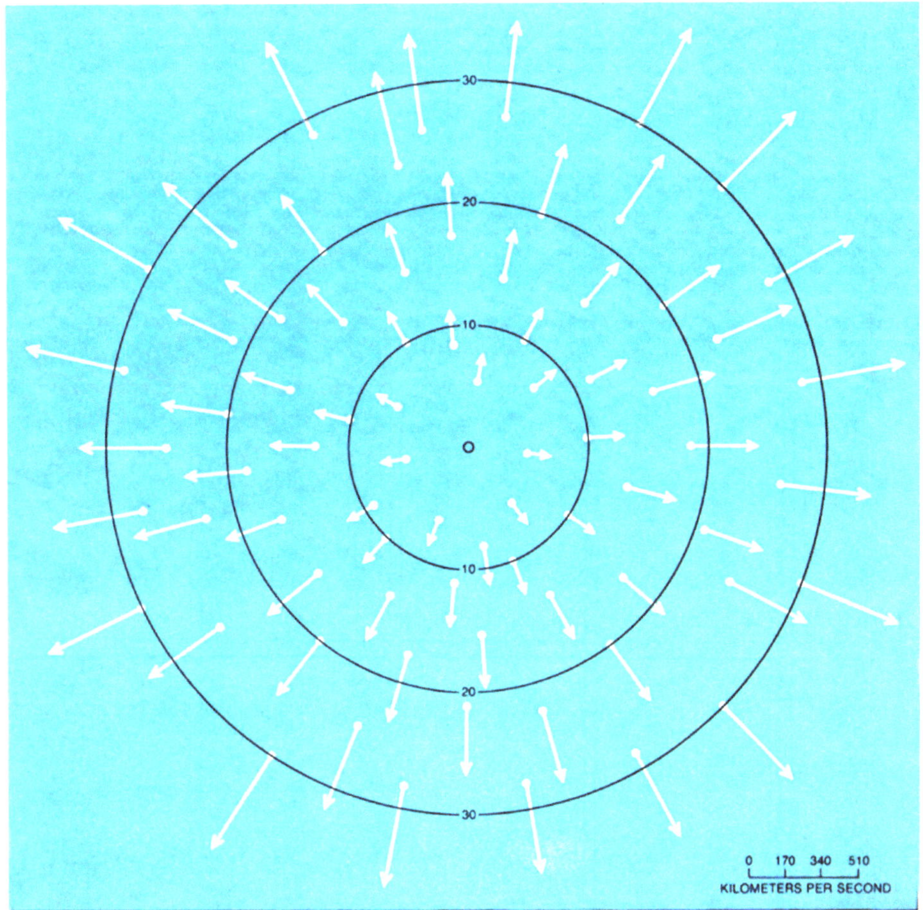

COSMIC EXPANSION seems to place the observer at the center of the universe, from which all distant galaxies are fleeing. The velocity with which a galaxy recedes is proportional to its distance from the observer, a relation first established in the 1920's by Edwin P. Hubble from observations made with the 100-inch telescope at the Mount Wilson Observatory. The principle that the ratio of velocity to distance is constant has since become known as Hubble's law. It can be interpreted most simply as evidence that the expansion of the universe began with the big bang, since the relation implies that in the past all the galaxies were packed together with infinite density. Distances are given here in millions of light-years; velocities are represented by the lengths of arrows, measured on the scale at lower right.

FIGURE 347

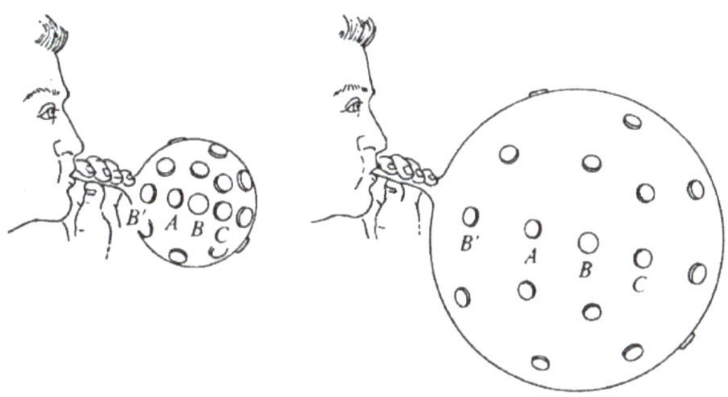

Inflation of a balloon covered with pennies as a model for the expansion of the universe. Each penny *A* may well consider itself to be the center of the expansion because the distance from *A* to any neighbor *B* or *C* increases the more the more remote that neighbor was to begin with ("the Hubble relation"). The pennies themselves do not expand (constancy of sun-Earth distance, no expansion of a meter stick, no increase of atomic dimensions). The spacing today between galaxy and galaxy ($\sim 10^6$ lyr) is roughly ten times the typical dimension of a galaxy ($\sim 10^5$ lyr).

Gravitation
Charles W. Misner, Kip S. Thorne, John Archibald Wheeler
W. H. Freeman and Co.,
San Francisco, California

FIGURE 348

Scientists use the words *dark energy* to express their dismay of how an *unobservable* Universal force of accelerated *expansion* can exist. It seems incomprehensible because our *mass* point of view *interprets* the force of gravity's *contraction* to be *Universally* the **same**. If all *mass* is ordered the **same** *Universally* under gravity, it is difficult to think that *mindless* nature would need yet another Universal force to order the Universe. Two such **opposing** universal forces would seem to cancel each other out. But the conceptual interactions of *mindless* nature's Universal forces don't interpret, they are the **same** at every place at any time. Therefore nature *interacts* **both** gravity **and** "antigravity" more complexly as the **Same** force with more energy than the sum of its parts, too great to be believable. Without the logic of interactions scientists only know that most of the force that governs the Universe is missing, but without a clue as to why.

The *timeless* **Sameness** of "antigravity" is too weak to be perceived close to the *gravitational* force of the solar system. The twin Pioneer spacecrafts, both launched at the **same** time and traveling at the **same** speed from the earth in 180° complementary directions, show a slight discrepancy from where they should be according to the predictions of gravity. **Both** twin spacecrafts are slowing down too much. After thirty years, the two Pioneer spacecrafts, now approaching the edge of the solar system, are about four million kilometers *closer* to the sun than its gravity predicts, as seen in "Solving the Pioneer Anomaly," by Constance Holden, *Science*, June 16, 2006. Also read, "Support Sought to Investigate Sluggish Pioneers," by Jim Giles, *Nature*, September 30, 2004.

Perhaps the force of "antigravity" has a very slight *expanding* effect on the mass of our solar system so that it is *pushing* slightly against the sun's *contraction* of gravity.

Read "Modern Cosmology: Science or Folktale?" by Michael J. Disney, *American Scientist*, September–October 2007. Also read "The Quest for Dark Energy: High Road or Low?" by Adrian Cho, *Science*, September 2, 2005. Also read "The Vacuum Energy Crisis," by Alexander Vilenkin, *Science*, May 26, 2006 that verifies what I have said.

The *complexity* of **both** Universal *contracting* **and** *expanding* forces is still **opposed** by the concept of *time* while conserving the *simple* interaction of **<u>opposite/same</u>**. An even more complex concept of **SAMENESS**—a force that excludes all **opposition**, even the **opposition** of *time*—is needed to complete such an omnipotent force. We can create this force of **SAMENESS** by simply *doubling* the *time* dominated megalogic on the inside front cover., Then place one of the temporal megalogics 180° to the other so that we get two *contradictory* (irreversible) arrows that cancel each other out. This *contradiction* then becomes the ultimate and complete *timeless* Universe of **SAMENESS**, forever bringing logical order without any **opposition**.

Don't confuse time-antitime with <u>time/timeless</u>. The *former* is a contradiction not subject to scientific thought or mathematical formulae, while the *latter* is an interaction that is subject to scientific proof.

The <u>time-antitime</u> universe is like the *collision* between the *negative* electron and the *positive* electron (positron) where both absolute parts of *mass* form a *massless* contradiction. Read "Alliance of Opposites: Electrons and Positrons Make New Molecules," by David Castelvecchi, *Science News*, September 15, 2007.

By adhering closely to the laws and forces of nature we have gone from the *simple interaction* of <u>contraction/expansion</u> between atoms to most complex interaction of <u>contraction/expansion</u> with <u>gravity/anti-gravity</u>, we have now proven that the *simple* interaction of **<u>opposite/same</u>** is *conserved on the horizontal* at each successive increase in complexity. Even further, that the **similarities-differences** of *temporal* mass on the *horizontal* **oppose** the *timeless* concept of **sameness** on the *vertical*. This 90° interaction *is only part of nature's square of* **sameness**, exemplified *by the antinomy in inference.*

With the *dominance* of deduction in inference and our **specific** mass point of view, it is only natural that scientists have chosen the "big bang theory" with its *temporal sequence* **from** *beginning* **to** *end* conclusively. On the other hand, the *perpetual* point of view as seen in nature's *interactions*, *constants*, and *conservation* laws, although *not obvious or dominant*, is just as valid. Read about the conceptual conflict between the *temporal* universe and the *timeless* universe as <u>sequence/simultaneity</u>, <u>big bang/steady state</u> respectively in "Going On from Observation," by E. Margaret Burbidge, *Science*, May 13, 2005, a review of two books about Fred Hoyle's failed attempt to convince other cosmologists of his theory of the "steady state" universe. How could Hoyle win an argument with scientists whose minds are *dominated* by deduction in inference and thus **opposed** to the concept of **sameness**? *Time changes* all *mass* **specificities**, but if *matter* can be **neither** *created* **nor** *destroyed*, how can there have been a beginning of the universe of mass and how can it come to an end? The two most basic and simplest parts of mass, the electron and the proton, are acknowledged by scientists to be "forever." Forever having a beginning is an oxymoron.

Despite the *mind's* perpetual *division* between deduction and induction in inference and the dominance of **similarities-differences** in *mass* that **oppose sameness**, scientists seek just one formula that will *unify* all the forces of nature. This, of course, will obviate the interaction of division/unity as **opposite/same** respectively. Scientists however are so enamored with the concept of *symmetry* that they overlook the more important concept of *paradox* that is the essence of nature's interactions. As we shall see, every mathematical formula that reveals a law or force of nature is the paradox of a *divided unity*, a **sameness** (=) with two **opposing** halves. *Absolute* concepts such as *unity* or *symmetry* will always shows their seemingly self-contradictory characteristics in our *relative* world of interactions, dominated by the *asymmetry* of *time*.

Because any **specific** *mass*, no matter how complex, can be only in one point in space at any point in time, our **specific** brain/minds *interpret* light *sequencing* through *time* and *distance* at a *constant* speed. By interpreting light at a *constant* speed of 186,000 miles per second in a vacuum, we can calculate the **different** *times* and *distances* it takes for a light wave to reach the earth from distant stars or galaxies. We can look *back* in *time* through *distance*. See Figure 319.

An observer at the point of a distant star from which light was emitted—say, 12 billion light years ago—could also do the **same** thing in **reverse** – see the **same** light years in the *past* to our star, the sun. So we have two **opposing** cosmic observers seen the **same** thing. Neither observer however can be in **both** places *simultaneously*. As with any mass point of view, the brain-mind of an observer would have to travel *alternately* in time between two cosmic points. But if one could, like gravitational **sameness**, be at **both** cosmic points at the **same** *time*, light would be *timeless*, having no concept of *distance* or *time*, which proportional interaction defines *speed*. Einstein said that if he could ride a light wave, *time* would stand still. If the clock of a light wave does not tick, how can light sequence in *time?* It can from any *time*-dominated **specific** mass point of view.

This non-intuitive phenomenon is explained by the fact that there is no relationship between the speed of any point of *mass* and the speed of any *massless* beam of light. For example, if a mass point and a beam of light collide head on at 180°, regardless of the speed of the mass, the speed approach between the two would be *constant* at 186,000 miles per second in a vacuum. The universal force of electromagnetic radiation remains *timelessly* the **same**, a *constant* for all the *variations* of mass.

That cosmic light has no concept of *distance* or *time* is also confirmed by the inescapable "entanglement" phenomenon where **both** cosmic photons seem to remain *inseparably* the **same** no matter how *distant* or how much *time* it takes for them to *speed apart* from each other at 180°. To the *mind*, they seem to move away from each other faster than the *constant* speed of light. **Both** such cosmic photons however view each other as the **same** phenomenon because their *speeds* are in 180° contradictory directions, **both** creating the contradiction of *speedless*.

With the *perpetual* universe the concepts of **both** *past* **and** *future* interact *simultaneously* the **same,** while the *temporal* universe *sequences* from past to future in **opposition**.

Each cosmic observer's point of view has his own isotropic universe, *each* one of which looks the **same** as any other. All such isotropic (**general**) universes however are more complexly the **same** for *mindless* nature. From any *cosmic* point of view, a **specific** *mind* would **opposes** nature's universal forces of **sameness**; otherwise one could not have *time* to move any *distance* **from** here **to** there, or distinguish between the **similarities-differences** of mass.

At any speed, all clocks tick at the **same** *rate*. But if the mass *changes* its speed, time will tick at a **different** rate. *Time* makes a **difference** with any mass **specificity**, including its *speed*. **Different** cosmic travelers with **different** rates of time frames led Einstein to propose that there is no single or *constant rate* of universal *time*. Since any point of *mass* always **opposes** Universal **sameness**, Einstein's **differences** in the rates of *local* times is consistent with **opposing** the **same** Universal time. We cannot escape the *timeless* interaction of change/constant as **opposite/same** any more than we can extricate ourselves from the gravitational acceleration of *constant change* in *time*.

Returning to the relationship between mass and time, if a ruler sped in the direction of its length and *increased* its speed, it would *shorten* at the ruler at **same** *rate of increase*. If the ruler *slowed* down, it would *lengthen* the ruler proportionally. In the former, if the ruler were a **specific** observer, he would see the **generality** of space *expand*, "warp," and its depth *contract*. At the absolute speed of light, spatial **sameness** also would become *absolute* to the exclusion of mass. There would be nothing. There would be no relationships and no such thing as relativity. No mass **specificity** therefore, no matter how simple or complex, can travel at the *absolute* speed of light. Any point of mass, even an electron, must travel less than the *constant* speed of light. It would take all the *energy* in the universe for an electron to reach the *absolute* speed of light. This can never happen because it would leave *mass* with no *energy* with which to *interact* as in the formula $E = mc^2$. Read "That Famous Equation and You," by Brian Greene, the *New York Times*, September 30, 2005.

Perhaps Einstein's greatest insight was that light is a universal *constant* with which all *varieties* of *mass* and *changes* in *energy* are related. Thus, mass is *relative* to the *absolute* speed of light. Thus, relative/absolute and variable/constant correspond to **specific/general** and **opposite/same** respectively.

Einstein introduced *mind* experiments to explain his theory of relativity. Many of these *conceptual* experiments were interactions of **sameness** from **opposing** interpretations, such as his famous "equivalence" principle. These *mind* experiments were later proven to be correct by *physical* experiments because mind/mindless *interact concepts* by the **same** logic. Even though any **specific** concept of *mind* is always an *incomplete*, its *concepts* are still *part* of *mindless* nature's interactions.

Einstein's "special theory" of *relativity* based on the *inertial* interpretation of mass states that because the greater the *speed* of any **specific** mass, the slower its *rate* of *time*, if **either** human twin traveled at a rate of speed faster than the other, the faster twin would travel into the *future* in comparison to his sibling. Since *time* goes only one way, into the future *irreversibly*, such a time traveler can never return to the **same** time as his twin who would have long since been dead when he returned to the point from which his travel began.

Einstein's general theory of relativity is based on the *accelerated* interpretation of mass using the *curvature* of space as its geometry, in contrast to his *inertial*, *straight-line* interpretation of distance and time that characterizes special relativity. In his general theory, *space* and *time* became the **same** thing. Space/time becomes the bases of measurement, a more complex metric unto itself that characterizes the *inseparable cosmic* **both/and** function. Einstein had to *alternate* in *time* between his two conceptual *interpretations* of *relativity*, but **either** one is *valid* because **both** are more complexly the **same** for *mindless* nature. Read, "Anthropic Reasoning," by Mario Livio and Martin J. Rees, *Science*, August 12, 2005.

The mathematical formula seen in Figure 320 appears to be a *doubling* of Einstein's mass/energy equation that a small group of physicists think would *unify* all the forces of nature. More to logic, $E = mc^2$ seems to be formed more in *simultaneous double reversal* to *complete* the **same** but more complex *simultaneous* interaction of **both** accelerated universal *contraction* **and** accelerated universal *expansion*.

To again see that there is **no difference** between **specific** and **general** from multiple scalar views of the universal mass, read the News and Views section and article on "The Evolution of the Universe," the cover story of *Nature*, June 2, 2005.

For our mass point of view on earth, our isotropic universe is *dominantly deductive*, **from** the **specific to** the **general**. Because *induction* is *part* of inference, one can get an increasingly accurate fix on a distant star by getting as many deductive conclusions as possible. The greater the **general** collection, the more accurate the star can be pinpointed. This deductive/inductive interaction is known as the "Gaussian paradox," seen in Figure 10. Again we see that induction's **generalized** conclusions enhance the mathematical accuracy of the deduction's **specific** conclusion.

Read the works of David Hume to understand the paradoxical nature of induction. Also see Figures 255a and b to see how a **general** set of inductive conclusions are *constrained* within the **same** circle with each successive point *concentrated* around a **specific** straight line sequencing in *time*.

More scientific evidence for the conceptual *paradox* of universal **sameness** is suggested when we look twelve or more billion light years away from our Earth's point of view. At these great *cosmic* distances, we see galaxies that are as "old" as our Milky Way. The universe of mass seems to looks the **same** *then* as it does *now*. There is **no difference** between the *present* and the *past*, or the *future* for that matter. I quote from *Nature*, July 8, 2004, under News and Views, "Old Before Their Time," by Gregory D. Wirth. His abstract reads, "The discovery of a massive evolved galaxies at much greater distances than expected—and hence at earlier times in the history of the universe—is a challenge to our understanding of how galaxies form." The two papers that Wirth describes in the same issue, along with other recent evidences, confirm that *late* is much too *early* than the predicted age of the universe. Also read "Crisis in the Cosmos?" by Ron Cowen, *Science News*, October 8, 2005.

Another paper that supports a *timeless* universe is seen in "Young and Near: Baby Galaxies Roam Our Backyard," by Ron Cowen, *Science News*, January 1, 2005, from which I quote:

> That's the finding of an ultraviolet-detecting satellite that has identified 36 such galaxies. Since 1995, astronomers have detected more than 2,000 baby galaxies, virtually all of which reside more than 10 billion light years away and therefore hail from the distant past . . . . Now, scientists have for the first time detected galaxies that resemble these youngsters but reside just 2 to 4 billion light years from earth . . . ."It's sort of like finding a dinosaur in

> your back yard," says Tim Heckman of Johns Hopkins University in Baltimore. "We thought that this type of galaxy had gone extinct." . . . Because newborn stars radiate most of their light in ultraviolet wavelengths, their host galaxies appear to GALEX like diamonds against a black backdrop. Each galaxy emits about 10 times as much ultraviolet light as does the Milky Way, and they are "dead ringers" for the much more common, young and massive galaxies found in the early universe.

Stars from the *early* universe shortly after the big bang have also been found among the *late*-universe stars of our own galaxy, the Milky Way, as described in "Cosmic Primitive," by Ron Cowen, *Science News*, April 16, 2005. Also read "A Large Population of Galaxies 9 to 12 Billion Years Back in the History of the Universe," by O. Le Fèvre, et al., *Nature*, September 22, 2005.

Early/late, young/old each interact universally the **same** that any *time* frame would **oppose**. Is the universe *temporal*, or is it *timeless*? Which one is true? **Neither** one is true, but **both** *simultaneously* are true, allowing the **specific** human *mind* to make **either** choice alternately in time as a valid interpretation. But in the *cosmic* realm, we run headlong into *duality*. The universe seems to become ever more *complex* and incomprehensible as information is accumulated. Read "Oh, for the Simple Days of the Big Bang," by George Johnson, the science writer for the *New York Times*, October 8, 2006, in which Johnson laments the seeming contradictions that cosmologists face. Worst of all, scientists can't seem to escape the anthropic principle that *mind* and *mindless* nature somehow interact.

Also read a book reviewed by Yadin Dudai, *Nature Neuroscience*, December 2004, of a book titled *The Physiology of Truth: Neuroscience and Human Knowledge*, by Jean-Pierre Changeux, Harvard University Press, 2004. I quote Dudai:

> Most experimental scientists do not have a theory of truth despite being confident that they are diligent truth seekers. At most the question "is it a true fact?" evokes an answer of the type, "Sure, I used ANOVA" (or what ever statistical favorite test was unearthed for the occasion). Those few scientists who contemplate the issue a bit further usually end up using what philosophers might call a "recursive correspondence theory." This is the belief—itself a loaded term—that our knowledge about an event or process is true if the behavior of that event or process in the world corresponds to the predictions of that knowledge, and if that knowledge integrates properly with other knowledge. Because this correspondence might break down when the analysis is pushed to its limits, research programs and academic careers are generated in an attempt to resolve the discrepancies. The new knowledge, alas, will generate new discrepancies; hence the recursiveness . . . yet there is much to say in its favor. First, it paved the way for intercontinental flight, e-mail, open heart surgery and Prozac, to name just a few dividends. It is difficult to argue with success. Second, it pleases scientists because they appreciate that the story will never end . . . . If Changeux and his colleagues are correct, we shouldn't worry too much about the validity of our basic picture of the world, because the world and our brain struck a deal, zillions of years ago, to enable them to comprehend one another correctly. An important caveat, though, is that this conclusion refers to the capabilities of our perceptual and cognitive faculties as a species, not to the veridicality of recall of specific episodes or of culturally entertained religious and political beliefs.

It seems unfortunate that we can never know truth without running headlong into paradox. But accepting conceptual paradox would be *enhancing* instead of *diminishing*. As of now, it is blatantly evident that *chaos* results from our minds conflicting ideologies. But since paradox is particularly evident at *cosmic* distances where the **both-and** *function* dominates the *conceptual* interaction of time/timeless as sequence/simultaneity, and where Einstein's theories of *relativity* reveal the reality of our *macro*-mass, and where the *interaction* of **opposite/same** is *conserved* at all levels of mass complexity, all of which are non-inferential, there is little doubt that conceptual paradox is mentally enhancing. Read, "Cosmic Bigness," by Ron Cowen, *Science News*, August 26, 2009, from which I quote, "Astronomers have found the largest structures ever discovered the universe—filaments of galaxies 200 million light years in length that date just 2-billion years after the Big Bang. The filaments contain 30 giant pockets of gas, each 10 times the mass of the Milky Way."

Also read "Spiral Galaxy in the Young Universe," by Ron Cowen, *Science News*, September 2, 2006, from which I quote, "The Milky Way is a spiral-shaped disk of swirling gas orbiting a central hub of stars—signs of a mature galaxy. Researchers have assumed that most younger galaxies are misshapen and full of chaotically moving gas. Now, astronomers have identified a galaxy that has already begun to resemble the modern Milky Way when the universe was only 3 billion years old, one-fifth of its current age."

The discovery of the 3°K cosmic microwave background seen in Figure 318 is not only evidence of the *temporal* universe but also evidence of a *timeless* universe. This universal signal clearly displays the universal **sameness** of **both** accelerated *contraction* **and** accelerated *expansion*, particularly at the mid-point– early-late –of cosmic *time*, as shown in, "From

Dark Matter to Light", by Tom Cowen, *Science News*, March 22, 2008. The **sameness** of cosmic background radiation is **opposed** by the mass of super-accelerated protons puncturing holes in the cosmic void. Read "A Singular Conundrum: How Odd Is Our Universe?" by Adrian Cho, *Science*, September 28, 2007.

For the human *mind*, creativity/destruction, genesis/entropy, chance/design, mass/energy, all interact as **specific/general** respectively. For *mindless* nature, all these conceptual interactions are more complexly the **same**. Read "Is Our Universe Natural?" by Sean M. Carroll, *Nature*, April 27, 2006. Also read "The (Long and Winding) Road to Reality," by Tim Folger, *Discover*, June 2005. This cover story explains the theory of Roger Penrose, Nobel laureate, for the need of a logical form for *simultaneity* to *unite* the various levels of mass complexity.

Returning to the scientific quest for a "theory of everything", or "string theory", if *everything* were *simultaneously* the **same**, there would be no *sequence* of inference—**from** data **to** conclusion—in *time* to **oppose** it, thereby nullifying their own idea of the big bang. *Absolute* concepts such as *unity* of *symmetry* will always show their seemingly self-contradictory characteristic in our *relative* world of interactions *dominated* by time. Moreover, a mathematical formula of *absolute* **sameness** would reveal *nothing*, *contradicting* their scientific quest to know *everything*. Also, this purely mathematical effort excludes *experiments* by induction the **generality** of which enhances the **specific** mathematical proofs of deduction. Excluding induction from deduction nullifies inference, the very logic of scientific discovery. From our **specific** mass point of view, we can only know the relationship between *some* things to each other, not everything. All knowledge is incomplete.

To show the self-contradictory nature of trying to *unify* everything, "string theory" has *divided* these scientists into two **opposing** camps. The idea of everything being based absolutely on the deductive reasoning of *mathematics* has produced so many **different** universes that many scientists fear that string theorists are becoming philosophical, contradicting the very idea of deductive predictability. Read "A 'Landscape' Too Far," by Tom Siegfried, *Science*, August 11, 2006, from which I quote:

> Physicists have long heaped scorn on anyone who tries to explain features of the universe by pointing out that had they been otherwise, life would be impossible. This "anthropic principle," many physicists charged, abandoned the long-standing goal of finding equations that specify all of nature's properties. Most preferred the notion that a comprehensive theory would account for everything the universe has to offer. Ironically, however, the favored candidate for that approach—string theory—may be exacerbating the very problem everybody hoped it would solve. Far from disposing anthropic reasoning, string theory has reinvigorated its advocates, leading to a schism within the physics community. The dispute has touched off sharp exchanges both within and outside science journals.

Read "Physics Ain't What It Used to Be" by George Ellis, who reviews of a book titled *The Cosmic Landscape: String Theory and the Illusion of Intelligent Design* by Leonard Susskind, *Nature*, December 8, 2005. To see the *incomplete* and self-contradictory nature of the *mind's absolute concept* of *unity*, I quote from Ellis, "As a philosophical proposal . . . the challenge facing cosmologists now is how to put on a sound basis the attempts to push science beyond the boundary where verification is possible—and what label to attach to the resulting theories. Physicists indulging in this kind of speculation sometimes denigrate philosophers of science, but they themselves do not yet have a rigorous criterion to offer for proof of physical existence. This is what is needed to make this area a solid science, rather than speculation."

Another good read about string theory that exposes its *unrealistic* concept of *absolute symmetry* that no *mind*, no matter how great its intelligence, can logically comprehend it is found in "A Crisis in Fundamental Physics," an essay by the theoretical physicist Lee Smolin, *Update*, January/February 2006 (www.nyas.org), a publication of the New York Academy of Science. I'm not going to quote the essay because you must read it all to understand that the *inconclusive* concept of **sameness** should be *dominantless* if the brain/mind is to determine anything conclusively. Also read *What We Believe But Cannot Prove: Today's Leading Thinkers on Science in the Age of Certainty*, edited by John Brockman, Harper Perennial. Also read *Not Even Wrong*, by Peter Woit, Basic Books, 2006, and *The Trouble with Physics*, by Lee Smolin, Haughton Mifflin, 2006—both of which explain that string theory predictions cannot be experimentally tested and therefore cannot be falsified. More to *logic*, deduction is absolute to the exclusion of induction, thereby denying the logic of inference that corresponds to the very logic of *mindless* nature and thereby has made scientific *minds* so highly predictive of *mass* behavior.

Finally, read "Unburdened by Proof: String Theorists Are Setting a Worrying Trend by Downplaying the Need for Experimental Evidence," *Nature*, October 5, 2006, by George Ellis, who reviews a book titled *The Trouble with Physics:*

the Rise of String Theory and the Fall of a Science, and What Comes Next by Lee Smolin, Houghton Mifflin, 2006. A rebuttal to the above book by Lee Smolin is, "The Universe on a String" by Brian Greene, the *New York Times*, October 20, 2006, who pleads, "Don't discard the most promising theory in physics." I quote further:

> Researchers worldwide are still working toward an exact and tractable formulation of the theory's equations. And without that final formulation in hand, the kind of detailed definitive prediction that would subject the theory to comprehensive experimental vetting remains beyond our reach . . . . Some critics have taken the lack of definitive predictions to mean that string theory is a protean concept whose advocates seek to step outside the established scientific method. Nothing could be further from the truth. Certainly, we are feeling our way through complex mathematical terrain, and do doubt have much ground yet to cover. But we will hold string theory to the usual scientific standard: to be accepted, it must make predictions that are verified.

The human brain cannot even remember *everything* it learns. The **opposing** concepts of *remembering* and *forgetting* are proportionally constrained in the brain within the **same** modular system. Read "Remember This: Forgetting Can Be Beneficial," from *National Geographic*, November, found in the *Wall Street Journal*, October 23, 2007, from which I quote, "'If we remembered everything,' writes journalist Joshua Foer, 'we'd be drowning in irrelevant information.' In the short story 'Funes in Memorious,' Argentine writer Jorge Luis Borges described how a character's inability to forget anything meant that he couldn't prioritize the events of his life or make generalizations. 'To think is to forget,' wrote Mr. Borges. Science is backing him up. Harvard University psychologist Daniel Schacter tells National Geographic that forgetting is the price individuals pay for being able to interpret the world around them."

The human *mind* can never find a *one* formula that encompasses *all* formulae of *mindless* nature, else we would be as gods ordering *everything* the **same** without **opposition**. String theory will never capture the *unity* of universal **sameness**. We will always be in **opposition** to the *logical* God of **SAMENESS**.

What God *is* can never be *determined* because to the *mind's concept of* **sameness** is not only *incomplete but* inconclusive. *How do you conclusively* **specify** a "spirit"? Yet the imperative to **specify** by deduction sets up a "the perfect storm" of perpetual conflict between the reciprocally **opposing** ideas of *faith* and *doubt*. The *question* of God can never be *answered* conclusively because doubt/faith interact, as explained by Peter Steinfels in the *New York Times*, September 18, 2004, whom I quote:

> Despite much talk of postmodernism, for many people the major arguments about belief in God's existence are the same today as they were a century ago, arguments pitting faith and religious experience against the philosophical naturalism that accepts only claims passing the test of scientific verification . . . the deep ethical impulses that, beneath the surface of logical argument, often drives people toward either atheism or faith in God . . . the big question that everybody cares about: . . . the origin, nature, and destiny of the cosmos and of human consciousness and intelligence. "Whether we realize it of not," Dr. Nicholi says, "we make one of two basic assumptions. We view the universe as an accident or we assume an intelligence beyond the universe who gives the universe order, and for some of us, meaning to life."

Read "Faith and Science Not Mutually Exclusive," by Eric Mink of the St. Louis Post-Dispatch, found in the *Tampa Tribune*, September 6, 2005.

If the megalogic on the inside front cover is the guide to the logical order of any mass **specificity** then, if we want to communicate with intelligent *life* on other cosmic points in the universe, then we should assume that they too have discovered inference and used it to advance science to a high degree of complexity. It would seem then that the megalogical form on the inside front cover would be the message along with Figures 5, 20, 105, 106, 107, and 205 by which **either** cosmic point could recognized intelligent life and thereby confirm that conscious life is not *unique*, at least not forever.

Because **either** *cosmic* sender **or** receiver cannot escape the *time* measured in light years for information to traverse the great *distance* **either** way between them, we may never know the answer. As with us, their mass point of view is *dominated* by *time* in **opposition** to **sameness**.

MUSIC

The foundation of different languages and of distinct species, and the proofs that both have been developed through a gradual process are curiously the same.

—Charles Darwin

Science is the search for resonance between mind and natural patterns.

—Niles Eldredge

Musical harmony takes the **same** logical *pattern* as the *inverse square law* of nature's universal forces. What the human *mind* hears as musical harmony is the harmony of the nature's universal **sameness**. Let us see the correlation.

We begin with the simple geometric concept of *frequency*. In Figure 119, you will see successive increases in frequency of a string's vibrations in whole natural numbers. For this to occur, the frequencies must be held *constant* between two *parallel* fixtures so that meaningful *change* can naturally occur in the numbers. Under this *constraint* change/constant interact at 90° the frequency of the string *quadruple* with every *doubling* of tension. Each *doubling* of a tone creates simultaneous *set* of overtones that takes the **same** form as the inverse square laws.

Now go to the piano keyboard seen in Figure 121 that shows sets of eight adjacent tones called octaves. Each octave has the **same** harmonic structure as all the others. The **sameness** of the octaves occurs with each successive *doubling* or *halving* of distance on the keyboard. Figure 122 shows that with the simple *linear* progression of overtones there is a *nonlinear* interactions between *distance* and *frequency*. With each *doubling* of distance, there is a *quadrupling* of frequencies, the **same** as nature's inverse square law. This occurs with *each* successive octave *up* the scale. There is **no difference** between the force of gravity in Figure 105, and the overtones of music in Figure 122 **both** have the **same e**nergy scale. Furthermore, **both** create *harmony* and *order*, instead of *noise* and *chaos*. The very definition of music is that a **specific** note is able to self-create the **general** overtones of *harmony*, the **same** as nature's *accelerated* forces. Without the *interaction* of **specific/ general**, there is no *harmony* between *mind* and *mindless* nature.

Figure 122 also shows that the *mind* is limited to overtones *up* the scale, whereas *mindless* nature's universal harmony is structured **both** *up* **and** *down* the **same** scale *simultaneously*. Our brain-minds sense only *half* of nature's squares of **sameness** from any tonal scale, yet **both** *mind* **and** *mindless* nature use the **same** logic of interactions. Our mind's harmony therefore is only half of nature's square of **sameness**.

The **sameness** of logic between *mind* and *mindless* nature allows us to naturally construct the language of music, just as our *minds conceptual* capacity also enables us to construct the language of words. The fundamental *logical* structure of these two languages comes *naturally* to human minds regardless of the **similarities-differences** of race, culture, faith, or environment. The logic of music has the **same** form as inference as seen in Figure 125.

To see how the *simultaneous* interactions of musical harmony can be transformed into a *sequence* with a conclusion the **same** as a verbal sentence, look again at Figure 122. Select any key of an octave—say, C—where the **sameness** of **both** the *first* **and** *last* notes of each octave sound the **same** as **either opposing** end of the *two* adjacent octaves. C, then, becomes the concept of **sameness** that is **no different** than a magnetic bar that *conserves* its **sameness** no matter how many times another magnet is added or no matter how many times a single magnet is cut in two.

Now, let's substitute numbers for letters to show that whatever key we select among *various* keys, their harmonies will remain *constant*. With the first note C, now numbered 1, sit down at the piano and strike the fifth note *above* 1 and the fifth note *below* 1 together *simultaneously*; and you will hear *conflicting* notes of *equal* distance on **opposite** sides of note 1. Now add the **opposing** notes 4 and 5 to note 1 and strike all *simultaneously* and you will hear conceptual conflict.

In order to make the language of music *sequential* and create a **specific** theme, we have to transform these *simultaneously* interacting notes in numerical *succession up* the scale only. Beginning with note 1, transfer note 5 *below* 1 the **same** distance *above* 1 so that note 5 from *below* substitutes for note 4 *above*. This is valid from the reciprocal interpretation of logic because we previously explained how the two **opposing** fifth notes **both** *below* **and** *above* note 1 sounded the **same**. By taking the geometric concepts of below/above and transferring *below* to *above* **both opposing**, fifth notes then can *sequence* as notes 4 and 5 within the *constraints* of octave C without making **any difference** to *mindless* nature.

The *dominantless left* hand can interact *simultaneously* with the *dominant right* hand so that **both opposing** hands can *sequence* the **same** theme *harmoniously* as seen in Figure 123. Observe that the notes of the *right*-hand *sequence* in whole natural numbers as the notes of the *left* hand also sequence **either** *back* **or** *forth* alternately, **either** *down* **or** *up* alternately, **no different** than the two **opposing** electric charges *alternating* to *sequence* a message in the electric current.

The *dominant sequence* of notes by the *right* hand deduces the *simple* theme to a **specific** *conclusion* when the last note 8 is struck. Play these eight numbers in *sequence* and you hear the simple doe, ray, me . . . theme. When you strike the next-to-last seventh note, it will *implore* you to strike the eighth note to bring the musical theme to a conclusion. Sit down at the piano, and hear for yourself this imperative to bring the numbers to a logical conclusion. Then read "Hitting the Right Note," *Nature*, September 7, 2006, by Petr Janata, who reviews a book titled *Sweet Anticipation: Music and the Psychology of Expectation* by David Huron, Bradford, 2006. I quote from Janata, "Imagine that just before sounding the final chord of a Beethoven sonata, a concert pianist abruptly interrupts the performance to answer the phone that has just rung in his tuxedo pocket. The audience is left pining for musical closure, livid that a sparkling performance should terminate with an insipid ring tone . . . . At the core of Huron's account is the premise that organisms derive an evolutionary benefit from being able to predict how their environments will behave. In other words, we internalize the probabilities with which certain events follow other events in everything we do." Although the author based this on statistical analysis, being able to bring the *sequence* of events to a predictable *conclusion* speaks more to the *logic* of inference dominated by deduction.

By *simultaneously* playing **both** the *dominant right* hand **and** the *dominantless left* hand, we get a more complex and enhanced musical harmony with our theme. This is obvious to the brain/mind if you first play all eight notes in numerical sequence with just the dominant *right* hand and then play the **same** theme with **both** *right* **and** *left* hands *simultaneously*. Again, regardless of the conceptual interaction, dominant/dominantless interact.

As for logic, the **specificity** of the *right* hand and the **generality** of the *left* hand interact the **same** as **both** the *deductive* conclusion **and** *inductive* conclusion in inference, thereby creating the *antinomy* that so greatly *enhances* our mind's logical capacity, **no different** than the two dominant/dominantless human *hands* interacting in Figure 125 that *harmoniously* perform a complex **specific** task conclusively.

Despite the *linear* dominance of inference in music and the language of words, it is customary to end them by *circling* back to the **same** idea as in the beginning to emphasize its meaning. In word language, from Sophocles to Cervantes to Shakespeare to Hegel to Kant, all great writers end their works with a deep and *inconclusive* irony. The *irony* of writing *concepts* in a *circular* pattern, along with the *linear* thoughts of *facts*, has a profound effect on the human *mind* that is *conserved* not only during the lifetime of the author but also generation after generation of readers.

The human *mind's capacity* for *conceptual* **sameness** in the **different** languages of words and music, if filled with adequate knowledge, creates *consciousness* of mind because its megalogical form becomes *part* of *mindless* nature's universal **sameness**. Read "Music, the Food of Neuroscience?" *Nature*, March 17, 2005. I quote the abstract, "Playing, listening to and creating music involves practically every cognitive function. Robert Zatorre explains how music can teach us about speech, brain plasticity and even the origin of emotion." Also read "Red in the Head," *Nature*, June 1, 2006, by Josh Weisberg, who reviews a book, *Seeing Red: A Study in Consciousness* by Nicholas Humphrey, Belknap Press, 2006.

To see how human *consciousness* is in logical harmony with *mindless* nature, refer again to Figure 125 and then contemplate Figure 105 where gravity's *accelerated* force is the **same** as the megalogical form seen on the inside front cover.

Now that we have made a *synthesis* between *consciousness* of *mind* and *unconscious mindless* nature, let us *analyze* the universal inverse square laws. Turn to Figure 105 where you will see that *time* squared in successive whole natural *numbers* always creates the **same** *geometrical* squares successively increasing in complexity the **same** as its inverse square law. Each successive *doubling* of *distance quadruples* the *acceleration* of gravity so that **both** *distance* **and** *acceleration* form a half/double proportion that is *conserved* in each square's successive increase in complexity. Also, the **opposing** numbers at the two ends of *each* diagonal of any square sum the **same** number no matter how complex, the **same** as in the squares of "magic numbers" seen in Figures 106, 107, and 108.

By contemplating Figure 105, you will discover that *geometry*, *concept*, and *number* are the **same** in the highly complex and enhanced universal force of gravity. *Mindless* nature interacts all three of these logical tools *simultaneously* the **same**. Gravity makes no distinction of any concept. The *minds* of scientists however have the logical capacity to discover **opposite/same** interactions, but this is only part of gravity's logic. Turn to Figure 175 to see how the geometric parts of *diagonal*, *diameter*, and *hypotenuse*, which must be viewed alternately, seem to be reciprocally the **same**. The **sameness** of these *geometric* parts *simultaneously* escapes the human mind, to say nothing of leaving out *number* and *concept*.

Since the human perception and conception are always **opposed** to **sameness**, we can never circumscribe the **sameness** of universal gravity. Knowledge of nature therefore is always incomplete. Despite its power of discovery, the megalogic on the inside front cover reveals only *parts* of nature's universal ordering principle at any given intervals of time. What we call "the laws of nature" all add up to more than the sum of any of its parts we can assemble: a **sameness** beyond the comprehension of any *mass* point of view no matter how complexly ordered. Einstein said that the laws of nature are comprehensible, but not all *simultaneously* the **same**.

A good rendition of our mental limitations is "Laws of Nature, Source Unknown: Which Came First: The Order or the Universe? And Can Science Supply the Answer?" by Dennis Overbye, the *New York Times*, December 18, 2007, from which I quote:

> Einstein hoped that the universe was unique: given a few deep principles, there would be only one consistent theory. So far, Einstein's dream has not been fulfilled. Cosmologists and physicists have recently found themselves confronted by the idea of the multi-universe . . . the alleged theory of everything, which apparently has 10^{500} solutions. Call it Einstein's nightmare. But it is [too] soon for any Einsteinian to throw in his or her hand. Since cosmologists don't know how the universe came into being, or even have a convincing theory, they have no way of addressing the conundrum of where the laws of nature come from or whether those laws are unique and inevitable or flaky as a leaf in the wind.

As reciprocal logic reveals, with *concepts* and their interactions, there is no answer to questions or solution to problems. The recursive and inconclusive *concepts* are not a problem, but instead they *interact* timelessly to *form* the very universal forces and laws of nature, revealed to the human mind by *dividing* the antinomy's *unifying* **sameness** that *forms* inference into its reciprocally **opposing** processes.

For the latest explanation of the concept of *simultaneity*, read "Concurrent Events," *American Scientist*, January–February 2008, by Jill North who reviews a book titled *Concepts of Simultaneity: From Antiquity to Einstein and Beyond* by Max Jammer, Johns Hopkins University Press, 2006. After reading this review and book, you will see how inference—the *sequence* from *data* to conclusion—is still hiding the *simultaneously* interacting *concepts* of reciprocal logic.

As with the **opposing** processes of inference, the *mind* must divide the **sameness** of the gravitational force *equally* into two reciprocally **opposing** parts, seen in Figure 105 bottom. Here we see the *strength* of Earth's gravitational force being **either** *weaker* **or** *stronger*. On Earth, we see that the force of gravity gets *weaker* as the distance *increases*—i.e., *inversely* proportional to the square of the distance, or stronger as distance *decreases*—i.e., *directly* proportional to the square of the distance.

From Earth or from any cosmic point of view, an observer would see the**se** "square laws" divided into **either** inverse **or** direct, dominated by the inverse interpretation of gravity. Mindless nature however places both inverse and direct in simultaneous double reversal to create a universal force that orders all mass **specificities** the **same**. This is **no different** than placing **opposing** processes of **both** deduction **and** induction in *simultaneous double reversal* to create, the **sameness** of the antinomy in inference.

Returning to the *accelerating* overtone that *simultaneously* interact, music can be transformed into a more complex and enhanced *interaction* of **both** *consonance* **and** *dissonance simultaneously*, as seen in "Calculated Tones," by Richard Webb, *Nature*, July 13, 2006.

That the concepts of *science* and of musical *art* are *parts* of the **same** harmony of *mindless* nature, and not just coincidental or a "dream of my imagination," read "A Genius Finds Inspiration in the Music of Another," by Arthur I. Miller, the *New York Times*, January 31, 2006, from which I quote:

> Einstein once said that while Beethoven created his music, Mozart's was so pure that it seemed to have been ever-present in the universe, waiting to be discovered by the master. Einstein believed much the same of physics, that beyond observation and theory lay the music of the spheres—which, he wrote, revealed a "pre-established harmony" exhibiting stunning symmetries. The laws of nature, such as those of relativity theory, were waiting to be plucked out of the cosmos by someone with a sympathetic ear. Thus it was less laborious calculation, but "pure thought" to which Einstein attributed his theories. Einstein was fascinated by Mozart and sensed an affinity between their creative processes, as will as their histories. As a boy Einstein did poorly in school. Music was an outlet for his emotions. At 5, he began violin lessons but soon found the drills so trying that he threw a chair at his

> teacher, who ran out of the house in tears. At 13, he discovered Mozart's sonatas. The result was an almost mystical connection, said Hans Byland, a friend of Einstein's from high school. "When his violin began to sing," Mr. Byland told biographer Carl Seelig, "the walls of the room seemed to recede—for the first time, Mozart in all his purity appeared before me, bathed in Hellenic beauty with its pure lines, roguishly playful, mightily sublime." . . . He also empathized with Mozart's ability to compose magnificent music even in impoverished conditions. In 1925, the year he discovered relativity, Einstein was living in a cramped apartment and dealing with a difficult marriage and money troubles. That spring he wrote four papers that were destined to change the course of science and nations. His ideas on space and time grew in part from aesthetic discontent. It seemed to him that asymmetries in physics concealed essential beauties of nature; existing theories lacked the "architecture" and "inner unity' he found in the music of Bach and Mozart. In his struggles with extremely complicated mathematics that let to his general theory of relativity of 1915, Einstein turned for inspiration to the simple beauty of Mozart's music . . . . In the end, Einstein felt that in his own field he had, like Mozart, succeeded in unraveling the complexity of the universe. Scientists often describe general relativity as the most beautiful theory ever formulated. Einstein himself always emphasized the theory's beauty . . . amazingly, the universe turned out to be pretty much as Einstein had imagined. Its daunting mathematics revealed spectacular and unexpected phenomena like black holes. Though a Classical giant, Mozart helped lay the groundwork for the Romantic with its less precise structures. Similarly, Einstein's theories of relativity completed the era of classical physics and paved the way for atomic physics and its ambiguities.

Also read *This is Your Brain on Music: Understand a Human Obsession*, by Daniel Levitin, Atlantic, 2007.

PARADOXES IN PHYSICS AND CHEMISTRY

I think it safe to say that no one understands quantum mechanics. Do no keep saying to yourself, if you can possible can avoid it, "But how can it be like that?" because you will go down the drain into a blind alley from which nobody has yet escaped. Nobody knows how it can be like that.

—Richard Feynman

How wonderful that we have met with paradox. Now we have some hope of making progress.
—Niels Bohr

The paradox is now fully established that the utmost abstractions are the true weapons with which to control our thought of concrete facts.

—Alfred North Whitehead

The only certainty is that nothing is certain.

—Pliny the Elder, ancient Roman author

From the ancient Egyptians to Pythagoras to Euclid to René Descartes to Galileo to Newton, and culminating in Einstein *concepts*, *number* and *geometry* have interacted in a synergistic relationship, accelerating the progress of science each step of the way. When Newton introduced the logic of inference with its *concepts* of **specific/general** interacting in *simultaneous double reversal*, modern science was created. This type of reasoning gave the human *mind* the capacity for discovering the universal laws and forces of nature. When Einstein with his mind experiments based on *relative concepts* revealed the paradoxical aspect of nature's interactions, *absolutes* were no longer applicable to our *macro*-mass realm. Our brain-minds dominance of deduction in inference with its ability to *quantify* **specific** conclusions *accelerated* the progress of science that continues to this day.

To show how *logically* enhancing inference is, Isaac Newton discovered the law of gravity by first using *deductive* reasoning to show that gravity's inverse square law applied to any **specific** mass. Then, he **reversed** this sequence with *inductive* reasoning to show that the force of gravity's inverse square acceleration applied to all mass in **general**. Looking at the force of gravity from these two reciprocally **opposing** interpretations *alternately* in *time* and getting the **same** result proved that nature *simultaneously* interacted **both** deduction **and** induction more complexly the **same**. These two **reversed** processes running *irreversibly* at the **same** *time* created a more complex and enhanced paradox in logic itself, the *inconclusive* **sameness** of its antinomy that the *conclusions* of inference **oppose**. The force of this logical paradox empowered the *temporal* human brain/mind to discover successive increases in the complexity of nature's timeless laws and forces.
Newton's method of *alternately* using **both** logical processes of inference convinced him and all other scientists that inference had to be the paradigm of scientific thinking. Nevertheless, scientists still overlook the enhancing value of the antinomy in inference, still thinking that this logical paradox is a mysterious defect, instead of having created a more complex and enhanced megalogic.

Having *synthesized* the concepts of nature, let us now *analyze* concepts in the hard sciences of physics and chemistry. At the same time let us understand that the imperative to find *meaning* in the information to be presented creates a conceptual paradox that is *meaningless*. So for the time being, let us ignore this paradox so that we can tease out the concepts embedded in the facts to be presented.

Read "A Search for Meaning," *Nature*, May 4, 2006, by Daniel Nettle, who reviews a book *The Happiness Hypothesis: Finding Modern Truth in Ancient Wisdom* by Jonathan Haidt, Basic Books, Britain, 2005. I quote from Nettle:

> What is the big idea of this book? In a sense there isn't one. Haidt quotes from the Monty Python Film *The Meaning of Life*, in which the answer is given as: "Try to be nice to people, avoid eating fat, read a good book now and then, get some walking in, and try to live in peace and harmony with people of all creeds and nations." This spoof formulation nonetheless contains an important point: there is, in fact, no single meaning to life. However, life has a definite form, a set of recurring emotions, interactions and experiences, and even if there is no ultimate solution to the dilemmas they pose, there are ways of understanding them better and navigating them with greater wisdom and purpose. And, as Haidt has shown, the ancient sages and modern psychologists often agree on these.

The conceptual paradox between quantum theory and gravity is conceptually logical in that the two *absolute* concepts of fermion mass—electron **oppositeness** and quarks **sameness**—*interact* thereby creating more complex *relative* mass. Relative/absolute then *interacts* as change/constant and **opposite/same** respectively. By *doubling* the interaction of **opposite/same** and placing *it in simultaneous reversal* we form the more *complex* and enhancing force of electromagnetism.

To transform *absolute* mass into *relative* mass, physicists *double* the electron's wave function. And to create the highly accurate theory of electrodynamics, physicists simply switch the *negative infinites* of the quantum world into the *positive* finites of the *real* world. This "quirky" mathematical stunt called, "renormalization", transforms *inconclusive concepts* of mass into *conclusive objects of* mass. This enables our *macro*-mass minds to interact the concepts of finite/infinite, conclusive/inconclusive, negative/positive, change/constant, hand/handless, as **specific/general**, dominant/dominantless respectively.

Despite Einstein's theory of *relativity* where he *interacted* geometric concepts with his *mind* experiments, he was so intent on using the paradigm of inference that he could not connect the *absolute* parts of *quantum*-mass with his own *relative* interpretations of *macro*-mass. Einstein's failure to *interact* relative/absolute contradicted his own idea of *relativity*. He could not *unite* the *division* between these two reciprocally **opposing** pillars of the **same** more complex idea of *mass* despite his "equivalence principal" that the **same** interaction could be *validly* interpreted in **either** of two **opposing** ways. Einstein's failure to use his equivalence principle for all pairs of *concepts*, including relative/absolute, is a classic example of how excluding **either** concept from its *interaction* always leads to self-contradiction. Read "One Hundred Years of Uncertainty," by Brian Greene, the *New York Times*, April 8, 2005.

With no *continuity* in the quantum realm, there is no **from** *cause* **to** *effect*, no **from** *data* **to** *conclusion*. Without any *constant change* in *time*, quantum "events" are *discontinuous* and *random*. Our *relative* minds can never *determine* which quantum concept will pop up at any given *time* because time is excluded from this realm. Baryons, *simultaneous* contradictory concepts, occur in pairs when two particles *collide*; but they quickly return these parts to their ground states, or lowest energy level.

Without *time* and without any *sequence* of events, there is *no determinism* in the *micro*-realm. Read "The Message of the Quantum," by Anton, Zeilinger, *Nature*, December 8, 2005, from which I quote:

> The discovery that individual events are irreducibly random is probably one of the most significant findings of the twentieth century. Before this, one could find comfort in the assumption that random events only seem random because our ignorance. For example, although the Brownian motion of a [*macro*-] particle appears random, it can still be causally described if we knew enough about the motions of the [*macro*-] particles surrounding it . . . . But for the individual events in quantum physics, not only do we not know the cause, there is no cause. The instant of a radioactive atom decays, or the paths taken by a photon behind a half-slivered beam-splitter are objectively random. There is nothing in the universe that determines the way an individual event will happen . . . . Most striking is the case of entanglement which Einstein called "spooky," as it implies that the act of measuring a particle [photon] can instantly [simultaneously] change the state of the anther particle [photon] no matter how far apart the two are.

The latter phenomenon is due to the *correlation* of the photon's *absolute* **both-and** function, while the former *randomness* is due to electron's *absolute* **either-or** function.

The more our mind's *relative* concepts try to find *logical* order with the *absolute* concepts of the quantum world, the more *illogical* they become. All this *illogical* behavior however is perfectly *logical* from the view of the *micro*-realm's *absolute* concepts but *illogical* from our view of *macro-relative* concepts.

Now let us see more scientific evidence that the *part*icles of mass are *absolute* concepts.

The German physicist Werner Heisenberg gave a *complete* description of the electron with his equation. About the same time, the Austrian physicist Erwin Schrödinger also gave a *complete* description of the electron with his equation. **Either** *complete* proof of the electron *absolutely* excluded the possibility of the other proof. The human mind is left with two *contradictory* proofs of the quantum world. In our *real* world of *macro*-mass this a *contradiction*, *nothing* that can be perceived or conceived, and therefore dismissed as *unrealistic*.

The *real* world of geometric concepts interact particle/wave, or mass/energy, as **specific/general** respectively. Einstein's formula, $E = mc^2$, elegantly shows that that mass/energy as **specific/general** respectively *interact* and that the formula itself *interacts* the *variable proportion* between mass and energy with the *constant* speed of light.

To show just how *unrealistic* quanta are to physicists and philosophers, let me quote part of a paper in *Science*, June 25, 2004, "Elite Retreat Takes the Measure of a Weirdly Ordinary World," by Adrian Cho:

> According to quantum mechanics, a tiny object does not follow a continuous trajectory. Instead, it must be described by oscillating quantum waves that reveal only the probability of finding it here or there. An object such as an electron may also be in two different places at once or spin in **opposite** directions at the **same** time, a condition known as superposition. But when one measures an electron it always spins either one way or the other. Even stranger, two separated electrons can be "entangled" so that both spin both ways at once, but in opposite directions from each other. If a measurement shows that one electron is spinning clockwise, the observer instantly knows that the other is spinning counterclockwise, even if it is worlds away.

Only when our *minds* attempt to *determine* something about the electron's *concept*, such as *measuring* it, does the electron concept reveal its *absolute* and *unrealistic* either-or function.

To reach the complex level of *object*-mass on the *vertical*, the concept of **sameness** must become a highly complex *collection* of **sameness** of *constants* each the **same** idea as all the others. This constant of constants, called alpha, is more complex than the sum of its parts. Alpha's complex **sameness** of *constants* enables the electron to interact with the increased complexity and enhancement in **opposition** by *alternating* its *negative* (-) and *positive* (+) charges in *time*. We will discuss the *creative* **sameness** of the constant alpha in more detail later.

The basic units of our *macro*-mass are atoms that are not only **specific** but each is unique. Each atom, conserves its uniqueness at any level of molecular complexity. Atoms are **similar-different** in logical patterns as explained below.

Each increase in the complexity of atoms, seen in the periodic table of the elements, is based on the weight of its protons that are counted in whole natural numbers. Hydrogen, the lightest element, has one proton; and each successive increase in atomic complexity is by unit 1 sequence. The proton, derived from quarks **sameness**, is a *constant* that conserves the sequence of whole, natural numbers, the intervals of which are also the basic unit of *time* that **opposes sameness**.

The electron also gets into the act of atomic complexity by **opposing the sameness** of the proton in the periodic table of the elements. Atoms increase in diversity and complexity by the number of electron pairs that "orbit" the nucleus. Each of these orbital rings forms the geometric concept of **sameness**, a *constant* platform upon which only two electrons spin in **opposite** directions. This **opposite/same** interaction is *conserved* in each successive increase in complexity. Atoms have **either** a *negative* **or** a *positive electron charge*, except that some atoms are *chargeless*. The **opposing** charges *attract* each other and *interact* **either** *weakly* **or** *strongly*. All of these *inclusive* interactions are derived from the Pauli *exclusion* principle. A few atoms are *chargeless* and thus inert. They are also *changeless*, and *constant* by remaining the **same**. These inert atoms that remain the **same** mark the end of each parallel line of **different** atoms in the periodic table. A *finite* number of atoms (92) combine to create an *infinite* number of *simple* and *complex* molecules. Read, *The Periodic Table: Its Story and Its Significance*, by Eric R. Scerri, Oxford University Press, 2007.

From the mathematical proof of electrodynamics to the more complex unit 1 whole natural numbers that **specify** successive increases in the complexity of atoms, we get the answer to the question raised by Albert Einstein, "How can it be that mathematics, being after all a product of human thought independent of existence, is so admirably adapted to the objects of reality?" Sir James Jeans said, "God is a mathematician." Even as far back as ancient Greece, the motto of the Pythagoreans was, "Number rules the universe." The quotation "All history involves the history of ideas, and this is especially true of the history of mathematics"—taken from *Numbers: Their History and Meaning* by Graham Flegg—says it all.

Also read *God Created the Integers: The Mathematical Breakthroughs that Changed History*, edited by Stephen Hawking, Running Press, Philadelphia, London, 2005. And read *How Mathematicians Think: Using Ambiguity, Contradiction, and Paradox to Create Mathematics* by William Byers, Princeton University Press, Princeton, New Jersey, 2007.

Could it be that the alternating *odd* and *even* numbers in the unit 1 sequence of whole natural numbers reflect the **either/both** function that is transformed into alternating *odd* and *even* numbers, like alternating (-) (+) electric charges in the electromagnetic force.

As with concepts, *numbers* to can become reciprocally ordered.. For example, *numbers* are not only *sequential* but also reversible. When the **same** sequence of numbes run **both** *forward* **and** *backward* simultaneously constrained within a square, the numbers become magically the **same** in **opposite** directions as seen in Figures 109, 110, and 111. This reinforces the **sameness** of *number, geometry, and concept* in nature interactions. Even prime numbers can form the geometry of squares, as seen in the Gaussian square of Figure 109.

Now let us circle back to the *timeless* pattern of the periodic table of the elements seen in Figure 152. Atoms with **similar** characteristics run *vertically*, while atoms with **different** characteristics run *horizontally*, showing that their **specificities** have *relative* characteristics. The periodic table is also divided into **opposing** parts, *active* on the extreme *left* and *inactive*, or *inert*, on the extreme *right*. There is also another set of **opposites** running *diagonally* between the upper right *nonmetals* and the lower left *metals*. There is also a **general** *division* among all atoms between *acids* that are *charged* and *bases* that are *chargeless*. These sets of **opposing** *concepts* bring logical order to the elements, and they *simplify* how the more *complex* molecules of mass interact. It serves the science of chemistry as a *qualitative synthesis* that enhances dominance of *quantitative analysis*.

However, there are two other atomic patterns of the elements that are more *qualitative* and *circular* instead of the above *quantitative* and *linear* interpretation. These two forms are more elegant but less efficient than the classical periodic table of the elements. One of these *qualitative circles* of periodic atomic relationships is by Dr. J. Franklin Hyde, seen in Figure 153-b. The other is by Philip Stewart—shown in FIGURE 349, taken from *Discover*, June 2005, by Susan Kruglinski—that divides the **general** circle into **opposing** parts of **similar-different** elements. One-*half* contains *donors*, and the other *half* contains *receptors*. One-*half* curves *inward*, and the other *half* curves *outward* with the most extreme **opponents** the most distant from the **same** dividing line the **same** as two complementary colors. Kruglinski's chart also looks like it has taken the transformations seen in Figure N of Addendum V where *half* of the mirrored *circle* makes the symmetry operations of translation and rotation to form two inside out circular halves that structure the diagonal of **sameness**. See also Figure 104 as well as Figure 175 where *diagonal*, *diameter*, and *hypotenuse* have all become more complexly the **same** geometric idea.

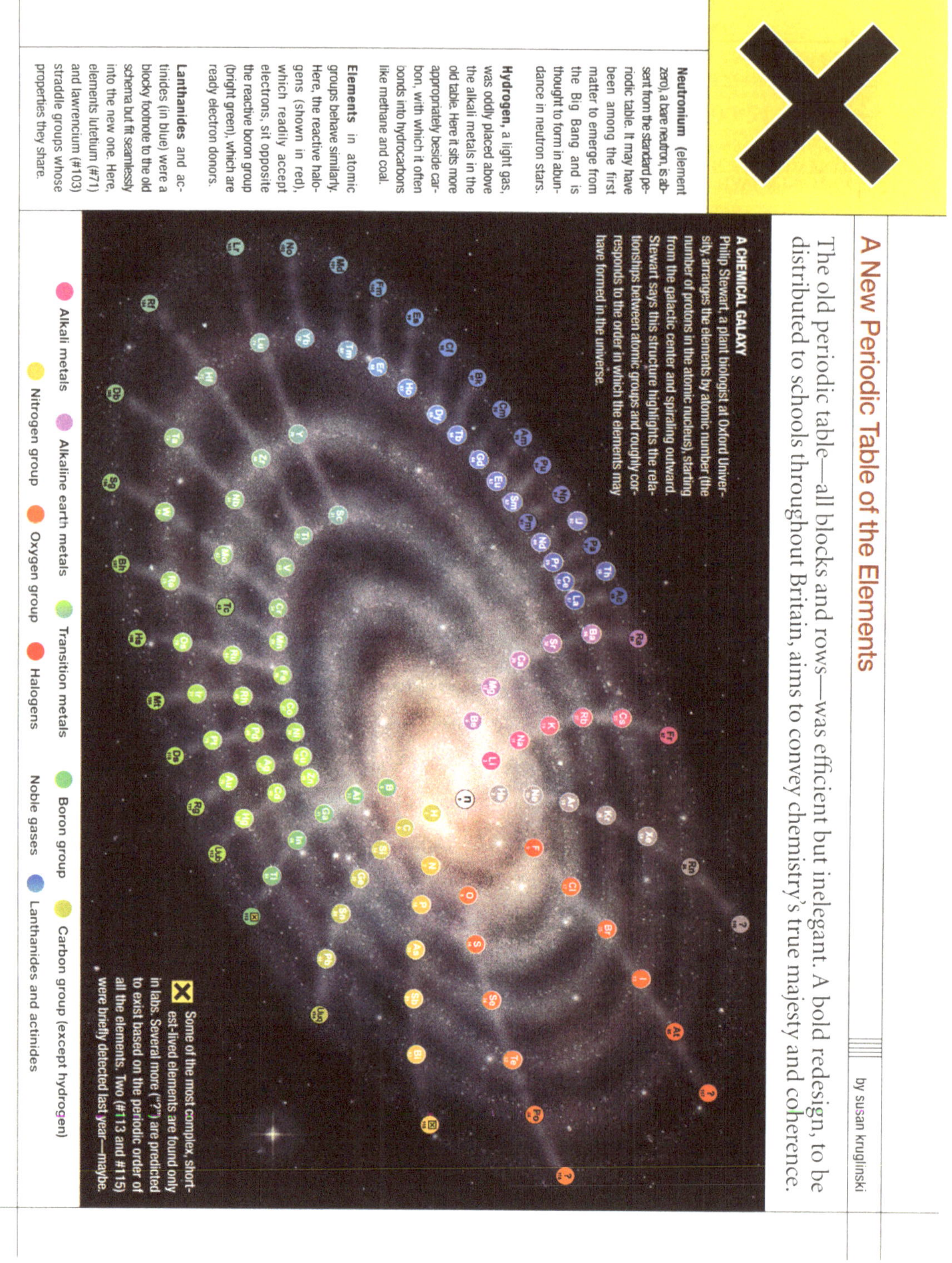

A New Periodic Table of the Elements

by susan kruglinski

The old periodic table—all blocks and rows—was efficient but inelegant. A bold redesign, to be distributed to schools throughout Britain, aims to convey chemistry's true majesty and coherence.

Neutronium (element zero), a bare neutron, is absent from the standard periodic table. It may have been among the first matter to emerge from the Big Bang and is thought to form in abundance in neutron stars.

Hydrogen, a light gas, was oddly placed above the alkali metals in the old table. Here it sits more appropriately beside carbon, with which it often bonds into hydrocarbons like methane and coal.

Elements in atomic groups behave similarly. Here, the reactive halogens (shown in red), which readily accept electrons, sit opposite the reactive boron group (bright green), which are ready electron donors.

Lanthanides and actinides (in blue) were a blocky footnote to the old schema but fit seamlessly into the new one. Here, elements lutetium (#71) and lawrencium (#103) straddle groups whose properties they share.

A CHEMICAL GALAXY
Philip Stewart, a plant biologist at Oxford University, arranges the elements by atomic number (the number of protons in the atomic nucleus), starting from the galactic center and spiraling outward. Stewart says this structure highlights the relationships between atomic groups and roughly corresponds to the order in which the elements may have formed in the universe.

Some of the most complex, shortest-lived elements are found only in labs. Several more ("?") are predicted to exist based on the periodic order of all the elements. Two (#113 and #115) were briefly detected last year—maybe.

FIGURE 349

All chemistry is divided into the **opposing** concepts of **either** *acid* **or** *base*, **either** *negative* ion **or** *positive* anion. The *dominance* of **similarities-differences** of **specific** compounds and molecules in chemistry hides the *conceptual interactions* of **opposite/same**. Again read, "The Past and Future of the Periodic Table," by Eric R. Scerri, *American Scientist*, January–February 2008.

To better understand this *division* between physics and chemistry, we find in chemical *processes* that regardless complexity of the components and regardless of their **similarities-differences**, the end product will be a **specific** *conclusion* that is **similar** or **different** than any one of its components. There is no paradox that would muddy or invalidate this *deductive* process, nor is there any **opposition** to the conclusion.

Although chemistry is more complex than physics, there is a *unity* in the *division* between them. To demonstrate the *vertical* **sameness** between *physics* and *chemistry*, consider the following two formulae for their respective forces:

For *physics*, we have the formula:

$$F_{(gravity)} = \underline{K\ m\ m1}\ r^2$$

where any two closely interacting weightless masses have reciprocally the **same** gravitational force.

For *chemistry*, we have two forces that are the **same** as the formula above for gravity except for their notations:

Chemical	Electro-charge
$F_{(coulomb)} = \underline{K\ i\ i1}$	$F_{(coulomb)} = \underline{K\ q\ q1}\ r^2\ r^2$

The fundamental formula on the *left*, called the "dielectric constant," is **specific** to a liquid medium with the inductive capacity being the *constant* K. Solvents with a higher dielectric constant encourage the ionization of the solutes which reduces the coulomb force between the charges, allowing the **opposing** ions to separate more easily toward the poles. The physical formula on the *right* also has a *constant* medium that instead of being a liquid is *air*.

The physical constant *unifies* the changing progression while the chemical constant *divides* the changing progression. However, **both** the coulomb force **and** the gravitational force *decrease* with distance.

From the point of view of the human *mind*, physics and chemistry are **different**; but for *mindless* nature, **both** have the **same** mathematical form, strong evidence that the *mind's* geometric concepts are *parts* of *mindless* nature's logic of interactions.

The *conceptual* functions of chemistry also look the **same** as those of physics:

BOND		IONS	
(+) --------- (—)		(+)	(—)
both --------- and		either	or
acid (pH)	base	charge	charge

The higher the pH, the lower the acid; and the lower the pH, the higher the base. Although each species of life has its own constrained ratio within the extremes of acidity and base, its health and indeed its life depends on striking a *balance* between its limited extremes. Substitute **sameness** for *balance* between the **opposing** concepts of acid and base and it is inevitable that healthier and more robust the **specific** species of life is, the closer its environment meets its criteria for **sameness**. As a former citrus grower, I had to test my soils each year for the pH required for my orange crops. If the ratio were *imbalanced* **either** way, the trees roots could not take up enough nutrients from the soils to feed them properly. I would have to *rebalance* the soils, or my orange trees would not produce enough boxes of fruit or enough solids per pound of juice for concentrate to be profitable. Again we see, even in this complex biological environment the necessity for **sameness** to enhance the **opposing** processing of nutritional information *from* the soil and its microbes up *to* the trees for efficient development of fruit.

As for *physical sequence*, when Caesar crossed the Rubicon, he did so conclusively; yet he could have re-crossed the river as many times as he desired just as conclusively without canceling out, or uncrossing, the previous event. Each crossing would be **similar** or **different**, but not alternately **same**

Chemistry is dominated by deduction in inference, the processing of data **from** the **general to** a **specific** conclusion. Regardless of the **similar/different** chemicals and their complexity, they all conclude a *unique* chemical with no seeming relationship to those from which it is made. There is no **sameness** in the processing of information, nor are there any *concepts*. All are *data* that are processed and whatever their conclusion, all the chemicals are **similar/different**. However, any process can be **reversed,** the **same** as the reciprocally **opposing** processes of inference.

The *concepts* of chemistry are also **reversible**. For example, *nontoxic* sodium + *toxic* chlorine creates common table salt, NaCl, that seems to have no relationship to either component. Nevertheless, table salt can be **reversed** back to its *separate* components of Na and Cl.

All chemical *interactions* require a *negative* electric charge and a *positive* electric charge, two **opposing** aspects of the **same** idea of *charge*. Also, all chemistry can be *interpreted* as processes that are **either** *endoergic* (heat absorbing) or *exoergic* (heat *releasing*). *Endoergic* processes are deductive, *contracting* heat to a **specific** conclusion that is *active*, such as **either** Na+ **or** Cl-. *Exoergic* processes are inductive, *expanding* heat as a **general** conclusion that is *stable*, such as table salt (**both** Na **and** Cl). See Figure 349.

These examples show that it is difficult to distinguish between *data* processing (*sequence*) and *conceptual* interactions (*simultaneity*) because they are **opposing** interpretations of the **same** idea—chemistry. Chemistry is more *complex* and seemingly messy than physics. Thus the value of the periodic table of the elements to simplify and synthesize *analysis*.

The *simplest* chemistry is *inorganic*. *Organic* chemistry is more *complex*, and the *biological* chemistry of life is even more complex. The *extravagant complexity* of *life's* chemistry interacts with *parsimonious* and *simple physics* of *lifeless* mass. The *concepts* of *physics*, *simple* and *parsimonious*, and the *concepts* of *chemistry*, *complex* and *extravagant*, show the *vertical* rise of complex systems.

The paradigm of inference *hides* the complex **sameness** of nature's **opposing** chemical interactions. For example, it is difficult to pin down the idea of *homeostasis*, the overall *balance* of an organism's interactions. In physics, the **sameness** of the vacuum in space has recently been found to be surprisingly profound. **Sameness** on the *vertical*, seen in the *conserved* set of Hox genes, is a profound concept to evolution. We see **sameness** even in most minute *interactions* of life's chemistry, such as in the junction between neurons called "synapses" that expedites and enhances the *sequence* of **similar-different** data throughout the neurological system. Read carefully "Matching at the Synapse," by Scott M. Thompson, *Science*, May 6, 2005, from which FIGURE 350 has been taken, that the conceptual **sameness** of the synapse has reciprocally **opposing** concepts that are proportionally constrained for *dynamic* fine-tuning the neurons' electrical pulses for distribution to their proper destinations. More on synapses later.

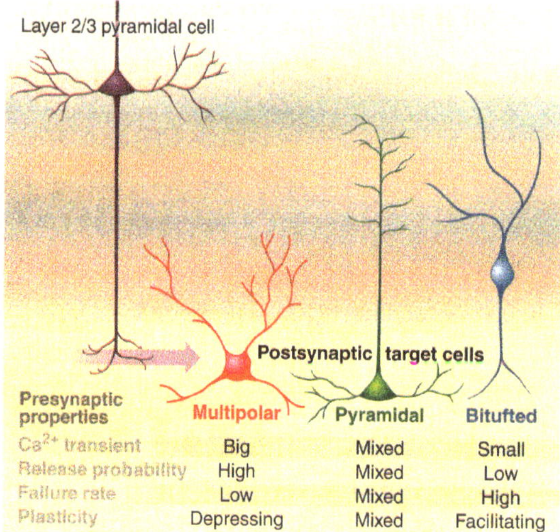

Making connections. Dependence of presynaptic terminal properties on the type of postsynaptic target cell. Presynaptic boutons formed by the axons of layer 2/3 pyramidal cells of the rat somatosensory cortex form connections with three different classes of postsynaptic target cell (*2*). The three postsynaptic cell types include two classes of inhibitory interneurons, multipolar and bitufted, and pyramidal cells.

FIGURE 350

In FIGURE 350, notice how the **opposing** perceptions all interact *simultaneously* the **same**.

Also read "Striking the Right Balance," by Heather Wood, *Nature Reviews Neuroscience*, March 2005.

FIGURE #351

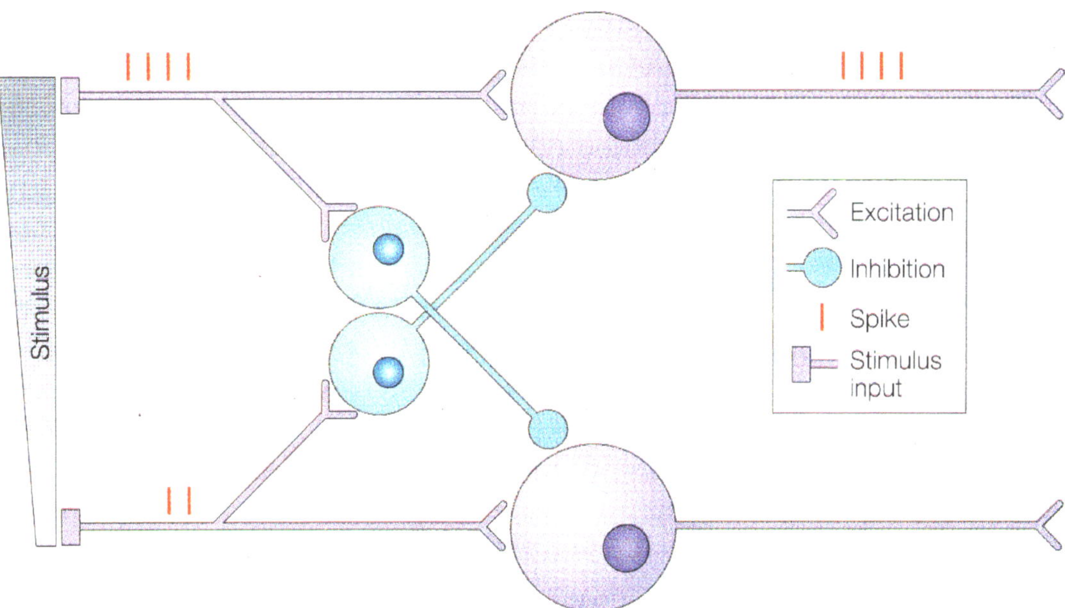

Figure 3 | **Neuronal connections that are mutually inhibitory can accentuate the differences in time and intensity.** Neuronal pathways (top) that are closer to a stimulus might fire earlier and produce more spikes than other pathways (bottom). Neuronal connections that are mutually inhibitory can enhance this difference by blocking transmission in neurons that receive later and less frequent excitation. Recent work has emphasized the importance of the latency to the first spike in rapid behavioural responses.

Box 1 | **Competitive, uncompetitive and noncompetitive antagonists**

An uncompetitive antagonist is defined as a drug whose inhibitory action is contingent upon prior activation of the receptor by an agonist. For example, a fixed, low micromolar dose of memantine blocks the effect of increasing concentrations of N-methyl-D-aspartate (NMDA) to a greater degree than it blocks lower concentrations of NMDA. This is characteristic of an uncompetitive antagonist. By contrast, a fixed dose of a competitive antagonist will manifest less effect as the concentration of agonist increases because the agonist will outcompete the antagonist, rendering it less effective. A third type of antagonist is a noncompetitive antagonist. Here, the inhibitor binds to an allosteric site whose availability is not affected by the concentration of agonist. A noncompetitive antagonist is therefore equally effective at various agonist concentrations.

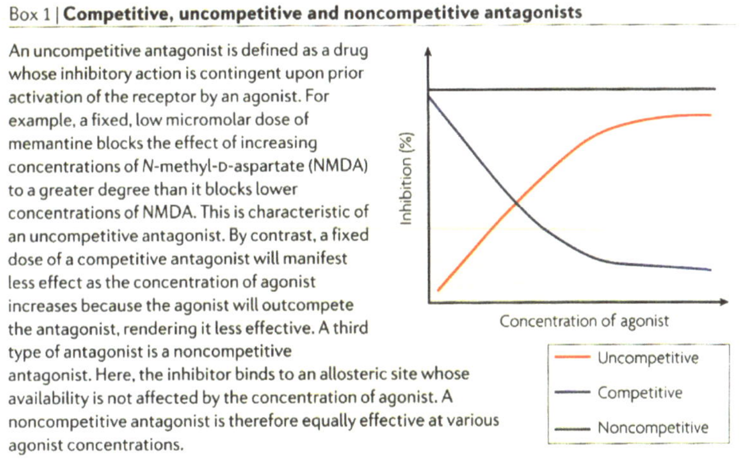

FIGURE #352

An example of <u>sequence/simultaneity</u> interacting <u>parallel/diagonally</u> respectively is seen in FIGURE 351, taken from “Neuronal Variability: Noise or Part of the Signal?” by Richard B. Stein et al., *Nature Review / Neuroscience*, May 2005, that also shows how the **opposing** geometric concepts of **both** *excitation* **and** *inhibition* interact with a **specific** *stimulus*. **Either** *excitation* **or** *inhibition* can exchange dominance to meet the *varying* needs and demands of the **same** more complex system. This work also suggests that <u>signal/noise</u> *interaction* is **<u>specific/general</u>** respectively, the **general** *noise* having been *constrained* to enhance the *dominance* of the **specific** signal. If the *noise* were *unconstrained*, there would be *chaos* instead of *order* because the signal would no longer be *dominant*. Read “Noise in Gene Expression: Origins, Consequences, and Control,” by Jonathan M. Raser and Erin K. O’Shea, *Science*, September 23, 2005, as well as, “No-nuisance Noise,” by Adi R. Bulsara, *Nature*, October 13, 2005, from which I quote, “When there is too much noise, it fails to lift the system over the threshold between the two states: the system cannot switch between them, as there is no flow of information . . . . Conversely, too much noise leads to a surfeit of switching events and a corresponding loss of information. But between these two information minima, the switchings acquire maximum coherence with respect to the applied signal, and an information maximum occurs. Thus we have an intuitive result from an otherwise counter-intuitive behavior—that applying noise can be helpful." This paper clearly shows the necessity for proportional constraint, a **sameness** of concept that enhances complex information processing.

Due to the dominance of the electromagnetic force in our *macro*-realms, on/off switching whether by alternating electric charges or light bulb switches, are ubiquitous even in the most complex systems such as the biological systems of life. If no switching mechanism is found, then something is missing from the system to maintain its logical order. Read, "Where Are the Switches on This Thing?" *Nature*, April, 27 2006, by Kevan A. C. Martin, who reviews a book titled *23 Problems in Systems Neuroscience*, edited by J. Leo, would be the promoter.

It should be understood that in biological chemistry *on* and *off* are mostly associated with molecules that signal other molecules to adjust the balance of an interaction. For example, a signaling molecule can tell one of the four interacting molecules that upsets the proportion to **either** *shut down* **or** *turn on*. When, say, a molecule of an interaction becomes too *strong*, the signaling molecule will instruct its diagonally **opposing** molecule to turn *on to rebalance its half of the interaction*. At the **same** time, the signaling molecule for the other half of the interaction will to the **same** thing in the **opposite** way to complete the *balance* of the interaction. Once the interaction is restored to the **same** proportion as before, the two signaling molecules also *balance* their **opposing** concepts so that they interact the **same**.
The switching of **either** on **or** off are the **same** as the concepts of *promoter* and *inhibitor* respectively. These biological interactions are the **same** as but more highly complex than the reciprocally **opposing** **either/or** electric charges that *alternately interact* with the **both/and** **sameness** of the magnetic field. These molecular interactions are more plastic in order to synchronize their proportion with other such interacting proportions so the whole system of interactions becomes harmoniously the **same**, in a word, "homeostasis."

A good example of fine-tuning a highly complex biological systems by switches is seen in FIGURE 352, taken from "Pathologically Activated Therapeutics for Neuroprotection," by Stuart Lipton, *Nature Reviews Neuroscience*, October 8, 2007, which shows only the *off* half of the system as "*inhibition*" that can, itself, fine-tune for *balance* (*noncompetitive*) **either** the *competitive* **or** the *uncompetitive part* of the system. What we have then is competitive/non-competitive interacting as **opposite/same**. With all this complexity, it is no wonder that the creative force of **sameness** is hidden.

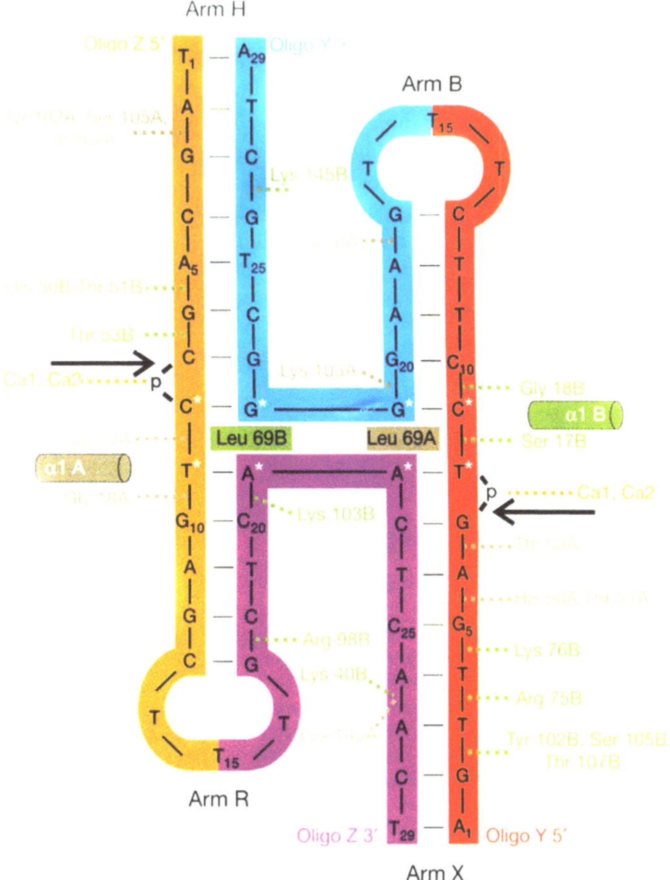

Figure 1 | Schematic representation of the Holliday junction in the crystal structure. The junction comprises four arms labelled B, H, R and X, as shown; arms B and R are 5-bp stem-loops. The constituent oligonucleotides are named Y (with its 5′ end in arm X) and Z (5′ end in arm H) and the sequences are coloured to match Figs 2 and 3. The continuous strands of the DNA junction are depicted in gold and red whereas the exchanging strands are in purple and cyan. The hairpin loops at the ends of arms B and R are not visible in electron density maps. Amino acids interacting with the DNA have been coloured according to their protein chain (protein chain A in fawn and B in pale green) (see also Supplementary Table 2). Each active site of the protein binds two calcium ions (Ca1 and Ca2), which also interact with the DNA backbone. Dipoles of $\alpha 1$-helices (residues 17–29; cylinders) of both protein subunits make additional interactions with the DNA. Scissile phosphates are shown (p), with scissile bonds indicated by arrows, and loss of base-stacking near the core of the junction by asterisks.

FIGURE 353

Let me show you a simpler and more obvious schema of the above found in Figure 154 where the *cold* and *hot* handles of a sink's water faucets substitute for the *inhibitor* and the *promoter*. This schema is largely self-explanatory in that we have all used *hot* and *cold* handles to fine-tune the *balance* of water temperature to suit our particular sensory response. The *top* represents only reciprocally **opposing** interactions of the *inhibitor*, while the *bottom* represents the *interaction* of inhibitor/promoter. To increase the ordered complexity of such a biological system, you would have to *double* the inhibitor/promoter interaction and place it in *simultaneous reversal*, forming the diagonaled square of **sameness**. All this is done with only a *right*-handed person. A *left*-handed person would get the **same** thing from his **opposing** interpretation. Mindless nature would place **both opposing** hands in *simultaneous double reversal* to get a more complex and enhanced **sameness**.

A schema that looks the **same** as a process that has been *doubled* and placed in *simultaneous reversal* is seen in FIGURE 353, taken from "The Structural Basis of Holliday Junction Resolution by T7 Endonuclease I," by Jonathan M. Hadden *et al.*, *Nature*, October 4, 2007.

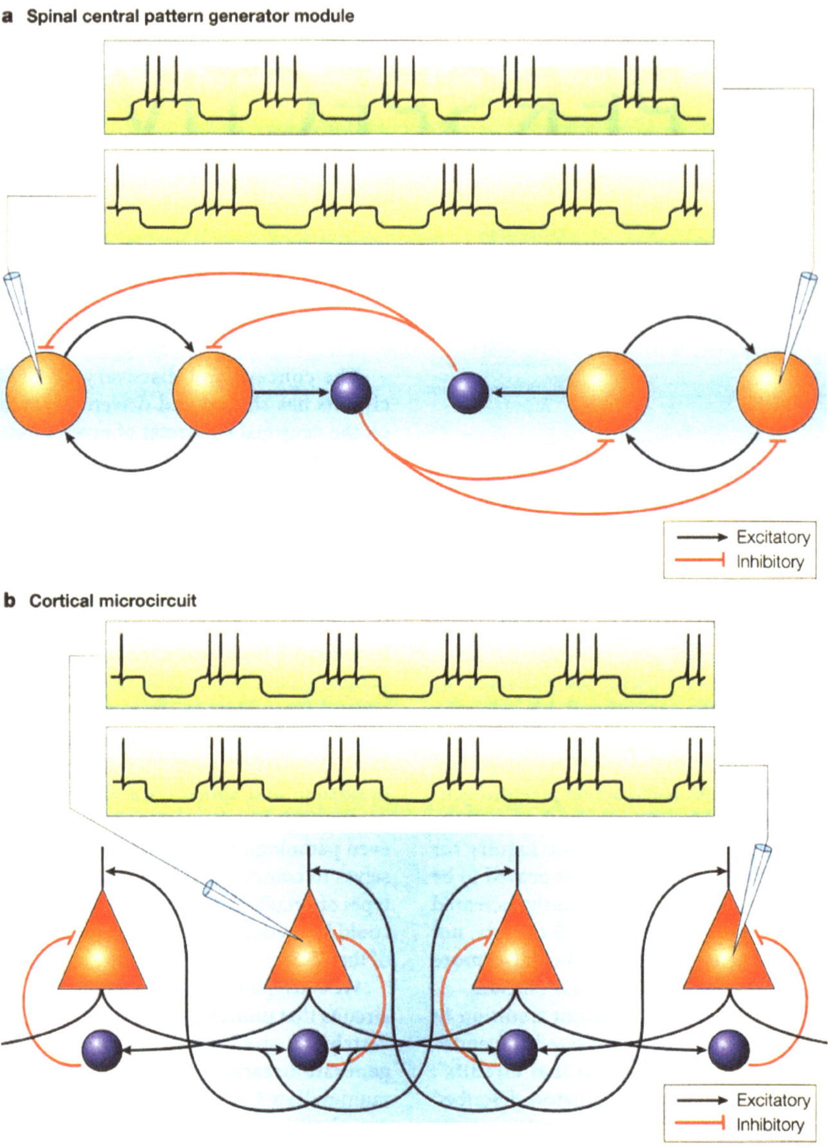

Figure 1 | **Similarities between vertebrate central pattern generators and neocortical circuits.** Central pattern generator (CPG) modules in both the brainstem and the spinal cord (**a**; schematically illustrated here with simplified architecture) contain excitatory interneurons that are functionally embedded within inhibitory networks. This is also true for the cortex (**b**), although a piece of cortex would contain many intermingled CPGs. In these recurrent circuits, a combination of intrinsic and circuit properties mediates neuronal bistability, which can lead to rich intrinsic dynamics, including oscillations (as indicated) and other forms of attractor dynamics.

FIGURE 354

If you refer to "PKA Type IIa Holoenzyme Reveals a Combinatorial Strategy for Isoform Diversity," by Jain Wu et al., *Science*, October 12, 2007, and look at the authors' Figure 3, you will see what appears to be a diagonaled square of **sameness** created with two **opposing** processes having been placed in *simultaneous double reversal*.

The simple **either/or** function in biology, **either** *on* **or** *off* switching, is found in the brain not only when we are *awake* but even in our *sleep*, as seen in "A Putative Flip-flop Switch for the Control of REM Sleep," by Jun Lu et al., *Nature*, June 1, 2006. Also read the importance of reciprocally **opposing** "switches" in "Flick the Switch," under Research Highlights, *Nature*, August 10, 2006, where, "NFAT may act as a molecular switch, allowing an organism both to launch an immune attack on a foreign protein, and to suppress a response to a 'self' protein."

Another switching of **opposites** is found within the **same** immunological system. Under Research Highlights in *Nature*, August 10, 2006, find "A Spoonful of Sugar," from which I quote, "The antibody immunoglobulin G (IgG;) plays an important role in activating inflammation to fight invaders but puzzlingly, it can also soothe inflammatory autoimmune diseases when injected into the bloodstream."

Another example is "The Cortex as a Central Pattern Generator," by Rafael Yuste et al., *Nature Reviews Neuroscience*, June, 2005. FIGURE 354 taken from this work displays how the two **opposing** *excitatory* and *inhibitory* geometric concepts formed in *simultaneous double reversal* alternate in time to maintain a *constant* proportion, within which *changes* in complex information can be deduced in both the spinal and cortical neurons. An insert in this issue by the company Sigma/RBI is a schema that shows a detailed *analysis* of the many *excitatory* and *inhibitory* concepts that hides the *synthesis* of FIGURE 354.

All of us, scientists included, have such a strong bias for substituting **different** for **opposite** that **opposite**, too, has been hidden from its interaction, although not to the extent of **sameness**.. Also, working scientists are so extremely *analytical* that most of what you read in scientific journals offers mere snippets and hints of any logical *synthesis*, and almost nothing of *concepts* and their interactions. The following are a few glimmers of *concepts* and their *interactions* dispersed among the proliferation of *data* that I have lifted from scientific journals and prominent science writers all of whom write under the paradigm of inference dominated by *deduction*:

Nature, June 26, 2003, under This Week, "In the developing embryo, neurons establish thousands of connections in the brain and peripheral nervous system, and with muscle cells throughout the body . . . the same axon guidance molecule can either attract or repel an approaching axon."

Nature Neuroscience, May 2000, "Anterior Cingulate and Prefrontal Cortex: Who's in Control?" by Jonathan D. Cohen et al. Their summary reads, "The results of Gehring and Knight may be explained by the possibility of two pathways for control, both of which respond to ASS conflict detection and are mediated by neuromodulations systems, but one of which is further mediated by PFC and has a more selective influence on task demands, while the other has a more general preparatory function . . . . Whether this simple account of the division of labor between PFC and ACC prevails or, as suggested by Gehring and Knight, more complex interactions occur, the authors are to be commended for their elegant and provocative study of the neural mechanisms underlying monitoring and execution components of cognitive control."

Science, July 11, 2003, "A Ménage è Trois in Two Configurations," J. Evan Sadler tells of two experiments by different authors. Sadler says that without thrombin, we would bleed to death; but with too much thrombin, we would die of thrombosis—blood clots. "To avoid these untoward consequences, thrombin is regulated by a bewildering array of other molecules . . . there are two 'exocites' that are distinct from the catalytic center exocite I and exocite II that are locked on opposite sides of thrombin, and both exocites participate in thrombin's interactions with platelets . . . . It is fascinating that two groups can study the same interaction with very similar regents, only to obtain different crystal forms . . . the astonishing disparity between these two structures is uncertain. In any case, the differences lead the two groups to offer strikingly different interpretations."

Science, May 12, 2000, under This Week in *Science*, "Shearman *et al.* show that the mammalian circadian clock, as was shown recently for the Drosophila clock as well, actually has two more 'gears.' The negative element is CRY, which interferes with the positive action of the heterodimer CLOCK, and BMALI. The authors now show that PER 2 is a positive regulator of BMALI and CRY is a negative regulator of PER 2. The next step is to understand how this intricate oscillator keeps circadian time." Also in the same issue is "Two Feedback Loops Run Mammalian Clock," by Marcia Barinaga, who states:

> The biological clock has two oscillations moving in counterpoint; the level of one set of proteins cresting while the others are low, and vice versa. The biological clock keeps those opposed oscillations in sync . . . the key to this regulation seems to be a pair of proteins that enter the cell nucleus together but apparently split their duties. One, called CRYPTOCHROM (CRY), turns off a set of genes, while the other, PERIOD 2 (PER 2), turns on a key gene. The work explains how genes can be activated by two opposite phases . . . findings build on work on the fruit fly clock, which features a negative feedback loop similar to the one in which CRY participates . . . . But this is only half of the story. A protein called CLK oscillates in counterpoint with PER and TEM; its level rises as theirs falls and vice versa. CLK is a positive regulator that pairs with a protein called CYC to turn the PER and TIM genes on. Indeed, PER and TEM shut their genes off by binding to and inactivating CLK and CYC . . . . Mammalian clocks use many of the same proteins.

Nature Reviews Neuroscience, February 2004, "Homeostatic Plasticity in the Developing Nervous System," by Gina G. Turrigiano and Sacha B. Nelson, I quote from the last sentence of the abstract, "Here we discuss evidence from a number of systems that homeostatic synaptic plasticity is crucial for processes." I now quote from the body of this paper, "It has become apparent that neural activity is itself subject to homeostatic regulation to prevent neural circuits from becoming hyper—or hypoactive. Without stabilizing mechanisms operating at the level of neural circuits, activity-dependent forms of plasticity such as long-term potentiation (LTP) and long-term depression (LTD) could drive neural activity toward run runaway excitation or quiescence. Similarly, without these mechanism operating at the level of single cells, the complex interplay of inward and outward conductances that subserve each neuron's unique pattern of electrical activity would be difficult to maintain in the face of morphological changes and protein turnover." Under the heading of Balancing Excitation and Inhibition, I continue to quote, "For highly recurrent cortical networks, tuning of excitatory synaptic strengths is probably not sufficient to maintain network stability. There are intensive positive feedback connections between excitatory pyramidal neurons both within and between layers, which are kept in check by feedback and feed forward inhibition mediated by complex networks of inhibitory interneurons. Even small changes in balance between excitation and inhibition in such networks can result in runaway excitability, disrupt sensory responses in primary visual cortex, and profoundly alter experience-dependent placidity, indicating that excitation and inhibition must be delicately balanced to keep cortical networks functional." Under Concluding Remarks, the authors state:

> Homeostatic forms of plasticity are ubiquitous in developing nervous systems . . . . Intensive study of these important phenomena has revealed a palette of mechanisms that contribute to the maintenance of excitability. Nervous systems seem remarkably clever at compensating for perturbations in activity of synaptic transmission, and the mechanisms engaged are likely to depend on these perturbations . . . . This flexibility might insure that firing rates are maintained within some functional range, regardless off which parameters of excitability are maladjusted . . . . One pressing issue is whether synaptic scaling is truly 'global' so that changes in activity . . . generates a cell-wide signal that operates on all synapses proportionally. Current experiments cannot distinguish between this and the alternate possibility that each synapse generates a local signal that regulates itself in a homeostatic way.

Nature, March 2, 2000, under New and Views, I quote Thomas Elbert and Andreas Keil, "In the visual system, several types of discrimination can be processes in the same small area of the cortex. A similar phenomenon has been seen in the motor system. Likewise, a single type of stimulation of two digits can produce two opposite use-dependent effects on the spatial relationship or the cortical representations of the digits. In other words, multiple maps, specific to different modes of discrimination or tasks, share the same region of the cortex. So the three-dimensional modular approach provides us with seeming conflicting results."

Science, August 11, 2000, "Long-sought Protein Packages Glutamate," by Laura Helmuth, I quote, "Among neurotransmitters, two stand out as stars, communicating most of the brain's urgent messages. These fast-acting, ubiquitous chemicals—GABA and glutamate—send the basic 'stop' and 'go' signals that most other transmitters merely modulate. Glutamate is called into action whenever rapid-fire excitatory signals are needed." Glutamate is the Go signal, while GABA is the Stop signal.

The protein serotonin has **both** a *promoter* signal **and** an *inhibitory* signal, while dopamine also has **both** a *promoter* signal **and** an inhibitor signal. All we seem to see is a pair of proteins *each* one with **opposing** concepts. These two sets of **opposing** proteins form a system, like that of Figure 154-a, which delicately balances the proportion between serotonin and dopamine. This proportion, in turn, is tilted in such a way that, along with other such proportions, the total system interacts all geometric concepts into the **same** proportion. Then, and only then, can these conflicting concepts *simultaneously* interact to enhance the highly complex *processing* of information within this conceptual constraint. Such various balancing acts, called "plasticity," occur within and between complex systems. The concept that describes such dynamic systems all interacting as the **same** more complex system is called "homeostasis."

Nature Reviews Neuroscience, March 2004, "Anterior Prefrontal Cortex: Insights into Function from Anatomy and Neuroimaging," by Narender Ramnani and Adrian M. Owen. I quote from the abstract, "Key features of existing models within a common theoretical framework. The results indicate a specific role for this region in integrating the outcomes of two or more separate cognitive operations in the pursuit of a higher behavioral goal." Surely, **specific/general** *interact* here.

Science, June 11, 2004, "A Dual Role for Hox Genes in Limb Anterior-posterior Asymmetry," by József Zákány, Marie Kmita, and Denis Duboule. I quote from the abstract, "Two collinear processes relying on opposite genomic topographies, upstream and downstream."

Science News, September 4, 2004, "Cancer Flip-flop: Gene Acts in Both Proliferation and Control of Growth," by Christen Brownlee, I quote, "These genes switch between halting and promoting cancer, depending on the presence or absence of a particular protein. 'What [these genes do] is put the brakes on cancer under one set of circumstances and [step] on the accelerator under another set of circumstances.'"

Nature, August 12, 2004, "The Principle of Temperature-dependent Gating in Cold—and Heat-sensitive TRP Channels," by Thomas Voets et al., from which I quote the abstract, "The mammalian sensory system is capable of discriminating thermal stimuli ranging from noxious cold to noxious heat . . . . Both channels are activated upon depolarization . . . . The chemical agonists menthol (TRPM8) and capsaicin (TRPV1 function as gating modifiers . . . . Our results suggest a simple unifying principle that explains both cold and heat sensitivity in TRP channels." From the body of this work, "recent studies have demonstrated that TRPM4 and TRPM5, two closely homologous monovalent cation channels, are dually gated by Ca2+ and membrane polarization."

Nature Neuroscience, July 2004, "Hebb and anti-Hebb meet in the brainstem," by Sacha B. Nelson. I quote only her abstract in which the accents are mine, "In a brainstem auditory nucleus, excitatory synapses form the **same** presynaptic neuron are shown to cause **opposing** forms of spike timing-dependent plasticity in projection neurons and GABAergic interneurons at *low* frequencies, but cause the **same** form of plasticity at *high* frequencies. The net effect is to *enhance* activation of the projected neurons by *relevant* stimuli."

Nature Neuroscience, July 2004, "A Mechanism for Homeostatic Plasticity," by D. James Surmeier and Robert Foehring. This paper shows that "homeostatic plasticity" is **no different** than the *interaction* of change/constant, or **opposite/same**, and that **either** concept to the exclusion of the other is fatal. I quote the authors:

> There is a lot of talk in neuroscience circles these days about "homeostatic plasticity." The term refers to the ability of neurons to maintain a preferred level of activity (spiking) in the face of sustained alterations in synaptic stimulation. This adaptive process lets neurons respond to rapidly changing signals even in the presents of slow changes in their synaptic drive that would otherwise make neurons spike near maximal or near minimal rates. Without homeostatic plasticity, neurons and networks run the risk of either "exploding" or "crashing." In spite of its importance we know relatively little about the molecular mechanisms engaged in this form of cellular plasticity.

The following example shows how our paradigm of inference creates needs to add reciprocal logic for *interactions*. I quote the abstract from, "Reciprocal Regulation of Haem Biosynthesis and the Circadian Clock in Mammals," by Krista Kaasik and Cheng Chi Lee, *Nature*, July 22, 2004:

> The circadian clock is the central timing system that controls numerous physiological processes. In mammals, one such process is haem biosynthesis, which the clock controls through regulation of the rate-limiting enzyme aminolevulinate synthase 1 (Alas 1). Several members of the core clock mechanism are Pas domaine proteins, one of which, neuronal Pas 2 (NPAS2), has a haem-binding motif. Indeed, haem controls activity of the BMALI-NPAS2 transcription complex *in vitro* by inhibiting DNA binding in response to carbon monoxide. Here we show that haem differentially modulates expression of the mammalian *Period* genes *mPer1* and *mPer2 in vitro* by a mechanism involving NPAS2 and mPER2. Further experiments show that mPER2 positively stimulates the activity of the BMAL1-NPAS2 transcription complex and, in turn, NPAS2 transcriptionally regulates *Alas1*. Vitamin B12 and haem complete for binding to NPAS2 and mPER2, but they have opposite effects on *mPer2* and *mPer1* expression *in vitro*. Our data show that the circadian clock and haem biosynthesis are reciprocally regulated and suggest that porphyrin-containing molecules are potential targets for therapy of circadian disorders.

Science News, September 18, 2004, "Flies 'R' Us," Christen Brownlee discusses a paper that appeared in *Nature*, September 16, 2004. I quote from Brownlee, "The common fruit fly . . . [has] cells that function much as those in the human pancreas do . . . . It contains two cell types that exert opposing effects. Beta cells secrete insulin, a hormone that prompts tissues to take in sugars and other nutrients, thereby lowering blood sugar concentrations. Conversely, alpha cells secrete glucagons, a hormone that triggers tissues to release sugar into the bloodstream. If functioning properly, both cell types sense blood-sugar concentrations, then together keep it within a healthy range."

Nature Neuroscience, September 2004, "From Neurotoxin to Neurotrophin," by Christian Mirescu and Elizabeth Gould. I quote their abstract, "Glucocorticoids are important for neuronal function, but their release in conjunction with a brain injury can intensify neuronal death, causing more harm than good. A new study shows that delivery of genes that have been modified to alter glucocorticoid signaling can block toxic effects of the hormone *in vitro* and *in vivo*, converting what was once a neuron's worst enemy into its best friend." Then, the body of the paper goes on to say, "Nature is conservative: the same signaling molecules exist across phyla and in different systems, where they may even have opposing roles. For example, adrenal steroids have many protective and beneficial actions in the brain—removal of the adrenal glands results in apoptosis [cell death] of dentgatengrus granule cells and decrements in synaptic plasticity and learning. But under conditions of stroke, seizures or brain injury, glucocorticoids are toxic for cell survival, in particular within the hippocampus."

Nature, November 23, 2000, "Local or Global?" by Shelley L. Berger, I quote, "Modification of histone proteins were thought to act only locally to control the expression of single genes. But the finding of such changes across the whole genome brings this view into question . . . . The US congressman 'Tip' O'Neil once famously said: 'All politics is local.' In the same vein, environmental groups have adopted the slogan, 'Think globally, act locally.' In biology, a similar sentiment is thought to be relevant to gene expression." **Neither** the **specific** conclusion of *deduction* **nor** the **general** conclusion of *induction* can adequately explain the "holistic" array of biological interactions.

Science, October 29, 2004, "The Evolutionary Origin of Cooperators and Defectors," by Michael Doebeli, Christoph Hauert, and Timothy Killingback. "The paradox of altruism, of which the tragedy of the commons is a celebrated avatar, is that although populations of altruists outperform populations of nonaltruists, selection will act to eliminate altruism altogether. Here, we have unveiled a different paradox of cooperation, which could be termed, 'tragedy of the commune'; in a cooperative system, in which every individual contributes to a common good and benefits from his own investment, selection does not always generate the evolution of uniform and intermediate investment levels but may instead lead to an asymmetric stable state, in which some individuals make high levels of cooperative investment and others invest little or nothing."

Science, October 29, 2004, "Symmetry Breaking and the Evolution of Development," by A. Richard Palmer, I quote, "Emerging from this diversity are two fundamentally different, yet easily distinguished, types: antisymmetry, in which dextral [right] and sinistral [left] forms are equally common within a species, and directional asymmetry, in which most individuals are asymmetrical in the same direction. These two asymmetry types differ in an important way. In antisymmetric species, direction of asymmetry is almost never inherited, whereas in directionally asymmetric species, it typically is."

Science News, "A.M. and P.M. Clocks: Fruit Fly Brain Has Double Timekeepers," by Susan Milius, I quote, "A fruit fly relies on a different group of cells to tick out the rhythm to perk up in the morning than it does to boost evening activity after daytime doldrums . . . . in the October 14 *Nature* . . . 'this is the first assignment of morning-ness and evening-ness to a specific cell' comments clock researcher William J. Schwartz of the University of Massachusetts Medical School in Worchester . . . . The team also found evidence that the two clocks interact . . . a view of circadian rhythm as a single feedback loop is giving way to a more elaborate vision that includes many interlocking cycles."

Nature Reviews Neuroscience, October 2004, "Interneurons of the Neocortical Inhibitory System," by Henry Markram et al., I quote, "Sensory stimulation results in a coordinated increase in excitatory and inhibitory conductances. Surprisingly, a balance between these two opposing conductances is maintained over a large dynamic range and for many stimuli. This means that regardless of the level of excitation, the inhibitory system can automatically scale its output to provide a matching opposition across a large dynamic range, analogous to a Yin-like inhibition opposing a Yang-like excitation."

Nature Reviews Neuroscience, October 2004, "Parallel Processing in the Mammalian Retina," by Heinz Wässle, from which I quote:

> Circuits that involve complex inhibitory and excitatory interactions represent filters that select "what the eye tells the brain." . . . The axons of OFF and ON cone bipolar cells terminate at different levels (strata) within the IPL: OFF in the outer half, ON in the inner half. However, superimposed on this ON/OF dichotomy, further bipolar cell types have been described and every mammalian retina that has been studied contains at least four types of OFF and four types of ON cone bipolar cell. We are just beginning to understand their functional roles. Double recordings from cone pedicles and ionotropic OFF cone bipolar cells in slices of the ground squirrel retina have shown that one type of OFF cone bipolar cell expresses ionotropic AMPA receptors, resulting in more plastic synaptic transfer, and therefore, in transient response to light. Meanwhile, two other types of OFF cone bipolar cell express kainite receptors, resulting in more tonic synaptic transfer and a sustained light response . . . . Axons that carry more transient OFF light signals terminate in the middles of the IPL; those that carry sustained OFF light signals are found in the more peripheral position. Although it has not been shown for the mammalian retina, there seems to be a comparable stratification of bipolar axons in the ON sublamina. Transient ON responses are transferred to the middle of the IPL; sustained ON-responses are found in the inner IPL. The preponderance of

> transient light responses in the middle of the IPL is further supported by the recent finding of voltage-dependent sodium channels at the axon thermals of these bipolar cells. Such channels would speed up the light response of the bipolar cells . . . each direction-selective ganglions cell receives signals derived from two neighboring image locations, one excitatory and one inhibitory. The inhibitory input is displaced toward the preferred direction of the ganglion cell and is delayed. So, a stimulus moving in the null direction would drive inhibition of the post synaptic sell and shunt the subsequent excitatory input, whereas movement in the preferred direction would result in a delayed and therefore ineffective inhibitory input leading to net excitation of the postsynaptic cell.

Nature Reviews Neuroscience, November 2004, under Research Highlights, "The Two Faces of Dopamine" by Rachel Jones, states:

> In mammals and other animals, dopamine receptors can be split into two groups—D1-like and D2-Like. Writing in **Nature Neuroscience**, Chase *et al.* describe a genetic analysis of dopamine signaling in the worm *Caenorhabiditis elegans* that sheds light on the antagonistic properties of these two types of receptor. There is strong evidence from mammalian studies that signaling through D1—and D2-like dopamine receptors has opposing effects on locomotor behavior, and that the two types of receptor can signal through different neuronal G proteins . . . . Worms in which the dop-3 gene was deleted showed a slowing of locomotion when a worm encounters a bacterial lawn . . . . By contrast, dop-1 or dop-2 mutants showed a normal basal slowing response . . . the antagonistic effects of D1-like and D2-like receptors can occur in the same cell . . . . Dopamine signaling in worms shows striking parallels with the dopamine system in mammals. In both types of animals, D1—and D2-type receptors are expressed at different level of the same neurons, and can have opposing effects on behavior . . . . If these parallels hold true for other details of dopamine signaling, results from studies of *C. elegans* could give important insights into how mammalian dopamine receptors function.

Reciprocal logic suggests that the concept of *signaling* also has its two **opposing** effects the **same** as the concept of *reception*, **both** pairs of **opposing** effects in *simultaneous double reversal* should form a more complex **sameness** of pathway that is plastic, allowing flexibility to orderly *changes* in the neurological system.

Nature Reviews Neuroscience, "Written in the Stars," by Heather Wood, refers to a paper in *Neuron* by Zheng and colleagues regarding electrical signaling in the development of the retinal system. I quote Wood:

> They find that the cells are reciprocally interconnected by GABA (y-aminobutyric acid)-releasing and nicotinic cholinergic synapses. Moreover, individual starburst cells can co-release GABA and acetylcholine in a calcium-dependent manner. Although GABA is best known for its inhibitory actions, it functions as an excitatory neurotransmitter in the developing nervous system. Therefore, in the early starburst network, the cells are wired together by two types of excitatory synapses . . . concomitant with eye opening and the onset of vision the nicotinic synapses are lost, and the GABA-releasing synapses become inhibitory, thereby transforming the hyperexcitatory network into an inhibitory network. The inhibitory activity is evoked by light, indicating that it has an important role in vision . . . . As the visual system matures and begins to process real vision signals, excitation switches to inhibition, generating a mutually inhibitory network that is better adapted for the processing of visual information.

Science, October 14, 2005, under Editor's Choice, "Converting Repulsion to Attraction," Elizabeth M. Adler states, "Inhibition of cAMP signaling converted Ca^{2+}—media attraction to repulsion, whereas pharmacological activation of protein kinase A converted Ca^{2+}—mediated repulsion to attraction . . . . Thus, a Ca^{2+} signal that triggers calcium-induced calcium release from intracellular stores simulates attractive turning, whereas a Ca^{2+} signal without calcium-induced calcium release elicits repulsion."

Nature Neuroscience, November 2005, "*Lbx1* and *Tlx3* Are Opposing Switches in Determining GABAergic Versus Glutamatergic Transmitter Phenotypes," by Leping Cheng et al. I quote from the abstract, "Most neurons in vertebrates make a developing choice between two principle neurotransmitter phenotypes (glutamatergic versus GABAergic)." From the text, "Dorsal horn neurons can be grouped as either excitatory neurons that utilize glutamate as their transmitter or inhibitory interneurons that use GABA and/or glycine. It is generally believed that a balance between excitation and inhibition within this dorsal horn neuron network 'gates' the relaying of the somatic sensory information to the CNS." These interactions occur dominantly at the dorsal, or *rear*, end of the development. The **same** complex interactions should also be found at the *front* end of the development, interacting **both opposing** ends in simultaneous double reversal as the **same** more complex and enhanced phenotype.

Nature Neuroscience, October 2005, "Synchronous Versus Asynchronous Transmitter Release: A Tale of Two Types of Inhibitory Neurons," by Shaul Hestrin and Mario Galarreta.

Nature, October 27, 2005, under Research Highlights, "Two Roles in Tumours," "The transcription factor KLF4 [in the p35 pathway] can suppress tumour growth in some situations, yet act as a tumour promoter in others . . . KLF4 regulates, in opposite directions, two genes central to growth control."

Science, October 21, 2005, "Interlinked Fast and Slow Positive Feedback Loops Drive Reliable Cell Decisions," by Onn Brandman et al.

Nature Reviews Neuroscience, December 2005, "More Than Mere Coincidence," by Rebecca Craven. In her discussion of the concepts of *excitation* and *inhibition* between the two **opposing** halves of a synapse, Craven said, "But why potentiate slow inhibition at the same time as fast excitation . . . ?" Reciprocal logic would suggest that this is but *half* of the **same** more complex interaction that also includes *slow excitation* with *fast inhibition*. This would account for the enhanced logic of complex interactions in the synapse that scientists call *plasticity*.

Science, November 11, 2005, "Separation of Conjoined Hormones Yields Appetite Rivals," by Rubin Noguerias and Matthias Tschöp, a perspective on an article titled, "Obestatin, a Peptide Encoded by the Ghrelin Gene, Opposes Ghrelin's Effect on Food Intake," by Jian V. Zhang et al. I quote from the abstract, "Two peptide hormones with opposing action in weight regulation are derived from the same ghrelin gene." Another read on this article in *Science* is "Appetite-suppressing Hormone Discovered," by Lauran Neergaard of the Associated Press, reported in the *Tampa Tribune*, November 11, 2005, from which I quote, "The same gene sparks production of the two opposing hormones . . . in the current edition of the journal **Science**. 'It is an unexpected but very, very intriguing finding,' said Matthis Tschöp of the University of Cincinnati, who reviewed the work. 'It seems counterintuitive that Mother Nature would press on the brake and one on the pedal at the same time.' . . . Scientists now know that dozens of hormones probably are involved in the balancing act of weight gain and loss."

Nature, September 28, 2006, under In This Issue, "Brassinosteroids and auxin are important regulators of plant growth, and the fact that the two pathways often regulate the same genes suggest that there must be some interaction between them."

Nature, September 28, 2006, "Different Paths to the Same End," subtitled, "Genetic dissection of yeast gene-regulatory pathways shows that the logical output of such a pathway can remain the same even though the molecular mechanisms underling the output have diverged remarkably," by Antonis Rokas.

Nature, September 28, 2006, "Evolution of Alternative Transcriptional Circuits with Identical Logic," by Annie E. Tsong et al.

Nature, September 16, 1999, "Antagonists on the Left Flank," by Tim King and Nigel A. Brown, from which I quote, "Several studies of left-right asymmetry in embryos . . . beautifully illustrate how direct and antagonistic signals can cooperate in concert."

Science, November 3, 2006, "Small RNAs Reveal an Activating Side," by Ken Garber, from which I quote, "The molecular machinery underlying RNAi appears to be involved in RNAa, raising the questions of how the same enzymes can sometimes turn genes off, and sometime on. 'What makes one siRNA [small interfering RNA] a silencer, and what makes the other one an activator?' asks Sontheimer. 'No clue.'"

Nature, September 28, 2006, "Balancing Regeneration and Cancer," by Christian M. Beausejour and Judith Campisi, in which the authors find that a protein "seems to act as a pivot of this delicate equilibrium." This suggests that the **opposing** functions of **either** *on* **or** *off* must be *balanced*, or the **same**, for normal cell duplication.

Nature, September 20, 2007, "Calcium's Double Punch," by Catherine Jessus and Oliver Haccard. I quote the abstract, "Fertilization promotes a calcium surge necessary to ensure the success of embryonic development. It seems that calcium activates apparently opposite molecular signaling pathways to achieve that end."

Science, January 19, 2007, "A Push-me Pull-you Neural Design," by William B. Kristan, the author asks, "How do spinal neurons control pattern motor output? A balance in simultaneous excitation and inhibition of neural circuits produces rhythmic activity that drives motor behavior." The author further states, "The very surprising result that the rhythmic input into motor neurons during scratching behavior in turtles excites and inhibits networks of motor neuron not alternately as expected, but in phase. This synchronization of simultaneous excitation and inhibition—pushing and pulling at the same time—appears to be counterproductive and wasteful. What possible function might it serve?" The answer to this is the interaction of **specific/general** in *simultaneous double reversal* that creates a more complex and enhancing **sameness** of concept shown in the megalogical form seen on the inside front cover.

Nature, January 18, 2007, under Research Highlights, titled "Opposing Forces," an experiment was a "surprise," in that **opposing** proteins had a *balancing* effect that was good for the immune system.

Nature, September 20, 2007, under In This Issue, titled "Balancing the Immune Forces," from which I quote, "Maintaining a balance between beneficial activation of immune responses and inappropriate immune attacks of normal tissues requires multiple checks and balances."

Science, August 3, 2007, "Mast Cells Show Their Might," by Mitch Leslie, whom I quote, "The upside of mast cells is bigger than anyone imagined a decade ago. But so is their downside."

Science, March 30, 2007, "Top-down Versus Bottom-up Control of Attention in the Prefrontal and Posterior Parietal Cortices," by Timothy J. Buschman and Earl K. Miller.

Science, October 6, 2007, online (www.stke.org), "A Unified Model of the Presynaptic and Postsynaptic Changes During LTP at CA1 Synapses," by John Lisman and Sridhar Raghavachari, whom I quote, "A structural model for long term potentiation accounts for seemingly contradictory results."

Science, July 27, 2007, "The Yin-Yang of Sirtuins," by Andrew Dillin and Jeffery W. Kelly, whom I quote, "Two related proteins protect cells form age-associated neurodegenerative disease, but one needs to be activated whereas the other must be inhibited." The title of the Figure depicting this interaction is, "Unity of Opposites."

Science, August 3, 2007, "Proteins That Promote Long Life," by Stuart K. Kim, whom I quote, "Quantitative analysis of proteins in a long-lived worm reveals changes that both increase and shorten life span."

Science, December 21, 2007, "Two Faces of miRNA," by J. Ross Buchan and Roy Parker, they explain a paper in the same issue by Vasudevan et al. I quote from Ross and Parker, "MicroRNAs can enhance or repress messenger RNA translation, depending on whether cells are proliferating or arrested in the cell cycle."

From the point of view of inference, all these examples raise more questions than answer, showing the dire deed for a *logical* method of *synthesis*, not to replace or exclude but to interact with *analysis* to enhance our understanding of complex systems. A *synthesis* (expansion) showing the **sameness** of interactions, however, is **opposed** by *analysis* (contraction), the *dominant* concept and method of working scientists.

THE MAGICAL FORCE OF SAMENESS

Truth is stranger than fiction; fiction has to make sense.
—Leo Rosten

The power of using abstraction is the essence of the intellect, and with every increase in abstraction the intellectual triumphs of science are enhanced.
—Bertrand Russell

It seems paradoxical that the truth predicate, for all its transparency, should prove useful to the point of indispensability. Truth is notoriously enmeshed in paradox, to the point of out-and-out antinomy.
—W.V. Quine

Scientists equate *symmetry* with Einstein's famous "equivalence principal." Theoretical physicists are enamored with the beauty of *symmetry* that in logical terms is **sameness**. Read "The Power of Symmetry," *American Scientist*, November–December 2007, in which David W. Farmer reviews a book titled *Why Beauty Is Truth: A History of Symmetry* by Ian Stewart, Basic Books, 2007. I quote from Former:

> In mathematics, symmetry has been given a more precise meaning. In his new history of mathematical symmetry, "Why Beauty is Truth," Ian Stewart gives this definition: "A symmetry of some mathematical object is a transformation that preserves the object's structure." So, a symmetrical structure looks the same before and after you do something to it . . . . In 1915, Emmy Noether proved that all conservation laws arise from a symmetry of the physical system. For example, conservation of momentum is a consequence of the fact that the laws of physics are the same everywhere. That is, Newton's laws are invariant under translation. Physics students use this principle every time they analyze a collision by switching to the center of the center-of-mass reference frame. Another example is the conservation of energy which is a consequence of the fact that Newton's laws look the same with time running in reverse . . . . Of course, modern physics provides wonderful examples of the power of symmetry. Special relativity is founded on the postulates that the laws of physics are the same in all inertial frames and the speed of light is the same for all observers. Both of those assumptions are symmetries in the mathematical sense of showing "invariance under an operation."

A good example of how the geometric concept of **sameness** can enhance the interaction between two conceptual **opponents** is by using the ancient mechanism of a *balance* or *scale*. This measuring FIGURE 355 is used to determine the **difference** in weights between two objects.

HORIZONTAL BALANCE

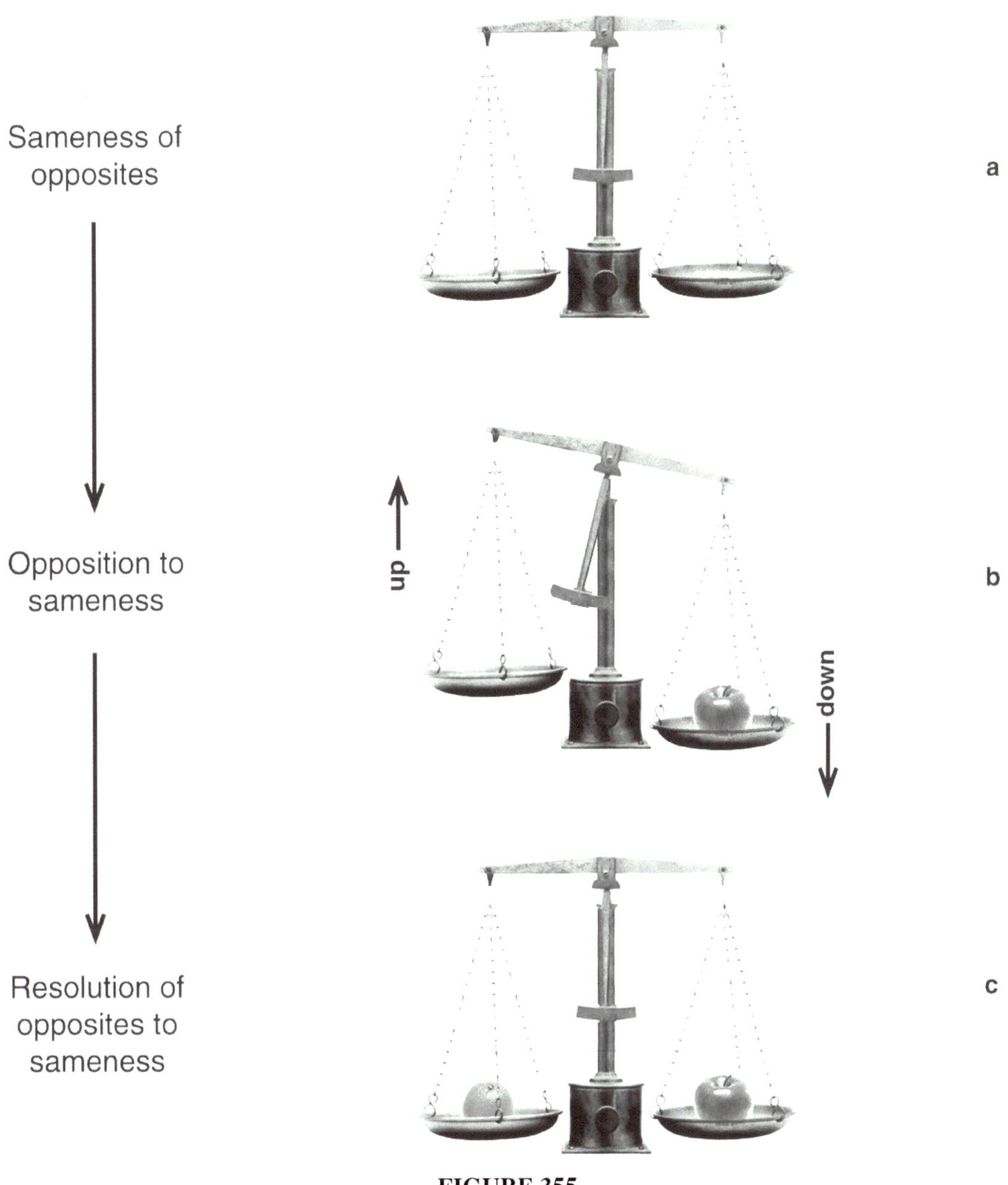

FIGURE 355

The two *balanced* arms with its two *balanced* pans for holding objects run *parallel* to the *horizontal* ground. If the two pans were *imbalanced* in any proportion, such *variations* would **oppose** the *constant balance* of the scale.

When two parties with **different** *weighted* objects want to negotiate an exchange the *interaction* of imbalance/balance, or change/constant, each as **opposite/same** respectively, allows both parties to *determine* the exact **difference** in weight between their respective objects by *rebalancing* the pans. This **sameness** between the two **opposing** pans creates mutual trust between the two parties, thus enabling them to negotiate a fair exchange. See the interaction between the buyer/seller in **http://www.reciprocallogic.com**.

The idea of *balance* in the scale of FIGURE 355 can be applied to the *balance* between *demand* and *supply* of goods or of services. In a free economic society, when these **opposing** concepts are proportioned close to *balance*—relative **sameness**, there is an exponential acceleration and enhancement of the *flow* of currency from hand to hand that powers a robust economic system.

The geometric concept of **sameness** in biology is much too complex to be so easily understood, particularly in the highly complex bodies and brains of life. The introduction to *Neuroscience: Systems-level Brain Development*, by Peter Stern and Pamela J. Hines, and *Language Acquisition and Brain Development*, by Kuniyoshi L. Sakai, in the Special section of *Science*, November 4, 2005, is adequate testimony to the awesome complexity of the brain's *processing* of an infinite *variety* of *data* that would be economically *inefficient* and *chaotic* without the *constant* constraint of **sameness**.

One would not readily think that a playing field needs to be *level*, like the *balance* of a scale. Without this geometric concept of **sameness** with its *constraining* boundaries, the **opposing** contestants could not **either** *win* **or** *lose* conclusively. The game would not be fair, like the **same** coin with its two *equally* **opposing** halves that when tossed will land **either** *heads* **or** *tails* regardless of the number of tosses. With the *constrained* **sameness** of the playing field and the **same** rules of the game enforce throughout the game, the contestants, whether individuals or teams, have the best chance for a fair and orderly outcome. To see the *logic* of two **opposing** teams under the *constrained* **sameness** of the field and of the rules enforce by the referee and judges, turn to the football game in Figure 327. This pattern of offence/defense alternating reciprocally in time is ubiquitous in all games, the **same** as the reciprocally **opposing** hands of **electro/magnetic** force alternating in time. The game of *life* should be played the **same** way for the **same** reasons.

Despite the awesome complexity of nature's complexity and in life's interactions, there are some exposures in the hidden *force* of *constrained* **sameness** as follows:

The **Sameness** (Symmetry) of Channel Pores

Nature Reviews Neuroscience, September 2001, "Shedding Light on the Opening of a Channel Pore," by Galen E. Flynn et al., from which I quote, "Ion channels are among the most efficient and selective biological molecules. To get a feel for their effectiveness, consider this analogy: if the entrance to your house were a channel pore, opening the door would allow the whole population of California to enter in a second. Moreover, the entrance would discriminate between men and women, or between people of different heights, despite the staggering flow. How ion channels achieve these tasks has captured our imagination for a long time, but only recently have we developed a clear picture of channels on an atomic scale."

Science, June 27, 2003, "The Puzzling Portrait of a Pore," by Greg Miller, whom I quote:

> A type of protein that is crucial for the generation of nerve impulses, the blips of electrical activity that underlie all movement, sensation and thought . . . charged molecules cross lipid membranes about as easily as people sprint up Mount Everest . . . the energy amount required makes it virtually impossible. Tiny ion channels . . . do the work of thousands of sherpas, allowing ions to skip through membranes with ease . . . voltage-gated potassium channels . . . are responsible for bringing a nerve impulse to an end so a neuron can prepare to fire again . . . . Without them, the electrical pulses that are the basic units of neural communication would be reduced to a meaningless buzz.

Science News, March 9, 2002, John Travis, "Channel Surfing," states:

> Because of **similarities** in the amino acid sequences of the proteins that comprise channels for positively *charged* ions, scientists suspect that sodium and calcium channels will prove to have **same** basic structure as the KesA channel . . . . Chloride and potassium channels are **similar**. **Both** channels contain helical structures that have a *distributed* electrical charge. *In contrast* to the helices in KesA channels, however, those in the chloride channel are *oriented* so that the positively charged faces of the helices *point inward*. "It's the **same** thing [as in KesA channels], but in **reverse**." . . . The ability of these ion channels to "shuttle up to 100 million potassium ions across a cell membrane in a single second, keeps out **similarly** charged sodium atoms, whose smaller size would seem to make passage even easier."

The emphases are mine. Notice the square of **sameness** through which the larger (**opposing**) potassium ions *sequence* as shown in Figure 279, taken from Travis's work is identical to that of the antinomy in inference.

Pictures of the *constraining symmetry* of some of the other channel pores within which *constraint* the messenger molecules flow with incredible speed and selectivity are seen in Figures 275, 276, 277, and 278, all of which have the geometric concept of **sameness** through which *simultaneity* the **opposing** ions *sequence* at 90°. Notice the "intercellular" symmetry of the *closed* and *open* pores at the bottom half of FIGURE 356 taken from "Towards a Structural View of Gating in Potassium Channels," by Kenton Swartz, *Nature Reviews Neuroscience*, December 2004. The logic of this *interaction, two* **opposing** *aspects of the* **same** *pore,* is the **same** as the logic of the electromagnetic force except that biological *processes* are much more complex.

Don't think of successive increases in the complexity of conceptual **sameness** as being like the **sameness** of a Russian doll, where successive sizes of the **same** shape allows them to nest as if they were the **same** doll. Any such doll is no more complex than any other doll no matter how many levels of *size* are added or subtracted. It is not intuitive to think that *doubling* the **same** interaction *quadruples* its complexity. If this were not the case, the quarks constrained force of **sameness** would not be able to "lift mass up by its own bootstraps" at each successive level of complexity while conserving logical order.

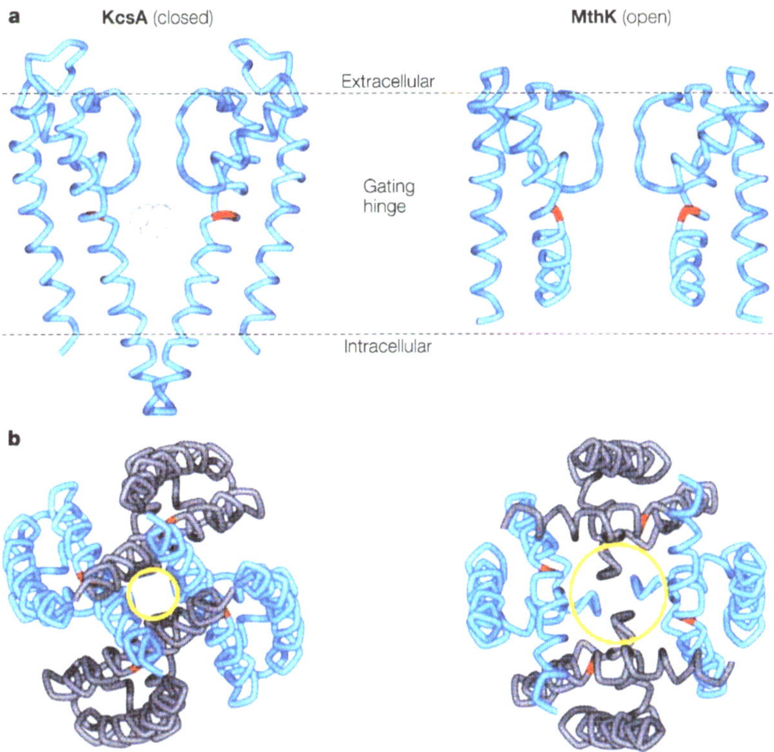

Figure 2 | **X-ray structures of KcsA and MthK potassium channels. a** | Side view of the two channels with the front and back subunits removed for clarity. The red residue in both channels (G99 in KcsA and G83 in MthK) is a glycine residue that has been proposed to serve as a gating hinge. The white molecule in the cavity of KcsA (closed) is a quaternary ammonium blocker. **b** | View of the tetrameric channels from an intracellular vantage point. The difference in aperture of the intracellular entrance to the pore in the two channels is outlined with a yellow ring. Protein database accession codes are 1J95 for KcsA and 1LNQ for MthK. Structures were generated using DS Viewer Pro (Accelrys).

FIGURE 356

FIGURE 357, taken from the cover of *Science*, August 5, 2005, pictures another potassium channel pore. As in FIGURE 356, notice the geometric square of the concept of **sameness** through which the **opposing**, yet logically enhanced, ion traverses at 90°.

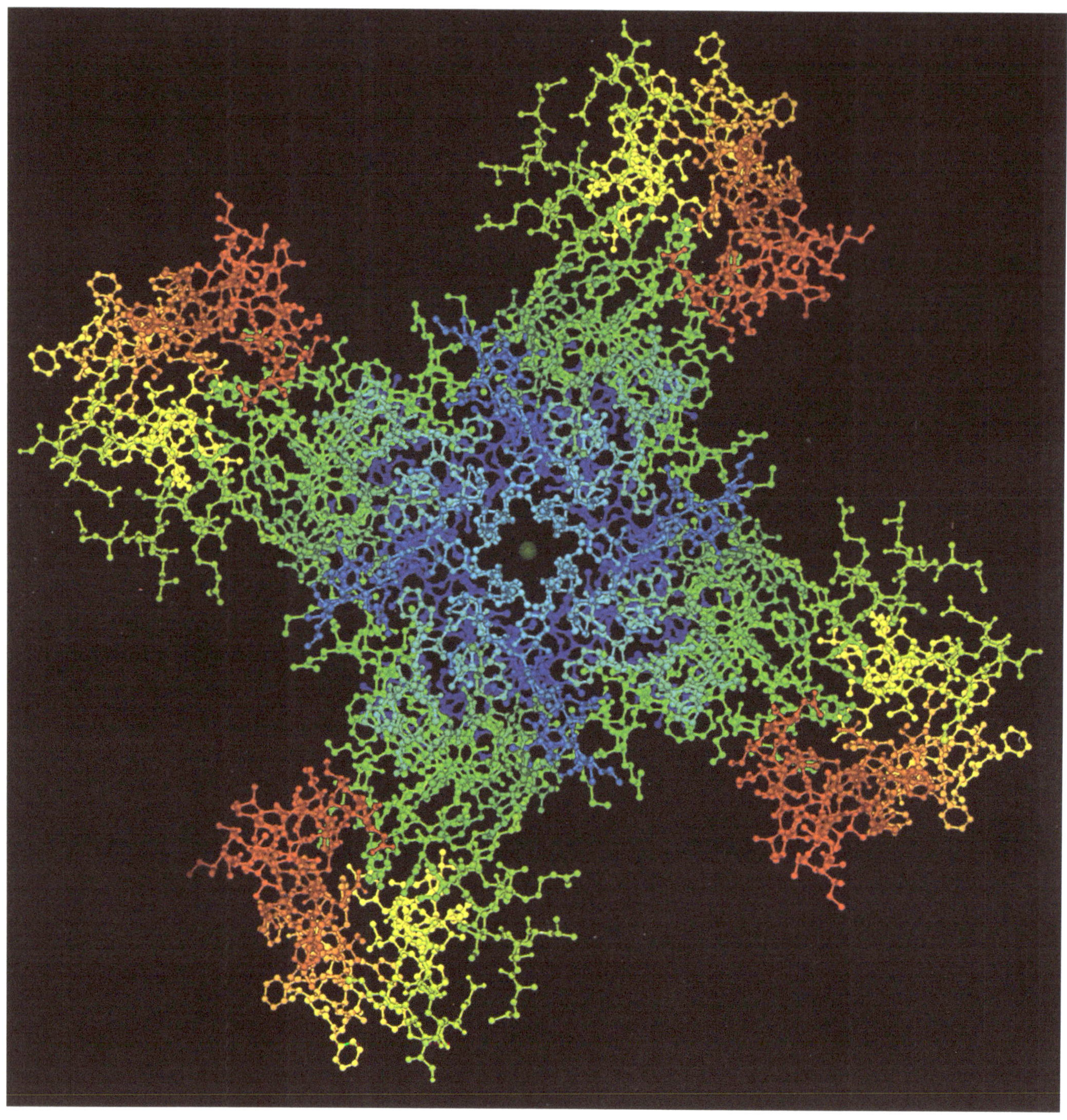

FIGURE 357

FIGURE 357 is taken from the cover of *Nature*, December 1, 2005. In News and Views under the title of "A Greasy Grip," Anthony G. Lee explains an article by Tamir Gonen et al. titled "Lipid-protein Interactions in Double-layered Two-dimensional AQPO crystals."

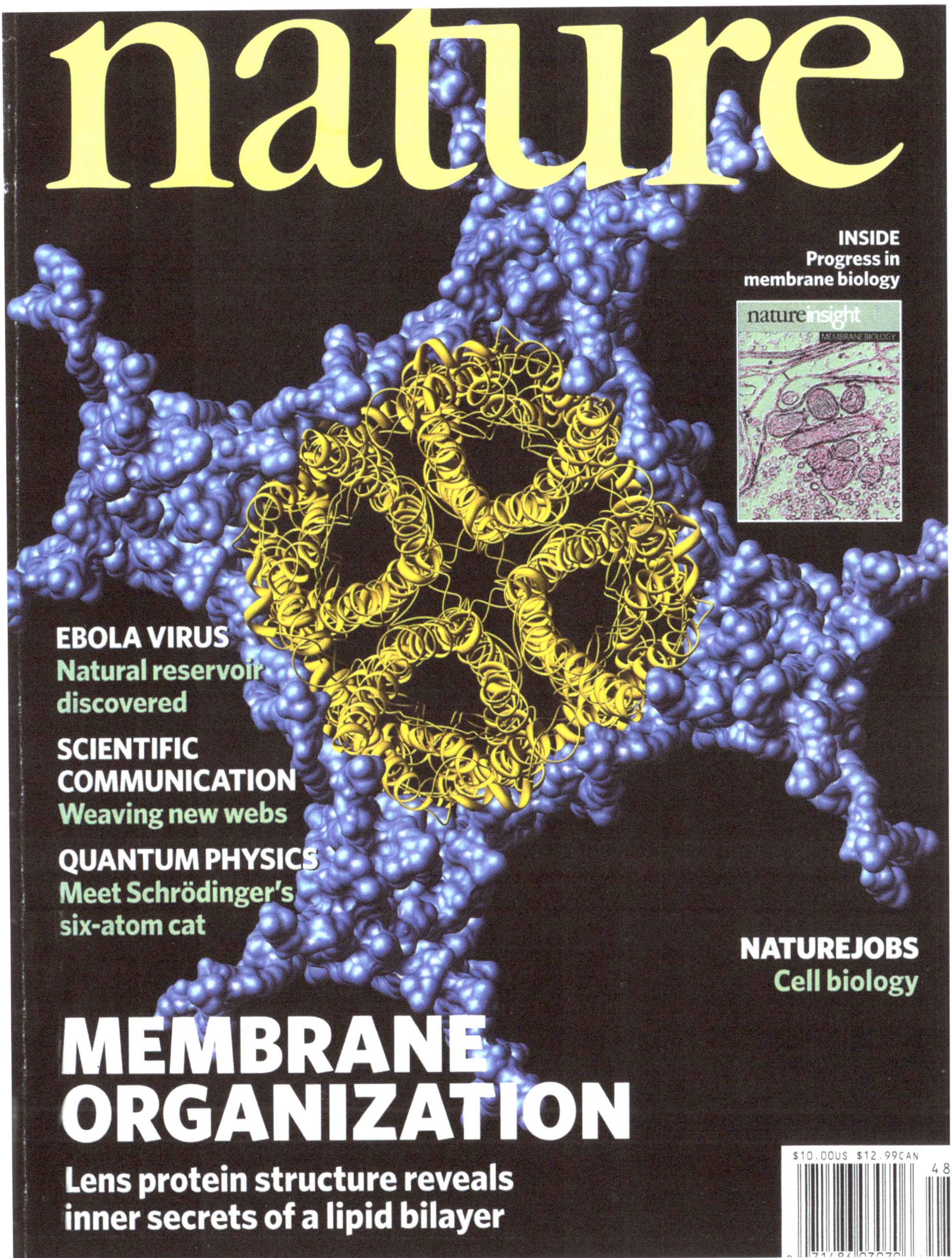

FIGURE 358

What is the **difference** between FIGURES 357 and 358? None. And they are the **same** as the square of **sameness** seen in the megalogic on the inside front cover that enhances the **opposing** logic of inference.

A FIGURE too late to be included in this book shows that pentameric ligand-gated ion channels have the **same** logic of geometric concepts as those above. See these FIGURES in, "X-ray structure of a prokaryotic pentameric ligand-gated ion channel", by Ricarda J. C. Hilf & Raimund Dutzler, *Nature*, 20 March 2008.

The Sameness of Geometry and Dimensionality

Geometry and *concept* interact to enhance the mind's processing of *numbers*; otherwise mathematics could not been discovered and successively increased in ordered complexity. The ancient Pythagorean theorem, based on the 90° triangle seen in Figure 29, is an example of the importance of *geometry* to form *mathematical* equations. When René Descartes later interacted *co-ordinates* and *number* at 90⁰ to create the *concept* of *dimensionality*, mathematics made another exponential leap in ordered complexity. Even Isaac Newton and Albert Einstein needed *dimensionality* to *mathematically* prove their respective *interpretations* of the force of gravity.

"The plane of complex numbers," pictured in Figure 303, is another example of how *number* (**specific**) interacts with the *geometry* of a square of squares (**general**). This **specific/general** *interaction* has given engineers the *mathematical* ability to solve complex problems far more quickly and easily than ever before.

What is so astounding about this Figure's conceptual perpendicularity is that all possible problems can be solved within the plane's constrained **sameness** of concept.

The *simplest* interpretation of whole natural number *interactions* is that the *mind alternately* **reverses** the sequence, but *mindless* nature interacts **both opposing** sequences, (-) = (+), *simultaneously* in a *straight line*. This seemingly 180° contradiction of numbers, **both** of which add the **same** quantity, is *nothing*, zero. This **sameness** (*equality*) between **both opposing** signs has given mathematicians an enhanced ability to solve problems that before this paradox was accepted were too complex to calculate.

The next most highly complex *interaction* between *number*, *concept*, and *geometry* is a straight line bent at 90°. This more complex *geometry* forms the right triangle, a quarter *part* of *mindless* nature's square of **sameness**, that *creates* the *constant* π, at the node or vertex of the 90⁰ angle, as seen in Figure 3. This *constant* is ubiquitous in all the basic *mathematical* formulae of nature, empowering mathematicians and physicists with an infinite *variety* of ways to discover the laws and forces of nature. The other two angles of the triangle **oppose** the *constant* 90° angle by being proportional, or *variable*, but this variable proportion is also *constrained* to 90°. This makes the *two variably* **opposing** angles the **same** as the *single constant* angle, all three of which total 180°, the **same** as the *simple straight line* from which is derived.

The four 90° triangles that form the square of **sameness**, seen on the inside front cover, is *static*. The more complex and *dynamic* **sameness** of the rhombus, seen in Figure 177, *conserves* the four 90° angles and is more *dynamic*, infusing geometric concepts with the ability to logically order highly complex interactions, such as life's protein folding seen in Figure 178.

A *third* coordinate added to the *two* coordinates of the plane of complex numbers, mentioned above, creates three-dimensional *perpendicularity*. This more complex **generalization** of space is *constantly directionless*, enabling physicists to calculate the *directional* motions of any **specific** object.

Each increase in the number of dimensions *quadruples* the complexity of **sameness, no different** than the power law of all nature's forces. This progression of spatial dimensionality cannot go beyond three for the human mind to envision. The infinite progression for gravity begins with 2 x 2, or 2^2—i.e., two multiplied by *itself*, equals four, then four multiplied by *itself* (4^2) equals sixteen, then sixteen multiplied by *itself* (16^2) equals sixty-four, etc. Multiplying the **same** *number* by the **same** *number* gives a product that if *constantly* repeated creates squares of squares that are all *conserved* the **same** but increasingly complex, as seen in Figure 105 top, but always **opposed** by *time*.

A good way to understand that our *perceptual* or *conceptual* point of view is but *part* of nature's squares of **sameness** is to refer to Figure 70. This *perceptual* two-dimensional page has been transformed *conceptually* into a three-dimensional *realistic* interpretation by adding the dimension of *depth*. The ability to transform two dimensions to three dimensions was discovered by artists during the European Enlightenment.

A three-dimensional building can be viewed from **either** 180° **opposing** end for a *valid* interpretation of the **same** more complex building. If one transforms a three-dimensional cube back to a two-dimensional "cube" as seen in Figure 340, *front* and *back* are each absolute to the exclusion of the other, creating a self-contradictory interpretation that is *invalid*. Stare for about half a minute at the vertex of such a cube in Figure 340, called a Neckar cube; and **either** the *front* **or** *rear* interpretation will *arbitrarily alternate instantly*, *unrealistically* contradicting itself in *no time*, the **same** as the isolated electron's *absolute* **opposing** interpretation.

Now visualize *parallel* railroad tracks viewed along its length from your *point* of view that in our *real* three-dimensional world always comes to a **specific** *point* in the *distance*, as seen in FIGURE 359. If you turn 180°, looking along the length of the **same** pair of parallel tracts, you see the **same** parallel tracts converging to a *point* the **same** *distance* as before in **opposite** direction. The brain/mind views the **same** thing from reciprocally **opposing** points of view. So we have two *reciprocally* **opposing** interpretations of the **same** thing. Without *mindless* nature *interacting* **both opposing** views *simultaneously* the **same**, one not choose **either opposing** view *alternately* at any *time*. **Either** 180° interpretation makes **no difference** to nature's **sameness** of **both** views. **Either** view of the parallel tracts is *dominantly deductive*, **from** the **general to** the **specific**. But we must keep in mind that we are also viewing **general** scenes from our **specific** brain-minds, thus conserving *induction's dominantless* part of inference. Dominant/dominantless also interact from *each* 180° complementary view.

Now turn back to Figure 70 again to see how the *dominance* of deduction with its **specific** point of view transforms a two-dimensional building into what seems to be a three-dimensional building, the **same** as with the railroad tracts.

Each of the two **opposing** eyes of humans also competes for *dominance*, rapidly *alternating* in time to *dominate* the **specific** part of the **general** scene. Read "Eye-specific Effects of Binocular Rivalry in the Human Lateral Geniculate Nucleus," by John-Dylan Haynes et al., *Nature*, November 24, 2005. If you place FIGURE 359 in *simultaneous double reversal*, you get *mindless* nature's more complex and enhancing **sameness** of concept that enables the *mind* to reciprocally choose **either** point of view dominant/dominantless.

FIGURE #359

The reciprocal interaction of **specific/general** as dominant/dominantless is seen in "Us and Them (Me and You): The Emerging Science of 'Tribal' Psychology," by David Berreby, *Update*, New York Academy of Science magazine, November–December 2005. Also read "Stuck in the Middle," by Dan Barber, the *New York Times*, November 23, 2005.

In all these cases, it is obvious that we must choose **either** a **specific or** a **general** conclusion of the **same** form of inference if we are to get any *meaning* out of the inescapable conceptual paradoxes of our *relative* world.

Despite our mind's *dominance* of *analysis*, we need the *dominantless* concept of *synthesis*, just as *deduction* needs *induction* in order to be logically realistic.

To again show how *chaotic* and logically confusing and *unrealistic* the *absolute* concepts of the *micro*-realm are to the *minds* of physicists, read "An Asymmetric World," an essay in *Nature*, December 15, 2005, by Oliver Penrose, from which I quote:

> Physicists want to explain the macroscopic behavior of matter in terms of the microscopic mechanical laws obeyed by its component particles. These laws are symmetric under time reversal: for any motion obeying the laws of mechanics, the time-reversed motion also obeys these laws. Newton's laws have this symmetry, as do Maxwell's electromagnetic equations, Schrödinger's equations and Einstein's general theory of relativity. So, naively, one would expect the matter made out of the particles obeying these laws to behave in a time-symmetric way. Yet it does not. Divers are not expelled from swimming pools by converging splashes even thought there is a dynamically possible motion of the molecules in the pool that would produce this effect. Nor do natural processes such as heat flow ever go in reverse. The question is, then, how to explain the irreversibility of the macroscopic behavior of matter, given the reversibility of the microscopic laws obeyed by its constituent particles.

The "answer" is that in the *micro*-realm, the geometric concepts of **opposite/same** do not interact as reversible/irreversible, or commutative/non-commutative respectively.

Returning to the *macro*-realm, any observer can be at **either** 180° point of view *alternately*, but not at **both opposing** *points* of view *simultaneously*, not only because this would *complete* nature's *viewless* and more complexly the **same** but because it would also violate the most basic law of physics that no two *points* of *macro-mass* can occupy the **same** place *simultaneously. Mass* is **opposed** to **sameness**. When two automobiles *collide*, they damage each other, creating *chaos* instead of order. Our *relative* interactions are the antipathy of the *absolute* conceptual *parts* of interactions in the *micro-*realm where the *conceptual* **sameness** of two protons spinning in **opposite** directions pass through each other *without colliding*, as previously seen in FIGURE 345.

Let me give you a couple of very *practical* examples of how **sameness** *enhances* its **opposing** concept. First, look again at Figure 205 where the *motion* of the *right* hand **opposes** the *motionless left* hand and the *motionless left* hand *resists* the *motion* of the *right* hand. One would think that the *resistance* of the *left* hand would not reward the dominant *right* hand for **opposing** it. But this is **no different** than the *resistance* of the closed gate of *magnetic* **sameness**, seen in Figure 20 where, instead of *blocking* the alternating **opposing** *electric* charges, this 90° *interaction enhances* the flow of the electric current.

An example of **sameness** *enhancing* the interactions of chemistry is seen in catalysts where the **opposing** geometric forms of two **different** molecules join to become the **same** more complex molecule to speed up their interaction. The **same** more complex molecule that a catalyst creates from two less complex molecules is **opposed** by **either** a *negative* **or** a *positive* ion. As we shall see, various forms of RNA not only catalyse highly complex biological interactions but also do the **same** thing to themselves; i.e., self-catalyse, to enhance their own *creative* force.

The *chargeless symmetry* of the carbon buckyball, C^{60}, seen in Figure 170, captures the *charge* of an *asymmetric* atom to create a more complex and enhancing molecule. *Symmetry*, as **sameness**, enhances the value of the **opposing** idea of *asymmetry*. Figures 285, 289, 328, 329a and b show even more complex molecules with *symmetries* that have the *magical* qualities of **sameness**.

As for *lifeless* self-organization toward increasing complexity, Wolfson's book, *A New Physics*, is a good example of how *changeless symmetry*, or **sameness**, attracts an *asymmetric* "agent." This *interaction* of asymmetry/symmetry, as **opposite/ same** respectively, replicates itself to become an increasingly complex pattern. Wolfson's many **similar-different** examples of *self*-organization always develop within the **sameness** of squares of squares that remain *constant* regardless of the *changes* in complexity that successively "emerge."

As for *life*, read "Undisciplined Science," by Brian Hayes, *American Scientist*, July–August 2004, in which the author suggests that placing all knowledge under a matrix, or lattice of **sameness**, should replace our current use of the "tree" of knowledge if we are to comprehend the highly complex **similarities-differences** that emerge in nature. As we shall see, the "quasi-**sameness**" of *fractals* with their "trees" is only *part* of nature's squares of **sameness**.

Read "Liquid Crystalline Networks Composed of Pentagonal, Square, and Triangular Cylinders," by Bin Chen et al. *Science*, January 7, 2005, from which FIGURES 360 was taken. Also, refer back to Figure 160.

Such simulations and experiments in *self*-organization could not possibly occur if *mass* was only **similar-different** *objects*. Without *conceptual* mass, there would be no *conservation* of **sameness** to be **opposed** by *temporal* **similarities-differences** of mass, no *interaction* of **opposite/same** on the *horizontal* to be *timelessly* conserved on the *vertical* rise in complexity of *temporal* **specificities**. Without this *logical* method of conceptual *interactions*, evolution would be impossible.

Fig. 2. The dual tilings of the topological class (3^2.4.3.4), representing the plane group *p4gm* (*8*). (**A**) and dashed lines: Laves tiling (*25*) by identical pentagons; (**B**) and solid lines: Archimedean tiling (*26*) by identical squares and triangles.

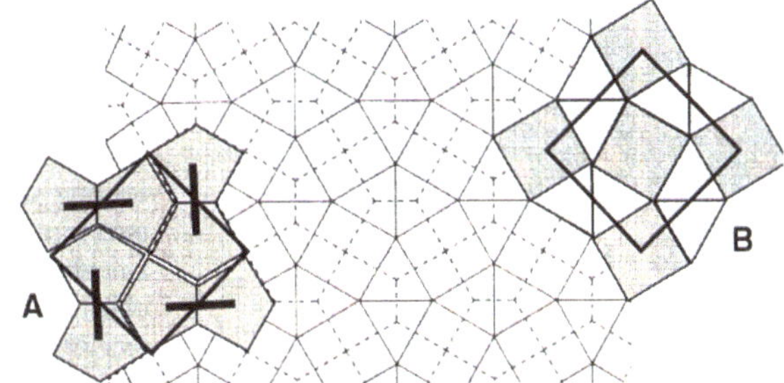

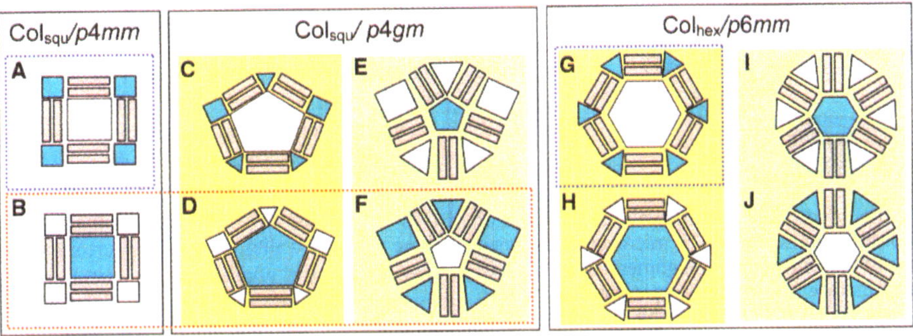

Fig. 3. (**A** to **J**) The topological isomers of the tiling of the plane in the different honeycomb-like network structures formed by different types of T-shaped molecules, facial amphiphiles (lower row), and rodlike molecules with two polar end groups and lateral nonpolar chains (bolaamphiphiles, upper row). Blue and white areas represent polar and nonpolar regions, respectively, whereas the rodlike cores are shown in gray. The arrangements in the lower row represent the color-isomers of the corresponding arrangements in the upper row. Topological duals are indicated by green and yellow backgrounds, respectively. Cross-relations between the duals [e.g., (C) and (F)] are characterized by a 90° rotation of the rodlike cores. The morphologies framed by a red dotted line are reported here, those framed by blue dotted lines have been found for bolaamphiphiles (*14, 15*), whereas the nonframed morphologies have yet to be discovered.

FIGURE 360

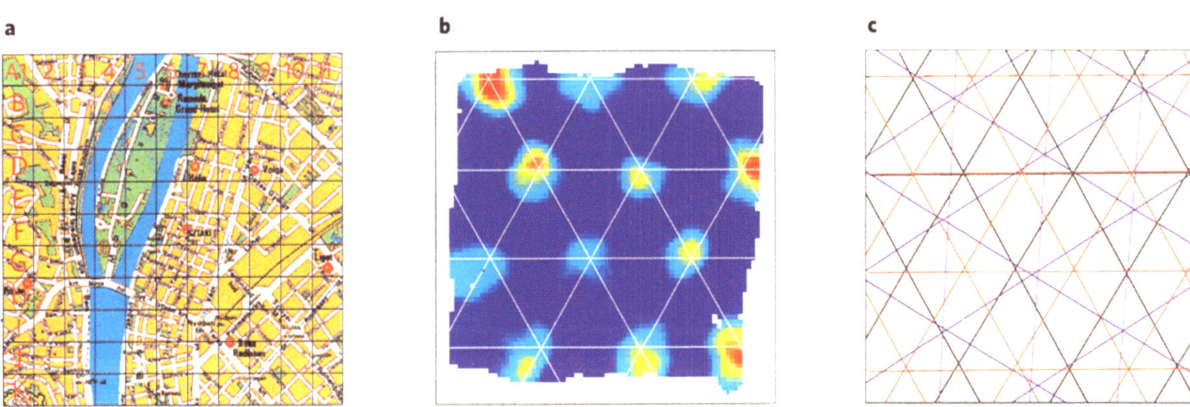

Figure 1 Tessellation and navigation. a, Tessellation of a city map by squares provides information about position, distance and direction, allowing specific places to be easily located. **b**, Hafting *et al.*[1] find that as a rat explores an experimental enclosure, the discharge rate of a neuron in the dorsocaudal medial entorhinal cortex increases at regular intervals corresponding to the vertices of a triangular grid. **c**, Integration of information from several grid components (that is, from the outputs of several neurons) can increase the spatial resolution of the environment. Three triangular grids are represented here, with red displaced and blue rotated relative to a neuron grid shown in black.

FIGURE 361

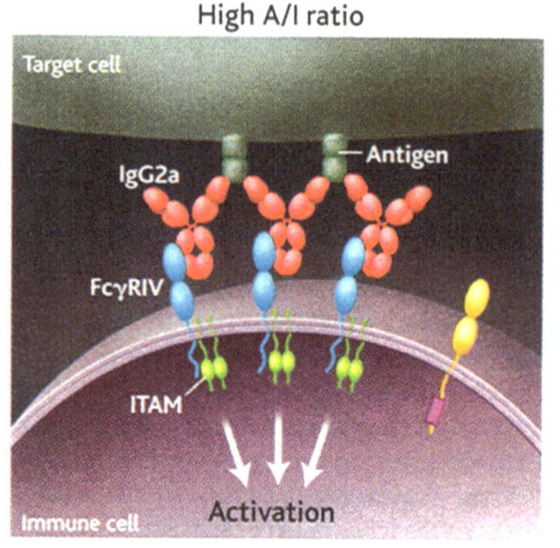

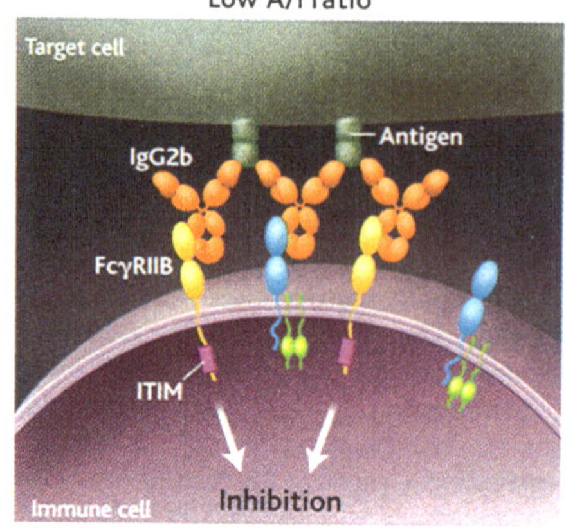

Activation versus inhibition. With a high activatory to inhibitory (A/I) ratio, IgG2a (red) antibody binds antigen (dark green) on a target cell surface and preferentially binds to activatory receptor FcγRIV (blue) rather than to inhibitory receptor FcγRIIB (yellow), and immune cell activation results. In contrast, an IgG2b antibody (orange), with a lower A/I ratio, will bind both FcγRIV and FcγRIIB and activation is dampened.

FIGURE 362

Figure 361—taken from "Neurons and Navigation," by György Buzsáki, *Nature*, August 11, 2005—shows that a city's squares of **sameness** enhances one's ability to navigate in a large city **from** here **to** there conclusively. Deduction needs *half* of inference's square of **sameness** to process **similar-different** data to a **specific** conclusion. Also read "The Map in the Brain: Grid Cells May Help Us Navigate," by Karen Heyman, *Science*, May 5, 2006, where it seems that *brain* has its own *grid* that does the **same** thing as the *mind,* **both samenesses** superposed to enhance our ability to navigate in **similar-different** directions.

FIGURE 362—taken from "Tipping the Scales Toward More Effective Antibodies," by Jenny M. Woof, *Science*, December 2, 2005—demonstrates the high/low ratio on each of the "*activation* versus *inhibition*" concepts, suggesting again that this pathway is enhanced by the *interaction* of activation/inhibition in *simultaneous double reversal*, the **sameness** of which creates a more complex and enhanced system of communication.

RNA and DNA

The master logical form of all life is DNA that has been *conserved* in deep time at every level of life's complexity. DNA is *divided* into two **opposing** parts that are formed anti-parallel, or in *simultaneous double reversal*, as seen in Figure 161. The two halves have the **same** two sets of **opposing** nucleotides as seen in Figure 163. *Along* the *length* of each strand, the nucleotides are **similar-different**, and their *unconstrained* sequence can be coded in an *infinite* number of ways. *Across* the two DNA strands, the four nucleotides are *constrained* to only two pairs, **either** A-T **or** G-C, as seen in Figure 161. **Both** pairs have the **same** *reversible*, thus each opposing pair reciprocally interact. The *geometry* of the *neuleotides* run *both horizontally and vertically* while *conceptually* interacting sequence/simultaneously unconstrained/constrained, infinite/finite all as **opposite/same** respectively.

What is the difference between life's DNA and the megalogical form seen on the inside front cover? None. They are *geometrically* and *conceptually* the **same** on the *vertical* rise of mass complexity.

Figure 161 also shows that **specific** bits of information are *constrained* to three nucleotides called "codons" that assemble in a variety of ways to create the more complex amino acids that in turn form the even more complex protein molecules that are the basic building blocks of life.

Most biologists agree that **either opposing** *half* of DNA, called RNA, was the precursor of more complex DNA. By RNA *doubling* and then re-forming in *simultaneous reversal*, these **opposing** RNA's formed the **same** more complex DNA that gave life a better *chance* to *create* increasingly complex forms of life,just as the reciprocally **opposing** "strands" of inference gives the human *mind* a much better chance to discover the Universal laws and forces of *mindless* nature. Perhaps primitive RNA was only *part* of life's logic, like that of the *micro*-electron, that by chance *doubled* and formed in *simultaneous reversal* to complete the megalogical form, Read the essay, "Stirring the Primordial Soup," by William R. Taylor, *Nature*, April 7, 2005.

Modern RNA alone has the ability not only to catalyze but also to self-catalyze RNA can **either** *silence* **or** *activate* transcription. RNA's enhancements also have been found in ancient immune-type reactions many of which are *conserved* in today's organisms. Read "The RNAi Revolution," by Carl D. Novina and Phillip A. Sharp, *Nature*, July 8, 2004. Read, "DNA and RNA Swap Roles," under This Week in *Science*, December 17, 2004. RNA is so versatile that it can do almost anything that the DNA requires, including self-assembly, *reading* bits of nucleotides and then *transferring* these bits of information from DNA to make increasingly complex molecules. RNA uses its many *catalyzing* modes to enhance an amazing number of complex molecular interactions.

DNA is *dominantly left*-handed, which if the logic of conceptual interactions is valid, there must be a *dominantless right*-hand of DNA. These two **opposing** concepts of the **same** DNA are seen in Figure 162 where **both opposing** hands enhance the interactions of the genome. Read "DNA Twists and Flips," by Richard R. Sinden, *Nature*, October 20, 2005, from which I quote:

> Nature's design of the B-Z junction maximizes helix integrity and base pairing, thereby minimizing the impact of the dramatic alteration of helix structure where Z-DNA forms in the chromosome . . . its static appearance in pictures belies the fact that DNA is flexible and dynamic. Accordingly, in a supercoiled molecule the two B-Z junctions either side of a Z-DNA tract will probably be mobile and flexible . . . . The structure of the B-Z junction comprises a canonical right-handed helix with a typical left-handed helix (Fig. 1) . . . the structure affords a remarkable conservation of energy and maintenance of the helical structure, with only a single base pair broken at the junction. An important feature of Z-DNA is that, compared with B-DNA, the hydrogen-bonded base pairs are turned upside down. How this structural transition occurs is unclear . . . the notion of a Z-DNA helix ensconced within a B-DNA chromosome punctuated by two pairs of extruded bases is a more reasonable, stable and energetically feasible view of this key DNA regulatory element.

To see how the great *complexity* of DNA can be *simplified* to the *interaction* of **opposite/same**, refer again to Figure 162 where the *constrained* **sameness** of the *circular* spirals at bottom are **opposed** at 90° by the *dominant* **similar-different** nucleotides at the top. So the **opposing** hands of DNA interact **specific/general** at 90°.the **same** as the dominant/dominantless conclusions of inference interact at 90^0, as seen in the megalogic on the inside front cover.

To see how the human *mind* uses the geometric concepts of **specific** and **general** alternately to increasingly *simplify* our understanding of the genome's great *complexity*, read from the *Tampa Tribune*, October 27, 2005, "New DNA Map to Help Cull Pins from Genetic Haystack," by Thomas H. Maugh II of the Los Angeles Times who explains a paper in the journal *Nature*, October 21, 2005, titled "Less Is More in Modeling Large Genetic Networks," by Stefan Bornholdt. I quote from Maugh, "In the past, researchers had to sift through the entire 3 billion individual chemical letters, called "nucleotides," that comprise the human blueprint in their search for disease-causing genes. But now it has become clear that each of those individual changes, called "single nucleotide polymorphisms" or SNPs, is linked to a large block of DNA, called a "haplotype," that is generally inherited intact. By focusing on haplotypes, researchers are able to reduce the number of sites that must be searched to a much more manageable 1 million, sharply speeding up the search for genetic change."

Another initiative to *simplify complexity* is to go from mapping the *genome's* nucleotides to mapping the proteome, the more complex proteins' one-on-one interactions. Read "Towards a Proteome-scale Map of the Human Protein-protein Interaction Network," by Jean-Francois Rual et al., *Nature*, October 20, 2005.

None of the above increases in simplifying complexity would be possible if *mindless* nature did not form all concepts and their interactions *timelessly* the **same** on the *vertical* so that the human *mind* can successively comprehend more and more complex *parts* of *mindless* nature's interactions.

Amino Acid Coding

As the conceptual *interactions* of life become increasingly complex, the **opposing similarities-differences** of mass overwhelm the *idea* of **sameness**. Where we can still see the *interaction* of **opposite/same** as **specific/general** in chemistry is in the logical coding of amino acids seen in Figure 248, taken from *American Scientist*, "The Invention of the Genetic Code," January–February 1998. Figure 248 highly suggests that each amino acid is formed with three nucleotides, *two* of which are the **same** while the other *one* **opposes** the **sameness** by being **different**. Just as RNA transcribes three nucleotides from *half* of DNA's two **opposing** strands to form a codon and then *doubles* it in *simultaneous reversal* to create an amino acid. The more complex amino acids do the **same** thing to create the even more complex and *diverse* **similar-different** proteins that are the basic molecules of all life's information processing seen in Figure 165.

Glial Cells and Synapses

Nine out of ten brain cells do not process information. These so-called glial cells—from the Greek word meaning *glue*—hold the wet gray matter of the brain together. But glial cells are more than just "glue." From a distance, glia are both *conceptually* and *perceptually* the **same**. On close examination, however, these cells have "astrocytes," round symmetrical "starburst" shapes with very short fine tendrils that are not absolutely the **same**. Their *relative* **sameness** allows them to *interact* with the two **opposing** halves of synapses where information is regenerated and interpreted for routing to various destinations along the neural system. Astrocytes are said to "pamper" the nerve cells electrical activity, but they do much more than just pamper synaptic parts of the neurological system. **Both** *electromagnetically* **and** *chemically* astrocytes create a highly complex geometric concept of **sameness** between the two **opposing** halves of a synapse that regenerates and enhances the processing of complex messages to their respective destinations in the brain. Read "Astrocytes, from Brain Glue to Communication Elements: The Revolution Continues," by Andrea Volterra and Jacopo Meldolesi, *Nature Reviews Neuroscience*, August 2005.

More evidence that **sameness** in the synapse enhances the **opposing** information processing in the neurological system is seen in "Traffic Jam at the Synapse," *Nature Reviews Neuroscience*, by Claudia Wiedemann, whom I quote, "Neurotransmission involves a tightly controlled balancing act between synaptic vesicle exocytosis and endocytosis." **Exo—**and **endo—**are obviously **opposing** interpretations of the **same** idea of vesicule.

Also read "Autism's Cause May Reside in the Abnormalities at the Synapse," by Ken Garber, *Science*, July 13, 2007, from which I quote, "Neuroligins and neurexins, proteins crucial for aligning and activating synapses, have now been implicated in autism, along with the Shank3 scaffolding protein. An altered balance between excitatory synapses and inhibitory could affect learning and memory during development." Logically speaking, **sameness** is required between **opposing** concepts if logical order is to be conserved. Read, "Making Synapses: A Balancing Act," by Natasha K. Hussain and Morgan Sheng, *Science*, February 25, 2005.

Each **specific** synapse interacts with other synapses in **general** as suggested in FIGURE 363, taken from an Insight Review article, *Nature*, October 14, 2004. Their collective **sameness** gives the brain its super plasticity to simultaneously infer a multiplicity of messages to their respective destinations.

The necessity for **sameness** for the logical ordering of information processing in the brain is again seen in From the Editors of *Nature Reviews Neuroscience*, September 2007, from which I quote, "Information processing in the brain relies on a careful balance of excitatory and inhibitory inputs."

That excitatory/inhibitory neurons interact in *simultaneous double reversal* to create a more complex and enhanced pathway for information processing is suggested in "Balancing LTP and LTD," *Nature Reviews Neuroscience*, April 2007, by Katherine Whalley, whom I quote, "Two major forms of synaptic plasticity, long-term potentiation (LTP) and long term depression (LTD) are cellular processes in learning and memory. Although they produce opposite effects on excitability, both LTP and LTD can occur in the **same** synapse n response to different patterns." It would seem that in addition to the dual role of *excitability* the **same** thing would occur with the **opposing** concept of *depression*.

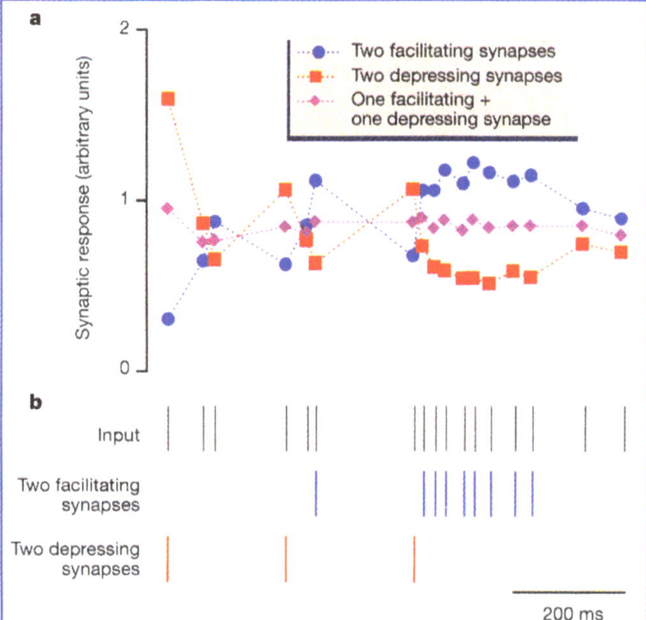

Figure 5 The ability of coactivated synapses to activate their targets depends on whether the synaptic inputs have the same use-dependent plasticity. **a**, The amplitudes of synaptic currents resulting from the simultaneous stimulation of two inputs: both facilitating, both depressing or one of each type. If the synapses share the same type of plasticity they reinforce each other's variations in synaptic strength. In contrast, if a facilitating and a depressing synapse are activated, their variations in synaptic strength tend to cancel each other out. This affects the ability of the synapses to fire their targets. **b**, The amplitudes of the responses cross a threshold and fire the postsynaptic cell during high-frequency bursts of two facilitating synapses (blue) and following pauses in presynaptic activity for two depressing synapses (red). The combination of a facilitating and a depressing synapse gives a relatively uniform input that is unable to evoke any postsynaptic spikes (not shown).

FIGURE 363

Read "Normal and Pathological Oscillatory Communication in the Brain," by Alfons Schnitzler and Joachim Gross, *Nature Reviews Neuroscience*, April 2005, from which I quote, "Importantly, evidence is emerging that a delicately balanced pattern of synchronization and desynchronization in space and time is fundamental to the functional consequences of synchronization . . . a balanced and temporally-precise pattern of synchronization and desynchronization is pertinent to cognitive function."

Note the paradox in this instance where the idea of *coordination*, or *synchronization*, cannot exclude the concept of *desynchronization*. These two concepts must *alternate dominantly* to reciprocally **oppose** the *dominantless* **sameness** of their perpetual interaction. This is another example of the inability of the mind to escape conceptual paradox at any level of complexity.

The authors then go on to say, "The insect olfactory system shows the importance of excitation and inhibition for efficient information processing." Such *alternating* **opposing** concepts interact within the overall **sameness** of neurological system to move **from** *here* **to** *there* in response to **either** *aggression* **or** *fear*, **either** *hurt* **or** *help*, **either** *rebel* **or** *cooperate*, etc.

FIGURE 364 shows how synaptic **sameness** by *simultaneous double reversal* is formed is taken from "Structure of a Synaptic γδ Resolvase Tetramer Covalently Linked to Two Cleaved DNAs" by Weikai Li et al., *Science*, August 19, 2005. The diagonaled square in this Figure is the **same** logical form as that of the antinomy in inference seen on the inside front cover that enhances and accelerates highly complex information processing.

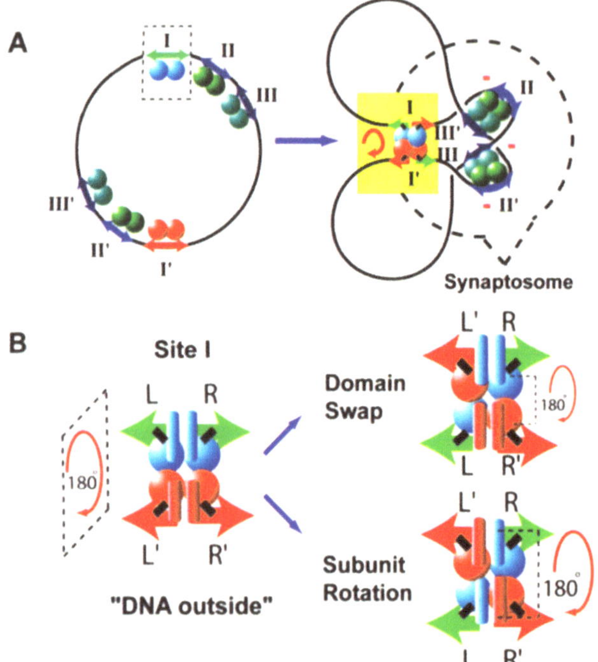

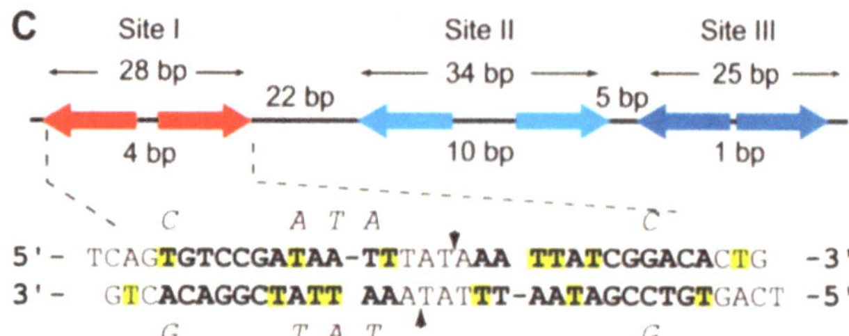

Fig. 1. Site-specific recombination by γδ resolvase. (**A**) Two *res* sites in negatively supercoiled, closed circular DNA bind wild-type resolvase as a dimer (blue and red spheres) at site I in the presynaptic state (dashed-line box). The synaptosome consists of the two *res* sites, each containing three resolvase dimers bound to sites I, II, and III (left) associating to form an assembly (right). During strand exchange, a synaptic complex at site I (yellow box) is formed by a resolvase tetramer that becomes covalently linked (black line) to four cleaved half sites (red and green arrows). (**B**) A tetramer of resolvase (left) recombines two site I DNAs [same as in (A)]. Two models of γδ resolvase strand exchange (right) are domain swap (top) and subunit rotation (bottom). The interfaces formed by E helices (blue and red sticks) are intact in the domain swap model but rotate relative to each other in the subunit rotation model. (**C**) (Top) γδ resolvase requires three sites to form the synaptic complex but performs recombination exclusively on site I. The length and spacing of these sites are shown. (Bottom) The sequence of the symmetrized site I analog used, with the bases of the original sequence of site I shown in italics above and below the mutated bases. The double-strand cleavage sites are shown in black arrows. Thymidines substituted by 5-bromo-2′-deoxyuridine in oligonucleotide derivatives are shown shaded in yellow. Nicks in the crystallographic substrates are opposite the dashes.

FIGURE 364

FIGURE 365—taken from "Multiple Routes to Similar Network Output," by Scott L. Hooper, *Nature Neuroscience*, December 2004—shows a *facilitating* (blue) synapse and a *depressing* synapse (red) interacting the **same** in **opposing** ways. **Both** of these reciprocally **opposing** concepts sequence information from a more complex square of **sameness**. The caption of FIGURE 354 clearly reflects my conjecture of reciprocal logic.

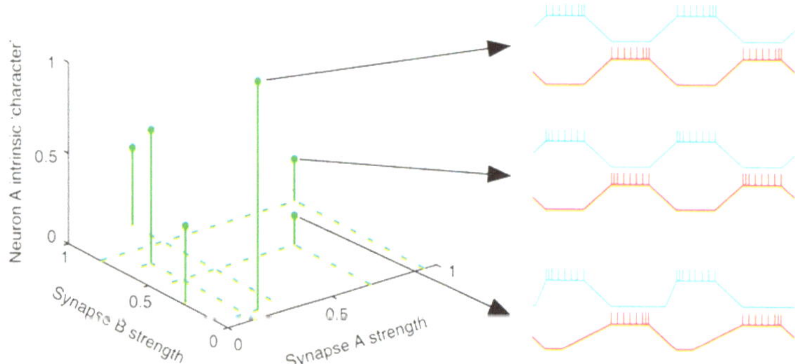

Figure 1 Example illustrating how different combinations of synaptic strength and intrinsic neuron properties could produce outputs that have the same cycle period, spike number and phase relationships (albeit with fine differences in action potential timing and slow-wave trajectories). The system modeled by Prinz *et al.* had nine dimensions (three neurons whose intrinsic properties could vary, and six synapses).

FIGURE 365

FIGURE 365 illustrates but a *part* of the overall more complex synaptic harmony in all the brain's interactions, seen in the **sameness** of 40 Hz vibrations, which **generality** enhances the *brain's* neurological system to send **specific** messages *conclusively*. It is the concept of **sameness** that allows millions of **similar-different** commands to interact between cells for the body/brain to accomplish **specific** tasks *deliberately*.

More information on the magic of **sameness** in synapses is found in, "Expression and Functions of Neuronal Gap Junctions," by Goran Söhl et al., *Nature Reviews Neuroscience*, March 2005.

Also read, "Connecting Spontaneous Neural Activity and Perception," under eBriefings, *Update*, New York Academy of Science, from which I quote, "General models of synaptic connections . . . argue that sensitivity to an external stimulus is a fine balance between the strength of connection within a neural network and the strength of the input signal."

Myelin Sheaths

The dendrite *asymmetry* of the neuron is surrounded by the *symmetry* of an "insulating sheath" interacting at 90° with the length of neuron electrical processing. The chemical composition of the sheath covering the axon "wire" is called "myelin," which is more than mere insulation. Healthy myelin appears under the microscope as dozens of spiral wraps with the **same** molecular pattern and with the **same** spacing. See Figure 225. Myelin is found in all large animals and has been proven to expedite and enhance the electrical pulses in the neurons. The larger the mass of the animal, the greater the thickness of myelin sheathes so that there is an increase in the force of **sameness** to further enhance the flow of information through the longer neural lengths of the greater mass. In addition, animals that move the quickest and fastest have much thicker layers of myelin **sameness** to increase their efficient use of energy.

In effect, myelin serves the **same** purpose as the **sameness** of the *magnetic* field surrounding the **opposing** *electrical* current of electromagnetism. In **both** instances, and many others in various guises, the *interaction* of **opposite/same** creates the *dynamics* of our *macro*-realm.

The brain's analog of myelin sheaths is "white matter" found in great abundance surrounding its neurological system. As with the **sameness** of gluon *gray* matter, the **sameness** of the brain's *white* matter enhances the *interactions* between neuronal connections. Read, "Why the White Brain Matters," by Christopher M. Filley, *Cerebrum*, the Dana Forum on Brain Science, Winter 2005. Also read "The Dark Side of Glia," by Greg Miller, *Science*, May 6, 2005.

For now let us turn to the ancient Greeks who correctly interpreted the **sameness**, or *symmetry*, of a square as being slightly too *static* and excessively *balanced*. Its *balance* seemed to need some degree of *imbalanced* to make it frame *realistic* and *dynamic*. This asymmetry/symmetry *interaction*—the ancient Greeks called the "golden mean," or the "golden ratio"—is seen in Figures 130 and 131. The framework of the golden ratio, therefore, is often used by artists to gives their *static* and *motionless* paintings the geometric concept of *dynamic motion*.

Another aspect of the golden mean is the Fibonacci ratio where a series of numbers have a special additive progression, seen at the *bottom* of Figure 27 and the *bottom* of Figure 28. The *middle* of Figure 27 shows that there is **no difference** between the geometry of the Fibonacci rectangle and the rectangle of the golden mean in Figure 131. The *top* of Figure 27 shows the *dynamic* "flying squares" of Fibonacci spiral. It is more than just interesting that **both** of these two interpretations are each based on the 90° angle, and that the dynamic of their spirals is seen in many forms of life.

All animal bodies, especially those with bilateral symmetry, need a sense of *balance*. Without this **sameness** of concept, there is no sense of direction, one's orientation becomes confused and chaotic. Read *Balance: In Search of the Lost Sense*, by Scott McCredie, Little Brown, 2007, who makes us aware of what seems to be hidden from us, like the **sameness** quarks.

Telomeres

There is a series of molecules at each end of a chromosome that link the chromosomes to each other. Instead of the **sameness** of **opposing** molecules *surrounding* myelin sheathes telomeres are connected with the **same** molecules *aligned* in **opposing** directions. As would be expected with the geometric concept of **sameness** connecting **similar-different** chromosomes, telomeres are more than just a linkage. They regenerate and logically interact all the chromosomes as the **same** complex system, as well as enhance the processing of information between the chromosomes and the more complex genetic system. As long as the telomeres retain their link of **sameness** between chromosomes, a healthy life is sustained. But as life ages, the length of the telomeres shorten; and once they become so short that they lose their geometric concept of **sameness**, life can no longer be *conserved*, terminal diseases set in and death occurs.

Pluripotent Stem Cells

The magic of the self-creative force of *constrained* **sameness** of the stem cell, seen in Figures 88 and 153a, is explained and discussed throughout the book, so I will not dwell on it here.

Alpha

This *constant* of nature with its **sameness** of concept creates the first glimpse of **opposing similarities-differences** in mass. How alpha created its own magical force will be presented in the last chapter.

Quarks: Constrained within the Proton

The magical force of quarks, with its **sameness** of concept presented in Figure 34, has already been discussed.

Gravity

As already explained, the **sameness** of the force of *gravit*ational *contraction* creates successive increases in complexity of mass **specificities**.

SPIN

As I have already emphasized repeatedly, no experiment has any meaning at all unless it is interpreted by theory.

—Max Born, physicist

The geometric concept of *spin* is the *sine qua non* of reciprocal logic.

The *spin* of the gyroscope convincingly demonstrates that its logical order is not inference with its **similar-different** *data* being *sequencing* to a *conclusion*. Instead, it reveals geometric *concepts* that are *inconclusive*, interacting **apposite/same** reciprocally at 90°. See Figure 298.

For example, when you *push* on the frame of a *spinning* gyroscope, instead of continuing along the *straight line*, the spin axis rotates in a *circular* motion *inconclusively* 90° to the force of *push*. The idea of *push* becomes *nonlinear* at 90° in a *circular* motion rather than *linear* at 180°. The idea of *pull* does not relate to *push*. Instead, *push* interacts with *pushless* at 90°.

You can do the **same** thing as the above in the **opposite** way—i.e., *pull* on the frame of the spin axis from the **same** point as before and its spin axis will rotate in a *circular* motion *inconclusively* 90° to the force of *pull* creating a pull/pull-less interaction. Now we see that the **opposing** ideas of *push* and *pull* have the **same** circular reaction. **Either** *push* **or** *pull* creates the *interaction* of **opposite/same** at 90°, and **either** one is the reciprocal of the other, the **same** reciprocity of **opposite/same** we saw with the **electro/magnetic** force.

Now circle the gyroscope so that you are positioned 180° from where you were before. Then do the **same** thing from this 180° **opposing** point of view. You will get the **same** results as before but in the **opposite** way.

New let's break this down into its four components. Let's assume that the first pair of interactions is by you and the second pair of interactions is by me. We stand at 180° **opposing** points of view, yet we each see the **same** thing. **Neither** one of us can be in both places *simultaneously, but* mindless nature *timelessly* interacts **both** 180° **opposing** points of view the **same**. If this were not so, **neither** one of us could switch points of view deliberately at any time or place. To see this schematized other than with the gyroscope, go to Figures 52 and 53.

The faster the *spin accelerates*, the greater the force of *inertia* that *conserves* the spin axis in the **same** place in space. **Both** the *straight line* of *inertia* **and** *circle* of *acceleration* interact at 90°. This is what makes your bicycle increasingly stable the faster you ride. The interaction of acceleration/inertia is the basis of the gyroscope that dominates all navigational systems. It is the *inertial constant* that enables the gyroscope to detect and accurately measure any *change* (*acceleration*) in direction.

Even more un-intuitively, if the spin axis of a fast-spinning gyroscope is positioned parallel to the ground, instead of falling *perpendicular* to the ground by the downward *pull* of gravity, the spin axis of the gyroscope rotates in a *circle parallel* with the ground, *conserving* the 90° interaction pull/pull-less. This interaction is called "precession" seen in FIGURE 366, taken from *The Laws of Physics*, Milton A. Rothman, Basic Books Inc., New York, New York.

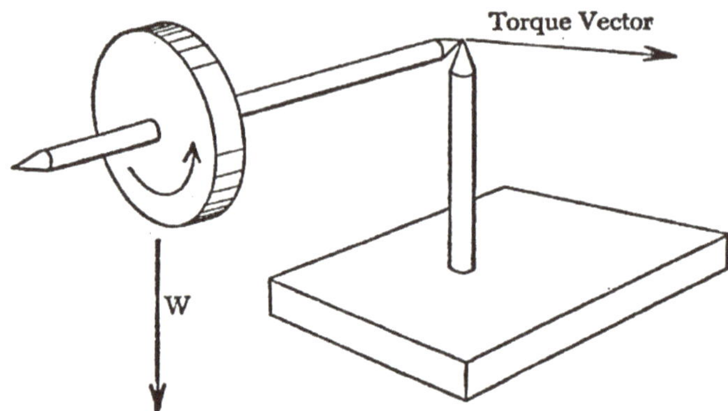

FIGURE 7. Action of gravity (W) on a spinning gyroscope.

Since the torque exerted by gravity goes on acting at right angles to the angular momentum, the axis of spin continues to shift around in the horizontal plane, so that the gyroscope slowly swings in a circle around the pivot. In other words, instead of toppling over, as an ordinary weight would if supported only at one end, the axis of the gyroscope moves around in a horizontal plane.

This motion is called *precession.* The angular velocity of the precession motion, *P*—that is, the rate at which the axis of the top turns in a horizontal plane—is given by the formula:

$$P = \frac{T}{I\omega}$$

where T is the torque resulting from the weight of the gyroscope, I is the gyroscope's moment of inertia, and ω is the angular velocity of the top around its own axis—its rate of spin.

This formula shows us that if the gyroscope has a large moment of inertia, and if it is spinning very rapidly, then the precession is slow. For a given amount of torque, the faster the spin, the slower the precession.

For an animation of FIGURE 366, go to http://www.reciprocallogic.com.

If one could *simultaneously* interact, **both** *left* hand **and** *right* hand *spins* along the **same** axis of rotation at the **same** time, the idea of *spin* would become more complexly *spinless*. The *simultaneity* of **both opposing** spins of any **specific** mass cannot occur in our *temporal* universe. *Mass*, including that of the gyroscope, is **opposed** to **sameness**. Gravitational **sameness** prefers **neither** *right*—**nor** *left*-handed *spin*, the **same** as fermion mass prefers **neither** *left* **nor** *right* ½-spin.

Our level of mass complexity *conserves* the ½-spin of electron **oppositeness,** but it takes *two* complete 360° spins of the isolated electron to complete the 360° spin of a *macro-object's* full spin. This demonstrates again that *micro*-mass ½-spin is only an *absolute part* (particle) our *relative macro*-mass conceptual interactions.

It is interesting to compare the 2/2-spin of the *photon* with the ½-spin of an isolated *electron*. When you place an *electron* within a magnetic field, its spin axis will align *parallel* with the lines of the field in **either** 180° direction. When you do the **same** thing with a *photon*, its spin axis forms *perpendicular* to the *parallel* lines of the magnetic field. See Figure 48. If **both** electron **and** photon could be superimposed, they would interact their geometric concepts would interact at 90°.

The *spin* of **opposing** complementary colors, those that lie at 180° to each other on the color wheel seen in Figure 21, also shows the interaction of **opposite/same** in our *macro*-world. In addition to the spin of **both** 180° complementary *colors*, the *spin* of **both** 180° *black* **and** *white*, and the *spin* of the three *additive colors each* create the **sameness** of *colorlessness*, as seen in Figure 26. All three **samenesses** occur with **either opposing** *left* **or** *right* spin.

More complexly, **both** the three *additive colors* **and** the three *subtractive* colors, seen in Figure 63, form the more complex *primary colors*. The spin of *each* of these **opposing** sets of *colors* creates *colorless* grey so that color/colorless interact. If **both opposing** sets of spin could be superimposed, a more complex and enhanced **sameness** of concept would be created.

We will have more to say on the subject of *spin* and *color* when we explain how the constrained acceleration of quarks **sameness** is conserved as mass becomes increasingly complex.

The gyroscope easily and simply demonstrates the futility of trying to completely *force* one's *interpretation* on an **opponent**. No single *idea* will continue along a *straight line* to a *conclusion*, as deduction in inference would have us believe. Instead, **either opposing** extreme will make the idea *circle* back upon itself *inconclusively*. Thinking that our *interpretation* is the way the world is and should be ordered and thinking of our conceptual **opponent** as being a problem to be solved only exacerbates the *chaos* of the *conceptual* conflict.

It is more than mere coincidence that arguing an *interpretation* is often called "spin." "Spin doctors" is a phrase often used to describe professionals that write speeches to argue a politician's point of view.

To show that even in the most complex interactions of life's molecular processes our macro-mass uses the reciprocal logic of *spin* to generate a force that enhances and accelerates deduction of inference. For example, mitochondrial DNA, the primary source of energy in each cell, has a molecule with all the characteristics of a generator, seen in Figure 195. This *spinning* molecular motor creates more energy than it needs for its own survival, enough to supply added energy to other molecules of the complex eukaryotic cell. I quote from a letter to *Nature*, dated March 20, 1997, by Hiroyinuki Noji et al. that describes this "molecular motor":

> A tripartite symmetry resembling the petals of a flower . . . . Two components: an electrical gradient and a pH gradient (Nernst potential), either of which can be tapped interchangeably for work . . . all three of the F catalytic sites are highly cooperative . . . all three of the participating sites were biologically equivalent . . . all three a-subunits were found to have an identical sequence, as did all three B-subunits. However, the sequence of the lone y-subunit did not show any tripartite character and yet all three of the enzyme sites needed to interact with it to function . . . the a—and B—subunits literally turned with respect to an axially located y-subunit, bringing the three sites into play, one after the other . . . it looked every bit like a piston motor, with a hexagonal ring of a-B pairs surrounding a drive shaft made up of the y-subunit (Fig1).

In Figure 195, note that its *force* has the geometric concept of **sameness** conserved on the vertical from the force of quarks **sameness** seen in Figure 34.

Also read "Nature's Rotary Electromotors," by Wolfgang Junge and Nathan Nelson, *Science*, April 29, 2005.

The molecule ATP cranked out by the spin of the miDNA motor carries more energy than the sum of its two complex parts. miDNA not only has the power to furnish the cell with all the power it needs to keep *alive* its particular task but also has the power to order its *death* when the cell is no longer able to make its **specialized** contribution to the **general** well-being of all the surrounding cells. This is an excellent example of the ubiquity of **specific/general** in that when a **specific** cell can no longer contribute to the well-being of the **general** cells surrounding it, they will signal the defective cell to commit suicide, which biologist euphemistically call "apoptosis." *Mindless macro*-mass will not tolerate the failure of **specific/general** perpetually interacting, and neither does evolution. Even at the cellular level of complexity, we see survival of the fittest and parsimonious economics *conserved*—something that philosophers, economists, medical science, and politicians should keep in mind if any system, including any socio-economic system, is to be *conserved* and successively increase in ordered *complexity*.

A spiral galaxy's *angular momentum* is transformed into *spin* when a black hole forms, as seen in FIGURE 367 and Figure 266. The black hole and its spiral galaxy take the **same** logical form as the *spin* of the gyroscope. In "Jet Set," *Science News*, October 7, 2006, Ron Cowen wrote, "Astronomers this week reported that they have finally identified the particles in the jets as electrons and protons . . . that carry much more energy than some astronomers had theorized . . . . Protons are about 1,8000 times as heavy as positrons." The great speed of these heavy positive (+) proton charges in complementary directions of 180° spin force the plane of the spiral galaxy to spin **either** *right* **or** *left* as if it were a gyroscope.

When a bright shining star's *accelerating* force of gravity exceeds the *constant* speed of light, it collapses into a black hole, leaving *nothing*. All information of *mass* is lost, leaving only the *energy* of *constrained* **sameness** ejecting in 180° this "polarity." The galaxy spins in **either opposing** direction in **opposition** to the **sameness** of the black hole 90° to its spin axis. This *force* generated by *spin* probably accounts for the "dark matter" that seems found around all *spiral* galaxies and their groups, the *spin* of which—instead of angular momentum—keeps the stars from escaping the spiral.

Read "Scientists Offer Proof of 'Dark Matter'" by Marc Kaufman, the *Washington Post*, August 25, 2006. Also read "Enlightened, Dark Matter Spotted after Cosmic Crash," by Eric Jaffe, *Science News*, August 26, 2006. In this work, the researcher Michael Turner of the University of Chicago says, "Its kind of like a cosmic centrifuge."

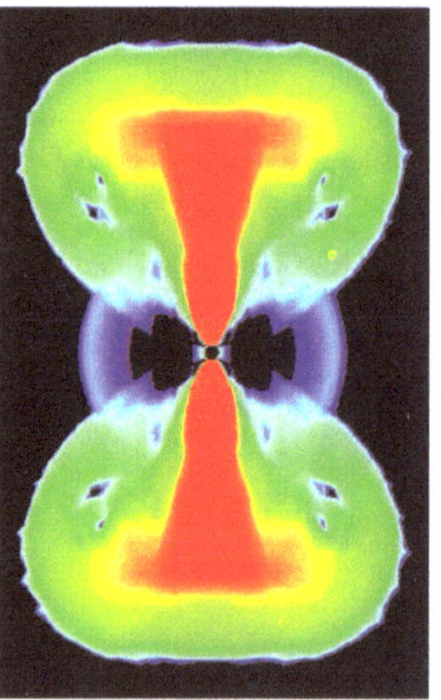

COOL IMAGES

Birthing a Black Hole

Two jets of gas (red) stream at nearly half the speed of light from an exploding star in this simulation of a collapsar, "a dying spinning massive star giving birth to a hungry black hole," according to University of California, Santa Cruz, grad student Andrew MacFadyen's home page. His and adviser Stan Woosley's collapsar model attempts to explain gamma ray bursts, mysterious flares billions of light-years away that emit more energy in a matter of seconds than our sun will in its lifetime. Observations earlier this year favor a collapsar scenario rather than explanations such as a collision between two neutron stars. MacFadyen has festooned his Web site* with images from his models and promises to add movies "after the holidays."

* www.ucolick.org/~andrew

<u>FIGURE 367</u>

Recent observations of the 90° interaction between the *massless* **<u>both-and</u>** polarity of the black hole and the **<u>either/or</u>** spin of the galaxy's *mass* is found in "The Hole Story," by Ron Cowen, *Science News*, January 22, 2005, from which I quote:

> Loeb suggested that Gebhardt compare the mass of each black hole with the average velocities of the billion-or-so stars that surround each black hole out to a distance of several thousand light-years. This swarm of stars—a major component of a galaxy—is known as the bulge, and the stellar velocities provide a measure of the bulge's mass. When Gebhardt made the comparisons, he was stunned. Regardless of their size, the bulges always turned out to be 500 times as massive as the giant black hole at the hub of their galaxies. What could be behind the ratio? . . . Laura Ferarese . . . made the same research suggested to her . . . . To understand how puzzling this numeric relationship is, Ferrarese notes, consider that a supermassive black hole can only suck in matter that reside less than a light-year from its own location a the center of the galaxy. Most stars in the bulge, which can lie as far as 20,000 light-years away from the center, aren't in the least affected by the black hole's gravity.

This indicates that when the *acceleration* of gravity exceeds the speed of light's *constraint* the geometric concept of *mass* disappear and all information along with it. Correspondingly, when **either** universal gravitational *acceleration* **or** universal anti-gravitational *acceleration* exceeds any **specific** observer's view of the *constant* speed of light, *knowledge* becomes *unknowable*. *Knowledge* of *time and space are* limited to the speed of light.

SUPERCONDUCTIVITY

We used to think that if we knew one, we knew two, because one and one is two.
We are finding that we must learn a great deal more about and.

—Sir Arthur Eddington

When atoms or molecules *collide*, they lose heat as they expand unconstrained. The *colder* the mass, the *fewer collisions* (interactions) occur between them. Less energy is generated and *entropy*, or disorder, *decreases*. The spaital environment becomes *colder* and inferential *order* becomes increasingly dominant.

Conversely, or should I say, reciprocally, the *hotter* the environment, the greater the number of *collisions* between atoms or molecules occur. They are *nonlinear* and less inferential. At a *balance* between these two extremes of temperature, but slightly on the colder side inference *dominates* and life has its greatest *chance* of emerging.

There is no such concept as **either** *absolute cold* **or** *absolute heat* because from our *relative mass* point of view, cold/heat *interact* as order/disorder, contraction/expansion, **opposite/same** respectively.

In extremely *cold* conditions, close to *absolute* zero, instead of **either** electric charge *alternating* **opposing** each other, **both** electrons *simultaneously* pair up and **oppose** themselves to become the **same** thing. So there is *nothing* to **oppose** their inferential *sequence*. These so called "Cooper pairs" can theoretically *sequence* forever without any resistance. In our *macro*-realm however, superconductivity continues only an long as it has an outside source of the electromagnetic current *During superconductivity, resistance* lies *outside* of the current in the form of a magnetic field. **Both** the *current* **and** the *magnetic field interact* the *simple* geometric concepts of inside/outside as sequence/simultaneity respectively. FIGURE 368, taken from "Superconductivity without Phonons," by P. Monthoux et al., *Nature*, December 27, 2007, shows the *complex supersimultaneity* of the magnetic field that has the diagonaled square of **sameness**, seen in the meglogic on the inside front cover, that enhances the logic of inference. Within this antinomy, we also see the **same** geometric concepts of **both** *contraction* **and** *expansion* that we found in the *cosmic* realm where the **sameness** of **both** gravitational *contraction* and the **sameness** of anti-gravitational *expansion* interact *timelessly* to create a **super-sameness** that also enhances the **opposing** *processing* of information in *time* throughout the knowable universe.

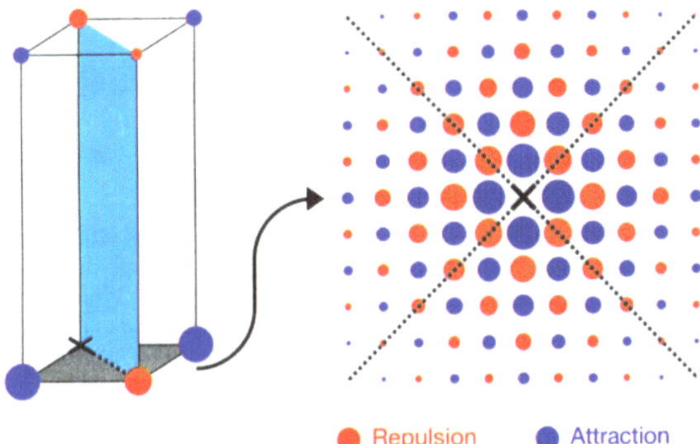

Figure 3 | Magnetic interaction potential in a lattice. Graphical representation of the static magnetic interaction potential in real space seen by a quasiparticle moving on a square crystal lattice given that the other quasiparticle is at the origin (denoted by a cross). The spins of the interacting quasiparticles are taken to be antiparallel, such that the total spin of the Cooper pair is zero. The dashed lines show the regions where the *d*-wave Cooper-pair state has vanishing amplitude. This is the state that best matches the oscillations of the potential, in that a quasiparticle has minimal probability of being on lattice sites when the potential induced by the quasiparticle at the origin is repulsive. The size of the circle in each lattice site is a representation of the absolute magnitude of the potential (on a logarithmic scale). This picture is appropriate for a system on the border of antiferromagnetism in which the period of the real space oscillations of the potential is precisely commensurate with the lattice.

FIGURE 368

Because the ideas of *conduction* and *insulation* are reciprocally ordered, an *insulator* can be transformed into a *conductor* and vice versa. Read "When Is a Metal Not a Metal?" by Steven C. Erwin, *Nature*, May 18, 2006. At room temperature, **neither** too hot **nor** too cold, a heavy coating of *insulation* must *simultaneously* surround the wire of a metal *conductor* so that an electric current will *sequence* information through the wire *conclusively*. Here, the **opposing** electric charges *alternate* in *time*, enhanced by the **sameness** of the insulator surrounding the wire, the **same** as the myelin sheath enhances information processing in more complex life's neurological systems. as previously discussed.

To show the pervasive and timeless nature of reciprocal logic, even superconductivity has its opposite. Read, "Opposite of a superconductor", by Rosario Fazio, *Nature*, 3 April 2008. Fazio explains a paper in the same issue of how *superconduction* transforms to *superinsulation*. They are reciprocally **opposing** interpretations of the **same** phenomenon that are more complexly the **same** to *mindless* nature's Universal ordering principle. For the *mind* therefore, **sameness** at any scale or level of complexity paradoxically *enhances* the very thing that **opposes** it.

FRACTAL AND DENDRETIC FORMS

Data, however fascinating, is not insight.

—Richard Kluge

If our minds *interact* order/chaos, then there should be some proportional *balance* between them. Benoit Mandelbrot discovered this "quasi-sameness" in the **similarities-differences** of mass. He called these patterns "fractals." They have a "self-similarity" at various scales of magnification such as seen in Figure 338, taken from *Sigma Xi Today*, May/June, 2002. Fractals are considered by scientists to be at the boundary between *order* and *chaos*. Reciprocal logic would also add that fractals are *are* parts *of* mindless *nature's* ***sameness*** *of concept that our minds* **oppose**.

Ideal fractals show dendrite type increases in complexity as seen in Figure 176. A comparison between the **opposing** *idealistic* and *realistic* forms is seen in Figure 339, taken from "A Brief History of Fractals." Even the *ideal* form is only *half* of nature's **sameness**—i.e., "quasi-**sameness**" overwhelmed by the **opposing similarities-differences** of mass.

Fractals have practical applications as seen in "It's a Rough World: Fractals Help Model Vexing Problems in Earth Science," by Sid Perkins, *Science News*, February 7, 2002, from which I quote:

> Not only do non-symmetry representations of objects show roughness and irregularity, they do so across the entire range of scale. A close-up look at a jagged non-fractal line would reveal that it's made up of small, straight segments. But for a fractal line . . . say, a 1.3 dimensional line—the view would be the same form afar as from close-up. Each segment would have the same degree of roughness no matter what scale at which it is viewed. Likewise, a microscopic look at a rock with a 2.7 dimensional surface would display the same texture as the telescopic view of a cliff face of the same fractal dimension.

Read *Fractals and Chaos: The Mandelbrot Set and Beyond* by Benoit B. Mandelbrot, Springer, 2004.

Fractal-like "scaling laws" are found in complex ecological systems and in the metabolic rates of animals. These rates are proportional to body size raised to the power of 3/4 and are the "same in all life from bacteria to whales and redwoods."

In *Science*, September 21, 2001, a paper titled "All Fired Up: A Universal Metabolic Rate," by Kathryn Brown, I quote, "All living organisms share roughly the same resting metabolic rate when body size and temperature are taken into account. The finding suggests that diverse species burn energy in predictable patterns. 'The basal metabolic rate of an apple is remarkably similar to a fish or a person.'" For a more detailed explanation of the fractal-like scaling laws, read "All creatures great and small" by John Whitfield, *Nature*, September 27, 2001.

If **sameness** and its quasi-form of fractals were not *dominantless* at any scale, we could not perceive **specificities** *dominantly*. The *inductive* logic of fractals is seen in Figures K-1, K-2, R, and 255, all of which take the **same general** form.

Later, we will see a dendrite neuron with its fractal form placed in *simultaneous double reversal* creates the **same** but more complex neuron than the sum of its parts, capable of enhancing and accelerating the neurological system's information processing. *Mindless* nature's highly complex *interactions* often need to be more efficient and enhancing than quasisameness, or *partly* the **same**. For example, *normal* blood vessels need to be distributed *evenly* throughout living tissue. The fractal forms of blood vessels therefore are placed in *simultaneous double reversal* to become the **same** more complex and enhanced system in **general**. When one of these *balanced* systems becomes *imbalanced*, the tissue becomes *abnormal* which can cause a tumor, as seen in FIGURE 369 taken from "Normalization of Tumor Vasculature: An Emerging Concept in Antiangiogenic Therapy," by Rakesh K. Jain, *Science*, January 7, 2005. Such *balance* is but part of the entire body's *conservation* of **sameness** called "homeostasis," itself *conserved* on the *vertical* from quarks geometric concept of **sameness**. Quasi-sameness cannot *create* a more complex and enhanced species ,that is more than the sum of its parts.

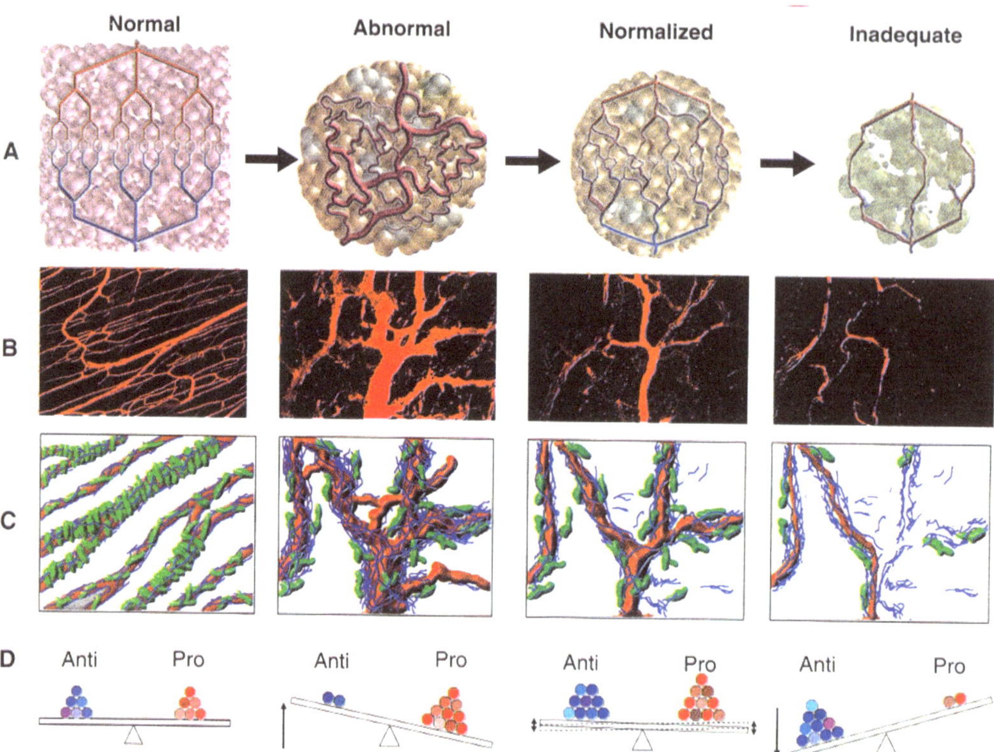

Fig. 1. Proposed role of vessel normalization in the response of tumors to antiangiogenic therapy. **(A)** Tumor vasculature is structurally and functionally abnormal. It is proposed that antiangiogenic therapies initially improve both the structure and the function of tumor vessels. However, sustained or aggressive antiangiogenic regimens may eventually prune away these vessels, resulting in a vasculature that is both resistant to further treatment and inadequate for the delivery of drugs or oxygen [reproduced, with permission, from (*11*)]. **(B)** Dynamics of vascular normalization induced by VEGFR2 blockade. On the left is a two-photon image showing normal blood vessels in skeletal muscle; subsequent images show human colon carcinoma vasculature in mice at day 0, day 3, and day 5 after administration of VEGR2-specific antibody [reproduced, with permission, from (*24*)]. **(C)** Diagram depicting the concomitant changes in pericyte (red) and basement membrane (blue) coverage during vascular normalization (*24, 29*). **(D)** These phenotypic changes in the vasculature may reflect changes in the balance of pro- and antiangiogenic factors in the tissue.

www.sciencemag.org SCIENCE VOL 307 7 JANUARY 2005

FIGURE 369

Even the food animals eat must be nutritionally *balanced* inductively—i.e., in **general**, to sustain life. Read This Week in *Science*, January 7, 2005, from which I quote:

> It is widely assumed in forging theory that predators cannot balance their nutrient intake, but instead maximize their energy intake subject to prey size, abundance, and time constraints. Mayntz *et al.* (p. 111) show that this is not the case, using three species of invertebrates (ground beetles, wolf spiders, and web spiders) with widely different feeding biology. When the diet of the predators was manipulated to render them either protein—or lipid-deficient, the animals adjusted their feeding to make good the specific defect. Compensatory nutrient selection occurred either by selecting among foods of different nutritional composition, by adjusting consumption of a single prey type, or by extracting nutrients selectively from within individual prey items.

The *self*-increase in complexity of scaling laws can be seen in the form of an isosceles triangle doubled and placed in simultaneous reversal, which form is the **same** as the *creative* **sameness** of quarks seen in Figures 34 and 36. This **sameness** of form is seen in crystals, such as snowflakes, that have been built up by repeating the **same** form over and over again along **opposing** spines of the **same** crystal. This is called a Koch fractal, but the fractals are limited to *each* of the spines of **sameness**.

The World Wide Web is a set of sets interacting between each other in **general** within the more complex **specific** Web. All sets have the **same** fractal characteristics as seen in "Sizing Up Complex Webs: Close or Far, Many Networks Look the Same," by Erica Klarreich, *Science News*, January 29, 2005. A work that explores the logical order of scale-free Web interactions is seen in a letter to *Nature*, July 1, 2004, titled "Evidence for Dynamically Organized Modularity in the Yeast Protein-protein Interaction Network," by Jing-Dong J. Han et al. The schemas in both of these works show the **same** *fractal* modularity and hub-to-node interactions as Figure 293.

FIGURE 370 shows the fractal form of any set of complex *mass* interactions at any scale, taken from "Self-similarity of Complex Networks," by Chaoming Song et al., *Nature*, January 27, 2005. I quote from the abstract:

> Complex networks have been studied extensively owing to their relevance to many real systems such as the Internet, energy landscapes and biological and social networks. A large number of real networks are referred to as "scale free" because they show a power law distribution of the number of links per node. This conclusion originates from the "small world" property of these networks, which implies that the number of nodes increases exponentially with the "diameter" of the network rather than the power-law relation expected for a self-similar structure. Here we analyze a variety of real complex networks and find that, on the contrary, they consist of self-repeating patterns on all length scales.

The modules of the brain can be likened to the hubs and nodes of the Internet, the hubs being the more complex sets and the nodes less complex sets. Each of these highly **generalized specificities** organizes and distributes highly complex messages to other modules of the brain with their respective dominance of concepts. All these modules interact more complexly the **same**, **opposed** by inference for animals to *move* instinctive, deliberately and conclusively.

FIGURE #370

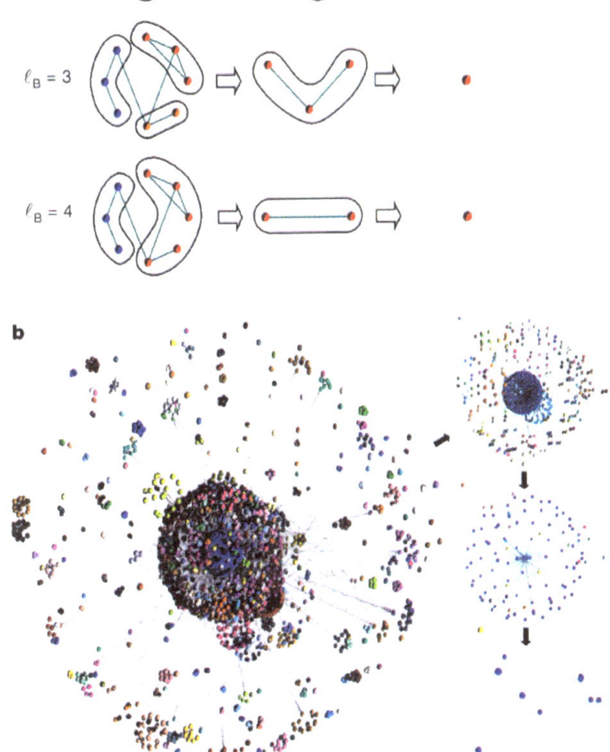

Figure 1 The renormalization procedure applied to complex networks. **a**, Demonstration of the method for different ℓ_B. The first column depicts the original network. We tile the system with boxes of size ℓ_B (different colours correspond to different boxes). All nodes in a box are connected by a minimum distance smaller than the given ℓ_B. For instance, in the case of $\ell_B = 2$, we identify four boxes that contain the nodes depicted with colours red, orange, white and blue, each containing 3, 2, 1 and 2 nodes, respectively. Then we replace each box by a single node; two renormalized nodes are connected if there is at least one link between the unrenormalized boxes. Thus we obtain the network shown in the second column. The resulting number of boxes needed to tile the network, $N_B(\ell_B)$, is plotted in Fig. 2 versus ℓ_B to obtain d_B as in equation (3). The renormalization procedure is applied again and repeated until the network is reduced to a single node (third and fourth columns for different ℓ_B). **b**, The stages in the renormalization scheme applied to the entire WWW. We fix the box size to $\ell_B = 3$ and apply the renormalization for four stages. This corresponds, for instance, to the sequence for the network demonstration depicted in the second row in panel **a**. We colour the nodes in the web according to the boxes to which they belong. The network is invariant under this renormalization, as explained in the legend of Fig. 2d and the Supplementary Information.

NATURE | VOL 433 | 27 JANUARY 2005 | www.nature.com/nature

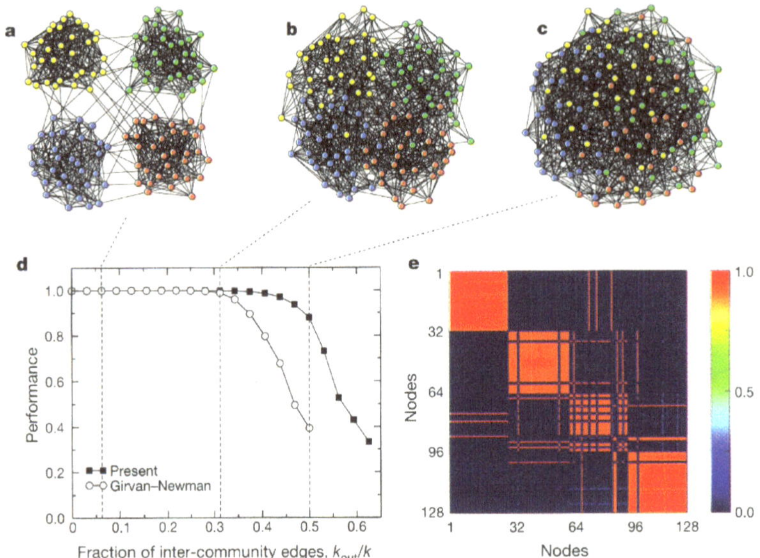

Figure 1 Performance of module identification methods. To test the performance of the method, we build 'random networks' with known module structure. Each test network comprises 128 nodes divided into 4 modules of 32 nodes. Each node is connected to the other nodes in its module with probability p_i, and to nodes in other modules with probability $p_o < p_i$. On average, thus, each node is connected to $k_{out} = 96\ p_o$ nodes in other modules and to $k_{in} = 31\ p_i$ in the same module. Additionally, p_i and p_o are selected so that the average degree of the nodes is $k = 16$. We display networks with: **a**, $k_{in} = 15$ and $k_{out} = 1$; **b**, $k_{in} = 11$ and $k_{out} = 5$; and **c**, $k_{in} = k_{out} = 8$. **d**, The performance of a module identification algorithm is typically defined as the fraction of correctly classified nodes. We compare our algorithm to the Girvan–Newman algorithm[5,18], which is the reference algorithm for module identification[11,18,19]. Note that our method is 90% accurate even when half of a node's links are to nodes in outside modules. **e**, Our module-identification algorithm is stochastic, so different runs yield, in principle, different partitions. To test the robustness of the algorithm, we obtain 100 partitions of the network depicted in **c** and plot, for each pair of nodes in the network, the fraction of times that they are classified in the same module. As shown in the figure, most pairs of nodes are either always classified in the same module (red) or never classified in the same module (dark blue), which indicates that the solution is robust.

FIGURE 371

Each of the fractal forms in FIGURE 370 is **specific** in a more **general** environment. FIGURE 371, taken from "Functional Cartography of Complex Metabolic Networks," by Roger Guimera and Luis A. Nunes Amaral, *Nature*, February 24, 2005, shows how a **general** set of four networks interacting as the diagonaled *square* of **sameness** can be transformed into a more complex and logically enhanced **specific** network.

COLOR THEORY AS CHROMODYNAMICS: HOW QUARKS ARE CONCEPTUALLY THE SAME

Geometry may sometime appear to take the lead over analysis, but in fact precedes it only as a servant goes before the master to clear the path and light the way.

—James Joseph Sylvester

Where does *self-creativity* come from? Color theory shows how and why *mass* can create successive increases in complexity, and that *each* such step is more than the sum of its previous parts. Increases in complexity originate with the quarks' geometric concept of **sameness** that is conserved timelessly on the *vertical*. Color theory, based on the colors of electromagnetic radiation with its **both-and** function, can accurately substitute for the scientific theory of quarks *mass* known as chromodynamics.

Let's start with the any two 180° **opposing** complementary *colors* that when mixed in *equal* amounts become exactly the **same** thing, *colorless*, as seen in the middle of Figure 58. They "mutually embrace," just as do two 180° complementary electrons that look exactly the **same** as **both** poles of the *chargeless* magnetic field. Let's take the complementary colors of **yellow** and **blue** that on the color circle are at 180°, as seen in Figure 21. When **both** are complements are mixed in *equal* amounts, they each embrace the other in diminishing intensities of their respective colors excluding only its 180° complement. **Yellow**, therefore, contains all colors of the color circle in diminishing intensities except **blue**; and conversely, **blue** contains all the colors of the color circle in diminishing intensities except **yellow**. **Both** complementary *colors* simultaneously interact as if they were *inseparably* the **same**—*colorless*.

The **sameness**—*colorless*—of just two 180° complementary *colors* in *equal* amounts are connected by a *straight line* that cannot complete the quarks *circle* of **sameness**. It takes all *three* sets of complementary colors to complete the quarks *circle* of **sameness**, as seen in Figure 21 bottom and Figure 65.

As each complementary pair of **sameness** is added in any order, each *addition* accelerates to more than the sum of its parts. If you *add* a third set to the first two, the three sets accelerate to more than the sum of the first two accelerations. One would intuitively think that any two sets of **sameness** would *double* the complexity of **sameness**, and the third set *quadruple* the complexity of **sameness**. This is merely a *numerical* progression, however, not one in which *each* pair *quadruples* the *acceleration* of the previous forces. This incredibly "strong force" of quarks is *constrained* within the proton's circle of **sameness**, as seen in Figures 34 and 36, conserved the **same** as the accelerated force of light.

By applying colors of light to chromodynamics, we get the **same** results is seen in Figure 34. Here the *additive* color of, say, *blue* is the **same** as the *subtractive* colors of **both** *cyan* **and** *magenta*; thus *blue* is more complex than the sum of its parts. The **same** can be seen for the other two *additive* colors, *red* and *green*. All three **samenesses** of **both** *additive* **and** *subtractive* colors *simultaneously* interact to complete the circle more complexly than their accelerating parts. This is demonstrated in Figures 34 and 36.

There is yet another interpretation that completes the quarks circle of *accelerated* **sameness**. Let us take the colors of *blue*, *magenta*, and *cyan* that together are the color *blue*; and then simultaneously interact this complex interaction of additive-subtractive with the complex subtractive-additive interaction of *yellow*, *red*, and *green*, *both of which* are *yellow*. Now superimpose these two complex complementary geometric concepts of **sameness**, and you get an even more complex **sameness** that completes the proton's *colorless* circle.

All three of the above interpretations of **sameness** *simultaneously* complete an even more complex **sameness** of the proton's *colorless* circle. No wonder we cannot *perceive* quarks since any way we try to interpret them we get the only the concept of **sameness**, *nothing*, the **same** as the more complex *nothingness* of three-dimensional *space* that mediates the **sameness** of the gravitational force.

Physicists speak of quarks as being able to lift its mass up by its own bootstraps, seemingly defying the laws of nature, thumbing its nose to the relentless concept of *entropy*. They haven't a clue that the quarks **sameness** of concept is forever the **same** and thus *conserves* its *constrained* force of acceleration *perpetually*, thereby giving electron a *chance* to choose **either** one of its **opposing** choices.

The quarks force of **sameness** is only *half* that of lights force of **sameness**. Quarks have 1/2-spin while the photons have a 2/2-spin. Quarks force has only the **sameness** of the *additive* colors, the other *half* being the anticolors of antimatter.

All these **samenesses** of mass and masses light amount to *nothing* perceivable, yet this *triviality* is a *profound* concept of physics. Read "An Emptier Emptiness?" by Frank Wilczek, *Nature*, May 12, 2005. Also, read "Generating Asymmetry," a side box in "Asymmetrical Threat Averted," by Eran Hornstein and Clifford J. Tabin, *Nature*, May 12, 2005, in which we see that the *creative* force in the pluripotent stem cell's **sameness** has to be "broken" so that the fetus can develop into the *complexity* **similar-different** body parts, terminating in **either** a *male* **or** *female* mammal.

The **sameness** of dark energy, the **sameness** of the gravitational force, the **sameness** of quarks, the **sameness** of cosmic light, the **sameness** of the pluripotent stem cells, the **sameness** of the brain's glial cells, the **sameness** of myelin sheaths are all *forces* of creativity that *mass* and *time* **oppose**.

At this point, in order to make sure that my reader is convinced that concepts of *mind* and *mindless* nature obey the **same** logic of *interactions* despite the brain/minds **opposition** to **sameness**, take a good look at Figure 342 where **both** *black* **and** *white—colorless—*form a grid, or *lattice*, of **sameness**. The 90° *angle* that *divides* this *colorless plane* into two equally **opposing** parts creates *color* in its depth. This visual experiment of the brain/mind demonstrates that the *reversible* interaction of color/colorless as **specific/general** respectively is a *part* of *mindless* nature's *timeless* matrix of Universal **sameness** that, to be knowable, is **opposed** by the *irreversible* arrow of *time*.
Now, let us go through successive increases in the complexity of *dimensional* **sameness** in color theory, seen in Figure 58, to show that concept and geometry are ordered by the **same** logic of interactions.

The *colorless* dimension of **both** *white* **and** *black* becomes medium gray at the exact center of the straight line between both extremes. Color theorists call this coordinate the *value* dimension, which corresponds to *quality* in quarks **sameness**.

The *intensity* dimension is created by connecting any pair of complementary *colors* in *equal* proportion *simultaneously* between both ends of a *straight line*. Again, **both** complementary *colors* simultaneously are transformed into the geometric concept of *colorlessness*, or the **sameness** of medium gray, at exactly the middle of the *intensity* dimension, the **same** midpoint of the *value* dimension. The *intensity* dimension gets its name from the extremes vividness of *color* at **either** end of the coordinate.

Place the *value* and *intensity* dimensions *perpendicular* to each other so that the two centers of **sameness** meet at the **same** point of medium gray where all four angles become *perpendicularly* the **same**, yet more complex and enhanced than the straight line of just the single dimension.

When we cross the *third* dimension of *hue perpendicular* at the **same** point as the *value* and *intensity* dimensions, all three dimensions interact simultaneously even more complexly the **same** than the former two dimensions. Conceptually this complexity of **sameness** is not a linear progression but instead is 2^2, plus 4^2, or sixteen times more complex, as seen in Figures 59 and 24.

Both the *value* **and** *intensity* dimensions are *constrained* to *colorlessness* while the *hue* dimension has an *unconstrained*, or *infinite*, number of *colors*. Instead of being dominated by *quality*, the *hue* dimension is *dominated* by *quantity* so that the **similar-different** *hues* can be **either** *added* **or** *subtracted* conclusively, as seen Figure 68. Here, the **same** *values* at the top do not add up, they remain the **same**. **Different** hues at the bottom do add up: 10 plus 20 equals 30, simple arithmetic. Substitute **similarities-differences** for the *colorful hue* dimension and substitute **sameness** for the *colorless value* dimensions, and you can understand how the *mind's* selection of a **specific** color **opposes** *mindless* nature's more complex **sameness** of all colors in **general**. A *colorful* picture is much more interesting and meaningful than the *colorless* canvas on which it is painted.

Substitute quantity/quality as **opposite/same** for color/colorless and you will see the link between the electron's concept of *quantification* by electrodynamics and the quarks concept of *qualification* by chromodynamics respectively.

To show how difficult it is for physicists and mathematicians to *quantify* the *value* concept of quarks, read "Starting from Square One," by Peter Weiss, *Science News*, August 7, 2004, from which FIGURE 372 is taken. In this FIGURE you can see that the **sameness** of these squares of squares make quarks so mathematically baffling. No wonder that *quantity* dominates *quality* else we could have no concept of *time* that *dominates* our lives.

FIGURE #372

QUARK JUNGLE GYM — By substituting a grid of discrete lines and points, or a lattice, for continuous space-time, physicists carry out many otherwise insoluble calculations of quark theory. The time dimension isn't depicted here.

Figure N, Addendum V reveals how the quarks *circle* of **sameness** is logically transformed into the diagonaled *square* of **sameness**.

Massless lights 2/2-spin *particle* of **sameness**, shown in Figure 75 top, interacts its *variable acceleration* at 90° in **opposed** to its *constant speed*. FIGURE 373 shows what light would look like if its **opposite/same** *interaction* were placed in *simultaneous double reversal* to create a three-dimensional form, more complex and enhanced than its *part*icle of **sameness**.

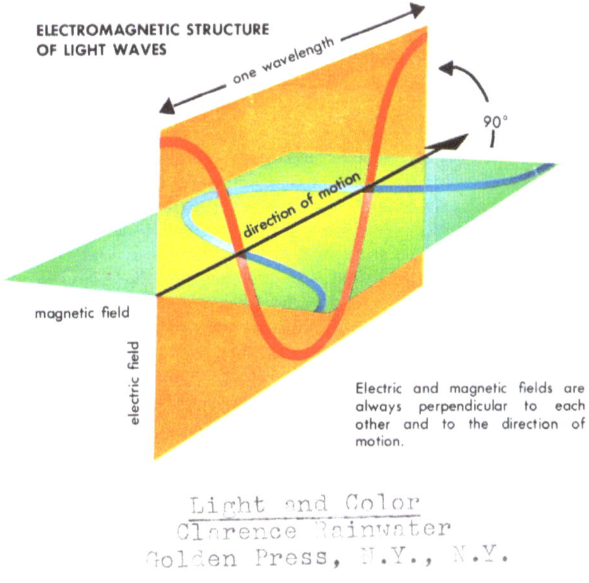

Light and Color
Clarence Rainwater
Golden Press, N.Y., N.Y.

FIGURE #373

When *cosmic* photons, *massless* particles of **sameness,** collide (interacts) with *mass*, it can transform the *imagined* square of **sameness** into the realistic FIGURE 360b. This occurs when light hits the chlorophyll complex in plant life. This highly complex and enhanced interaction creates a **sameness** of geometric concept in the "basic porphyrin ring" of the chlorophyll molecule, seen in FIGURE 374. The enhancing property of chlorophyll's **sameness** has *accelerated* the evolution of *complex* life while conserving the *simple* **opposite/same** interaction between mass/energy. Read *Eating the Sun: How Plants Power the Planet* by Oliver Morton, Fourth Estate, 2007.

FIGURE #374

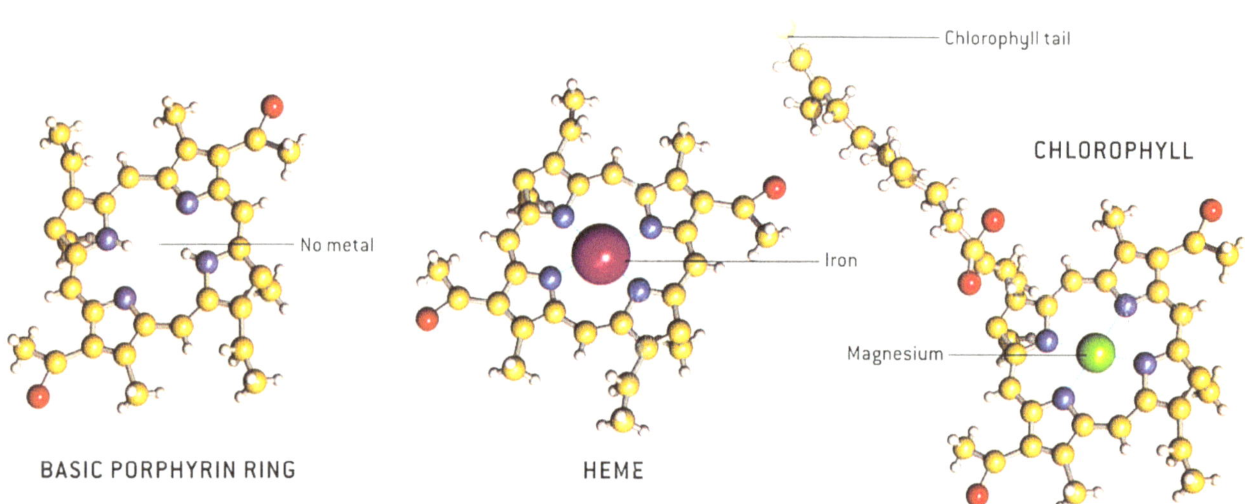

PORPHYRINS all have in common a flat ring, mainly composed of carbon and nitrogen, and a central hole where a metal ion can sit. The basic ring (*far left*) becomes caustic when exposed to light; molecules useful for photodynamic therapy also share this trait. Nontoxic examples include heme (a component of the oxygen transporter hemoglobin) and the chlorophyll that converts light to energy in plants.

From an article in *Scientific American* by Nick Lane, January 2003

A more complex system of photosynthesis, called "photosystem II", used by plants to convert light into energy came too late to include in this book, but can be found in "The photon trap", by Katharine Sanderson, *Nature*, 27 March 2008. You will see again that it is the geometric concept of **sameness**, **no different** than the magical **sameness** of the previously shown calcium pores, that creates and stores more fuel than plants require for their sustenance.

FIGURE 375 is another biological molecule with the geometric concept of **sameness**, taken from "Magnetic Blue," by Jeroen van den Brink and Alberto F. Morpurgo, *Nature*, November 8, 2007. This molecule's geometric concept is the **same** as that of a **generalization** or a magnetic field, the **sameness** of which when **opposed** by a **specific** metal at its *center* becomes an electrical conductor. I quote from the authors, "Perhaps their most interesting characteristics are its magnetic and electrical properties . . . . It is now clear that these compounds display a rich interplay of electronic, magnetic and structural properties with potential technological relevance . . . [theorists] have made some startling predictions. For example, they propose that, under appropriate conditions, these systems may become superconductors at high temperatures."

Figure 1 | Structure of metal phthalocyanines. Metal phthalocyanines are a class of molecule that comprises an organic, four-leaf-clover structure with a metal at the centre. Heutz *et al.*[1] show that the magnetic state of films of these molecules can be switched by controlling their crystalline structure.

FIGURE 375

Despite the relevance of **sameness** at any level of complexity, our *mass* **specificities** are always **opposed** to the *concept* of **sameness**. The whole gamut of our *factual* knowledge as well as our *concepts* and their interactions are only *part*s of universal ordering force of **sameness**. We are limited to the interactions of complete/incomplete as finite/infinite, change/constant and **specific/general** as **opposite/same** respectively. The great mathematician Kürt Gödel proved that no theory of knowledge is complete, not even a theory of mathematics. Even though our concept of *time dominants* every level of complexity it too is *incomplete* because it *asymmetry* must always **oppose** the *timeless symmetry* of Universal **Sameness**.

Read "The Incomplete Gödel," by Gregory H. Moore, in *American Scientist*, September–October 2005, in which Moore reviews two books about Gödel.

THE NEED FOR CONSTRAINT

The next great era of awakening of human intellect may produce a method of understanding the qualitative content of equations. Today we cannot.

—Richard P. Feynman

In certain particular problems that I have done it was necessary to continue the development of the picture as the method before the mathematics could really be done.

—Richard P. Feynman

Nature cannot be ordered about, except by obeying her.

—Francis Bacon

Equations, or formulae, are necessary for solving mathematical problems because without their *constrained* **sameness** of concept *simultaneously* divided into two equally **opposing** parts, *numbers* could not be *sequenced* to a logical conclusion and *quantification* by deduction would not be the *dominant* concept of physics and chemistry. With the proper notations substituting for the concepts, the mathematical formulae of science enable the human *mind* to solve the most difficult and complex problems of *mindless* nature.

If we think of the symbol for *equal* being the idea of **sameness** between the two **opposing** halves of any formula, we can write, say,

$$F = ma$$

which means that Force is the **same** as mass *multiplied* by acceleration. In reciprocal **opposition**, ma = F. Whichever *interpretation* is chosen is only a *part* of nature's more complex **sameness**. Our *interpretation* of nature's universal laws and forces are *constrained* to the formula's divided/unity or **sameness opposed** by the two **opposing** halves. Even the solution to the equation is but an interpretation of its conceptual point of view, such as F in the above equation. We cannot come to a **specific** *conclusion* with the concepts or numbers that perpetually *form* the equation. And the **specific** conclusion at **either** side of the formula is **opposed** to the **sameness** of the *equation.*

To get any *meaning* out of the **sameness** of an equation, we must *divide* it into two reciprocally **opposing** *parts* the **same** as we must *divide* inference into two reciprocally **opposing** *parts*, and even more complexly *divide* the megalogic on the inside front cover into two reciprocally **opposing** *parts.*

Equations are also *constrained* to the interaction of variable/constant. For example, in the formula F = ma, if mass is *constant*, then f/a is the *variable* proportion, written as m = a/F. *Conversely*, if mass *varies*, then the acceleration is *constant*; then mass and Force are *variables* written as a = F/m.

By the **same** token, the mathematical formula for determining speed, or velocity, is

velocity = (is the **same** as) length divided by time.

If the concepts of *length* and *time* are known to *vary* proportionally, we can determine the *constant* speed conclusively. Conversely, if *speed* varies, then *proportion* of distance/time must remain *constant* if the formula's *simultaneous* conceptual *interaction* is to give the mind the ability to *sequence* data to a conclusion. If you take **both** interactions of variable/constant as **opposite/same** respectively above that have been *reversed*, and superimpose them *simultaneously*, then they will become as complexly the **same** as *mindless* nature. However, we must *interpret* nature's **sameness** with **either opposing** formula to get any *meaning* from a *meaningless* universe.

Einstein's famous formula ($E = mc^2$) shows that **both** *mass* **and** *energy* can *vary* proportionally within the *constant* constraint of the speed of light. You can *interpret* this formula **either** from the point of view of *energy* ($E = mc^2$), **or** from the point of view of *mass*, written as $m = E/c^2$. In fact, *mass* was Einstein's original interpretation. **Either** reciprocally **opposing** *interpretation* of *mind* is a *valid part* of the **sameness** of the equation, which formula is also part of all the formulae of universal **sameness**.

Knowing that gravity is universally the **same** from all *cosmic* points of view, Einstein tried to find a *single* formula that encompassed *all* formulae. But like string theorists, his *mass* point of view could not encompass all points of view in the universe *simultaneously*.

Einstein's grasp of universal **sameness** however caused him to believe in a transcendental spirit that governs the universe, "Spinoza's God Who concerns Himself in the lawful harmony of the world, not a God Who concerns Himself with the fate and doings of mankind." Like many scientists, Einstein believed in an *impersonal* God that treats all mass **specificities**, no matter how complex, universally the **same**, including any *conscious* creature that can *conceive* of *itself* being part of nature's universal laws and forces.

Read "The Heretic Jew," in the Book section of the *New York Times*, June 18, 2006, by Harold Bloom, who reviews *Betraying Spinoza: The Renegade Jew Who Gave Us Modernity* by Rebecca Goldstein, Nextbook/Schocken. Bloom says, "Spinoza taught an intellectual love of God himself incapable of love." Reciprocal logic envisions an *impersonal* God, a **sameness** of *concept* that any **specific** *mass* **opposes**, even one as complex as a *person's* brain/mind.

Now let's again examine *geometry* to show the importance of its *constraint* on *concepts* to proportion their interactions. Within the *constraints* of just three geometric symbols, *point straight line* and *circle*, we can build successively increases in complexity to create not only the megalogic on the inside front cover but the logical form of gravity seen in Figure 105. Let's begin by taking two *points* and connecting them with a *straight line*. Now fit this *straight line* within the *constraints* of a *circle* so that it divides the circle into two mirror halves, as seen in Figure 2. This *constrained* straight line is called the "diameter." The interaction, or ratio, between the *diameter* and the *circumference* of the circle creates a more complex dynamic *point* that can lie anywhere on the circumference of the circle. This *variable* point on the circumference of the divided circle forms the vertex of a 90° triangle as seen in Figure 3. The idea of *variable* however needs a *constant* with which to *interact*. The *constant* is actually *created* at the *variable* point of the 90° angle. The constant *pi* is thought to be *transcendental* since it appears in all the basic formulae of science from seemingly *nothing*.

The formula for the circumference of a circle is:

$$C = 2 R$$

The *constraint* of this formula is the beginning of the accelerated complexity of geometric concepts seen in Figure N of Addendum V.

The ancient Greeks discovered the magic of this formula, but they had no logic of *geometric concepts* that *simultaneously* interact. Even later, the logic of inference could not explain why it could not solve its own *concepts*. Read, *Fearless Symmetry: Exposing the Hidden Patterns of Numbers* by Avner Ash and Robert Gross, Princeton University Press, 2006.

Equations with four numbers are proportioned the **same** as the diagonals of the antinomy in inference. Whatever you do to one of the numbers you must also do to its complement at the **opposite** end of the diagonal so that its proportion will remain the **same**. If the numbers are not *constrained* to the diagonaled proportion, the solution to the equation will be in error. Any three numbers of the four on the vertex of the diagonals have to be *known* if the *unknown* fourth number (x) is to be found. The problem is solved when the *unknown* number x *completes* the **sameness** of **both opposing** halves of the formula.

Read *Unknown Quantity*, by John Derbyshire, Plume, a member of Penguin Group, 2006. Also read "Naturalness in Theoretical Physics," subtitled, "Internal constraints on theories, especially the requirements of naturalness, play a pivotal role in physics," by Philip Nelson, *American Scientist*, January–February 1985.

The need for *constraint* occurs no only with the concepts of mathematics but with all concepts of the human mind. Whether *quantitative* or *qualitative*, all concepts must be constrained to a proportion to keep them from becoming *absolute*.

The *dominance* of inference in human language compels us to bring *concepts* to a *conclusion*. But this has had a devastating effect on our conceptual order. Throughout history, our *conceptual* engagements have been chaotic and diminishing.

An example of an *idea* that goes too far toward **either opposing** extreme of an interaction becomes *self*-contradictory and *unrealistic* is seen in FIGURES 376 and 377, by Wiley, taken from the comics of the *Tampa Tribune*.

FIGURE #376

FIGURE #377

The *unreality* of *self*-contradiction is what makes it so *comical*, except that in the *real* world it is usually *tragic*. The *tragedy* of absolute concepts now expressed in what is now called "political correctness," or PC for short, has turned the admirable concept of tolerance into intolerance. What was originally intended to promote the free expression of **opposing** ideas has contradicted itself into requiring everyone to think the **same** *absolute* idea, thereby muzzling the freedom to express **opposing** ideas. Just one of many examples of *self*-contradiction, the *intolerance* of *intolerance*, is seen in "PC Dangers Extend far Beyond UF's Doors," by Sheryl Young, the *Tampa Tribune*, December 20, 3007.

Another example of an *absolute* concept that has seemingly contradicted itself is in "Heil Woodrow!" the *New York Times*, December 30, 2007, by David Oshinsky who reviews a book titled *Liberal Fascism* by Jonah Goldberg, Doubleday, 2007. I quote from Oshinsky:

> Coming of age in the 1960's, I heard the word "fascist" all the time. College presidents were fascists, Vietnam War supporters were fascists, police that tangled with protesters were fascists, on and on. To some, the word smacked of Hitler and genocide. To others, it meant the oppression of the masses by the privileged few. But one point was crystal clear; the word belonged to those on the political left. It was their verbal weapon, and they used it every chance they got. Forty years have passed and not much has changed, complains Jonah Goldberg, a conservative columnist and contributing editor for National Review. Leftists still drop the "f" word to taint their opponents, be they global warming skeptics or members of the Moral Majority. The sad result, Goldberg says, is that Americans have come to equate fascism with right-wing political movements in the United States when, in fact, the reverse is true. To his mind it is liberalism, not conservatism, that embraces what he claims is the fascist ideal of perfecting society through a powerful state run by omniscient leaders. And it is liberals, not conservatives, who see government coercion as the key to getting things done.

We have already seen that Einstein, by making inference *absolute*, contradicted his own theory of *relativity*. So, too, did the ancient Greek Heraclitus of Ephesus. By making the idea of *change* absolute, he contradicted his belief of the *changeless* relationship between **opposing** ideas.

The unprecedented leisure time that science and technology have given us requires increased emphasis on *self*-constraint. Instead of using our spare time for the continuing development of the brain/mind, most of us take the least line of resistance by being entertained and satiated with pleasurable things that add little or no value to ourselves or to society. Read the Editorial in *Science*, January 21, 2005, that implores governments to do something about ignorance and self-defeating behavior.

The better emphasis is self-discipline to *constrain* one's own immediate emotions and temptations. Instead of blaming others for a lack of self-discipline and self-control, we should more often blame ourselves. The "blame game," however, is the quickest and easiest way to exonerate one*self* from *responsibility* so that blame will fall on some else.

The concepts of the human *mind* are evolution's greatest contribution to the complexity of mass. Whether one *interprets* the evolution mass **specificities** from *simple* to the great *complexity* of human consciousness by **either** *chance* **or** *design*, we take our gift of *conceptual* thought too much for granted. The logically ordered complexity of the human *mind* added to the complexity of our *mammalian* brain may be the most precious and valuable thing that has ever evolved, and may still be evolving. Yet we give this great mental capacity for logical order and wisdom short shrift.

Read "Future Perfect?" by Norman Myers, *Science*, June 23, 2006, who reviews a book titled *The Meaning of the 21st Century: A Vital Blueprint for Ensuring Our Future* by James Martin, Riverhead, 2006. I quote from the first paragraph of Myers, "The past 50 years have probably seen more change than the previous 500 years, and those 500 years witnessed more advances than the previous 5,000. The current century will surely pass the lot, with an onrush of scientific, technological, economic, social and political change, all of a character and on a scale way beyond anything we have experienced to date. The future will not be an extension of the past, but will mark a departure from much of our activities since the start of civilization 5,000 years ago."

If we spent just half of our *leisure* time exercising and developing our brain/minds as much as we do exercising and developing our physical bodies, our minds would be greatly enhanced. Knowledge is more powerful long term than physical strength and athletic ability. If we spent a reasonable portion of our nonworking time learning about logic and science rather than watching sports, reading gossip about celebrities, reading and watching pornography, and seeking the solution to absolute *happiness* or absolute *freedom*, we would not only get a much greater sense of fulfillment out of our own lives but also become more responsible citizens able to make wiser choices.

Substituting some of the thrills and excitements of our emotional brain that dumb down our minds with the lifelong learning of logic and science will be met with derision by those that need it the most: the poor, the youth, and the exploited. To them, it will seem ridiculous that mental exercise is more valuable and self-supportive than physical exercise, especially those who find that sports is their only hope to escape from poverty and achieve financial success. Due to the excessive popularity of competitive sports and the paucity of income for working scientists, it will be a daunting task to teach youth the enormous power of scientific thought.

Young athletes are willing to suffer endless and boring drills to achieve *physical* prowess but don't seem to understand that they must do the **same** thing to achieve *intellectual* prowess. Whether you are trying to learn to play a musical instrument, master computer technology, learn to be a surgeon, lawyer, doctor, mathematician, or scientist, you have to practice, practice, practice. Learning data by rote is essential to long-term and working memory, yet without the mind's *lateral* thoughts, students can't get any meaning from the data to understand why they are dong this or that. Read "Time for Hillsborough Schools to Bid Bon Voyage to Fuzzy Math," an Editorial in the *Tampa Tribune*, September 13, 2007, in which the editors state, "Florida's new math standards look more like Singapore's which is considered to have the best math instruction in the world. Instead of covering an average of 83 fast-paced concepts in a single school year, the curriculum will cover 18 with greater depth and mastery."

Because Muslim nations do not allow conceptual conflict within their culture, learning is almost entirely by rote. What they learn is their culture's *absolute* interpretation of the world. Read "Memorizing the Way to Heaven, Verse by Verse," by Michael Luo, the *New York Times*, August 16, 2006, from which I quote, "It's almost like a bank account for the afterlife," says an 11-year old who knows 6,200 holy verses. If Muslims spent just half as much time memorizing scientific data mathematical structures by rote as they do their holy verses, their *minds* would become much more in harmony with God.

If the Islam culture and religion would teach children science and logic the way they teach their *interpretation* of a universal God, the chaos and carnage now being suffered in the Near East would cease, their diminishment of mind would transform to enhancement not only for themselves but for all human minds. Read the section on "Islam and Science: Oil rich, science poor" *Nature*, November 2, 2006. If the *conceptually* brainwashed and *enslaved* minds of Muslims substituted scientific facts and the history of the difficult struggle for individual freedom, and the *conceptually free* nations copied the Muslims rigor of learning, human dialogue would be lifted to a higher level of ordered complexity.

Education in the United States has become excessively free in that young students too often set their own standards of learning. They select the easiest courses so they can get good grades with little effort of brain/mind. They get no sense of having to work to gain the power of knowledge. The result is that the very subjects that they reject are those that teach them how to comply with the forces and laws of nature—logic, mathematics physics, chemistry, biology, and evolution—and no less important, the history of the arduous struggle to free the human mind from conceptual tyranny. Without this knowledge, citizens become physical adults but remain mental children, unfit to vote in a free society.

That this is the actual situation here in America is seen in the following: "Most Students in Big Cities Lag Badly in Basic Science," by Diana Jean Schemo, the *New York Times*, November 16, 2006; "Freestyle Education Survived the '60s," in which, "Some Students Flee Rigid Schooling," by Nahal Toosi, the Associated Press, the *New York Times*, November 24, 2006; "As Math Scores Lag, A New Push For the Basics," by Tamar Lewin, the *New York Times*, November 14, 2006.

The best way to achieve enrichment of human interactions is to learn the logic of *mindless* nature's conceptual interactions and use this knowledge to create the laws for the *minds* part of conceptual **sameness**. An overarching human culture that understands and accepts the need for *self*-constraint due to our **opposition** to **sameness** at any level of complexity will lift us to higher level of logical order.

Since all argument takes the form of inference, it behooves us to teach the *logic* of inference as the key to understanding the need for not trying to infer the very concepts that form inference and thus do not sequence along with the *data* to a conclusion under the *conceptual* constraints of the antinomy. We will then understand the need to *interact* the geometric concepts of **specific/general** as dominant/dominantless. Once we understand that our *concepts* and their *interactions* are but an *interpretation* of nature's more complex constraining force of **sameness**, we can forge our human laws to comply with the *constraining* forces of nature and interact **opposites** with **sameness** despite our **opposition** to it.

As it is now, the greater our scientific knowledge and its accompanying technology, the more leisure time our minds have created for sensual pleasures and entertainment. We increasingly turn our minds away from the *interaction* of, say, individualism/collectivism as **specific/general** to choose **either** one as the "solution" to conceptual conflict. Anecdotal evidence for this is seen in the three references below:

"Technology's Future: A Look at the Dark Side," by Barnaby J. Feder, the *New York Times*, May 17, 2006. "Science and the Theft of Humanity," by Geoffrey Harpham, *American Scientist*, July–August 2006. "Art's Marginal Cost: The Ideal Frame," by Tracie Rozhon, the *New York Times*, June 4, 2006.

Paying *more* for the development of the brain-mind and *less* for the development of the body/brain would be a powerful incentive to enhance our cultural behavior.

For example, sponsors reward Howard Stern handsomely because so many people watch his dissemination of *unconstrained* vulgar thoughts and behavior, mostly in regard to explicit sex. Read Mark A. Stein under Openers in the *New York Times*, January 8, 2006, on how the more people listen to Stern's *unconstrained* trashing sex the more he is paid. This situation needs to be reversed. How much more uplifting and mysterious sex would be if it were not so lightly treated and considered to be merely a fun thing to do.

If parents understood the *science* of human development and evolution—which we will discuss later—and exposed their children more to such works of *art* as Richard Wagner's Liebestod from his opera Tristan and Isolde, Puccini's opera Tosca, Rachmaninoff's Concerto no. 2 in C Minor, Mozart's sonatas, the sculptures of the ancient Greeks, and the paintings and sculptures of Renaissance masters such as Leonardo da Vinci and Michelangelo, then love and sex would take on the deeper meaning.

Evolution will teach that the emotional thrill and sensual ecstasy of young sex has been selected to ensure reproduction and the survival of life. The consequences of casual sex can be deadly diseases and the creation of a child with our proper parental care. *Self*-constraint is by far the best choice for the couple and for society. Read Letters in the *Tampa Tribune*, September 22, 2007, headed, "Students Must Take Responsibility for Their Own Actions."

Read in the *Tampa Tribune*, September 25, 2006, what "the Legend of the Keyboard," Van Cliburn, has to say about the *intellectual* benefits of teaching children the art of music.

There is a relatively new encyclopedia of classical music on digital disks called *The NPR Listener's Encyclopedia of Classical Music*, by Ted Libbey, Workman. I quote from the first paragraph of "Readers Link to Site to Hear Classics," by Kurt Loft, the *Tampa Tribune*, August 22, 2006, "The latest entry in the cannon of great music takes readers on an interactive adventure, where background on Beethoven's 'Emperor' Concerto or Verdi's 'Aida' comes alive to the notes on a computer linked to the internet. With the flip of a page and click of a mouse, text and sound embrace in a way that makes even the most esoteric art educational and enlightening."

The reading of great books of Western and Eastern cultures is important for understanding the need of constraint, as suggested in "Books Are Great, But What Your Child Is Reading Does Matter," by Sheryl Young, the *Tampa Tribune*, Monday, January 16, 2006. Also read "Study Links Drop in Test Scores to a Decline in Time Spent Reading," by Motoko Rich, the *New York Times*, November 18, 2007. This study was done "by the National Endowment for the Arts, based on an analysis of data from about two dozen studies from the federal Education and Labor Departments and the Census Bureau as well as other academic, foundation and business surveys . . . . The results are alarming, a report's preface says."

The need for parental *dominance* and *constraint* in "Kids Gone Wild," with the subtitle, "Parents are more involved than ever before. So why do children today seem so rude?" by Judith Warner, the *New York Times*, November 27, 2005. The answer is the breakdown in parental authority. Children of the '60s generation who despised parental authority and are now themselves parents are much too permissive, allowing their children to do anything they wish. Without discipline, children will choose to do the most selfish, emotionally motivated things that are usually highly dangerous with little thought of the costly consequences to themselves and to society. Punishing disobedience and rewarding physical and mental achievement will always be the better way to mold character and integrity.

Read "This Is Your (Father's) Brain on Drugs," by Mike Males, the *New York Times*, September 17, 2007, in which the author states, "the real behavior crisis is among baby boomers, not teenagers." See FIGURE 378 taken from the *Tampa Tribune*, September 17, 2007. And read "When the Limits Push Back," by William Yardley, the *New York Times*, September 16, 2007. Also read "Right Stuff and Wrong in the Boys Who Dare," by Tamar Lewin, the *New York Times*, January 8, 2006.

FIGURE #378

As for concepts, a good way to get children of any culture or race out of their ignorance, superstitions, and tribal thinking would be for businesses, particularly large corporations with government assistance and tax incentives, to set aside some of their employees' time for the lifelong learning of logic and science. Since large corporations have as much to gain as their employees, such programs could be the catalyst for change to better cope with an increasingly complex world. Hopefully, this book will be helpful in this regard, a standard for inculcating the logic of inference and teaching the value of its antinomy human concepts reciprocally *interact*.

Read "The value of the Stick: Punishment Was a Driver of Altruism," by Jocelyn Kaiser, *Science*, June 23, 2006, and "Costly Punishment across Human Societies," by Joseph Henrich et al., *Science*, June 23, 2006.

Also read "How Does the Teenage Brain Work?" by Kendall Powell, *Nature*, August 24, 2006, subtitled "Changes in the structure of children's brains may account for the risky business of adolescence." There will be more on this when we get to the chapter on the long and difficult *development* of maturity in the human brain/mind.

If the wife as mother and husband as father understand that their relationship is *logically* ordered as two **opposing** parts of the **same** more complex idea of *sex*, the family circle's interactions can be properly *constrained*, thereby setting an example for their off-springs that will not just empower them to fulfill their own ambitions but also motivate them to *conserve* an orderly society.

Such a *self-constraining* culture will be much more efficient and economical, without costly enforcement from outside the immediate family. But outside enforcement of laws will always be essential to communal order, which is why religion may have played such an important role throughout the history of all human cultures. Read *Breaking the Spell: Religion as a Natural Phenomenon*, by Daniel C. Dennett, Viking, 2006, which states, "A community that believes in an immensely powerful enforcer gets the benefit of its norms being followed without paying the cost." Read also "Futile Campaign for One Man, One Woman," by Joseph J. Brown, the *Tampa Tribune*, January 12, 2006. Also read "Just Say No," by Peg Tyre,

Julie Scelfo, and Barbara Kantrowitz, *Readers Digest*, June 2005, taken from *Newsweek*, September 13, 2004. Also read "Yes, You Have Permission to Just Say No to Your Children," by Jennifer L. Boen, The News-Sentinel (Fort Wayne Ind.), the *Tampa Tribune*, October 8, 2007.

Gifts to many charitable organizations often end up going mostly to the employees of such organizations and then to the leaders of the countries that need the donations. Many leaders snatch away most of the money for themselves then dole out to the poor only amounts sufficient for their survival. Keeping their subjects poor and ignorant with the impression that their leader is their only hope of survival, the tyrants are able to stay in power. If the citizens of a country were well educated and had the *chance* to rise out of poverty, they would rebel and overthrow such a tyrant and his corrupt bureaucrats and determine their own fates. Without scientific education and opportunity, there is no hope for the masses. Read "Obama Urges Kenyans to Get Tough on Corruption," by Jeffrey Gettleman, the *New York Times*, August 29, 2006.

Well-meaning charity can be beneficial in the short term, but in the long run, like feeding wild bears, such beneficiaries grow increasingly dependent on their benefactors. The unintended consequence is that the *beneficiary* usually bites the hand that feeds it, as occurred in the French Revolution. The very culture that makes them poor and dependent reinforces their dependency on continuing handouts, thereby defeating the very purpose of the charity.

The gifts of medical research, building of schools, and training teachers from the Melinda and Bill Gates Foundation is a more enlightened charity for the poor, sick and deprived people of Africa and is more efficient because it bypasses a lot of bureaucracy. The most heartfelt and charitable donations of food and medicine however will not get primitive cultures out of their long-term predicament without formal education in science, logic, and the history of the long and arduous development of free and self-sufficient societies. Given the opportunity to learn these subjects through charitable donations is the only hope for individuals in primitive tribal cultures to become their own *masters* instead of the *servants* of tyrants.

Read "The Strongman's Weakness," the *New York Times*, September 10, 2006, by Jeff Turrentine, who reviews a book titled *The Wizard of the Crow* by Ngugi wa Thiong'o, Pantheon Books, 2006.

If the people of the world are to free themselves from dictatorial rule, they must be taught the history of how those that have freed themselves have done it. They must understand the intent of the Founding Fathers of the United States who designed the Constitution and were then elected to set an example for those to follow. While in office, they considered themselves to be *servants*, not *masters*, so that the people would be free to select others in future elections if they so desired. By *serving* the people, the Founding Fathers set an example that unfortunately has not been followed. Far too many seek public office in their lust for power and personal gain.

Read "Original Intent," by Jon Meacham, author of *American Gospel: God, the Founding Fathers, and the Making of a Nation*, who reviews two books in the *New York Times*, June 25, 2006. I quote from Meacham's reviews:

> "They sought, often unsuccessfully but always sincerely, to play a part, to what Jefferson called natural aristocrats—aristocrats who measured their status not by birth or family that hereditary aristocrats from time immemorial had valued but by enlightened values and benevolent behavior." Character for the founders was the most public of matters. "To have honor across space and time was to have fame, and fame was what the founders were after." It was this definition of character that . . ."made the founders different." "Fame resulted when they acted not as office seekers but as selfless statesmen, summoned from rural splendor to fight in the arena for a time, and then retire to their farms . . . for the contemplation of philosophy and the cultivation of the land and its virtue . . . . They did not conceive of politics as a profession and of office-holding as a career as politicians do today . . . . The founders were not intrinsically more virtuous than those that came before or after them. What they managed brilliantly was to marshal ambition in the service of benevolence . . . . Our task, as always, is to overcome our worst instincts . . . . The motives which predominate most human affairs," [Washington] once said, "are self-love and self-interest."

To get a sense of the intent of the Constitution, read "Plan of the New Constitution," the first British printing of the American Constitution. To capture the spirit of freedom of the American Revolution, read the collection of newspaper articles, *Common Sense*, by Thomas Paine, 1776.

Read also "Other Nations Catching Up to United States," by Justin Rattner, chief technology officers, Intel Corporation, *Science*, September 1, 2006. Also read "The New World of Global Health," by Jon Cohen, *Science*, January 13, 2006.

Many Oriental countries have been reluctant to give up being *masters* of the people because they would lose their *unconstrained* power to bribe and arbitrarily suck the wealth from their own people and from foreign investors.

The "Republic" of China is a case in point where the *paradise* communism (read, **sameness**) is a living *hell*. This is starkly revealed in, *Will the Boat Sink the Water?* by Chen Guidi and Wu Chuntao, Public Affairs, 2006. You can get a quick sense of the *chaos* in this book from its review in the *New York Times*, July 6, 2006, titled "Betrayed by the Revolution," by Emily Parker from whom I quote, "The tragedy, although whitewashed at the county level, captured the interest of Beijing when officials there noticed a discrepancy between the county's version and another (more brutally accurate) news report. The Central Committee launched an investigation that revealed the dire condition of the peasants in Ding's part of Anhui province. The head of the Supervisory Committee for Law Enforcement, an eyewitness recalls admitted tearfully: 'I never dreamed that so many years after Liberation, the peasants are still so poor, and their lives so hard, their tax burden so heavy, and they are so badly treated by cadres.'" Also read "Rumblings from China," by Nicholas D. Kristof, the *New York Times*, July 2, 2006.

Such tribal cultures are difficult to overcome because the *dominance* of *self*, and thereby *selfishness*, is *conserved* in our animal brains, along with the dominance of alpha males that rule the social and reproductive order.

The unique *concepts* of the human *mind*, however, can and should rise above such primitive *emotional* and *perceptual* behavior by using the logic of *interactions* where *dominance* and *dominantless* are **opposing** parts of the **same** more complex idea of *sex*.

A good way to counter extreme selfishness and corruption is to introduce economic competition, the hallmark of a relatively *free* society if its laws are enforced the **same** for all citizens. Competition makes the consumer, the populous, the master of society free to choose the *lowest* prices along with the *highest* degree of goods and services. Nothing instills more efficiency and quality in the economic system than enforced competition. In such an economic society, *self*, as **specific**, is dominant; but *selflessness*, as **general**, must be included to conserve a *healthy* market. An economy that is too extremely *free*, lacking enforced constraint on greed and exploitation, as history repeatedly reminds us, will *sicken* a nation's economy. The **same** thing will occur in the **opposite** way if there is too much constraint and excessive enforcement. For the two **opposing** sides of this **same** coin, read "The Free Market: A False Idol After All?" by Peter S. Goodman, the *New York Times*, December 30, 2007. Also read "Contrivances to Raise Prices," an essay on the history of business by John Steele Gordon, *Barrons*, December 31, 2007. For a logically ordered economy, human minds must interact greed/generosity, self/selfless, exploitation/conservation, as **specific/general** dominant/dominantless respectively.

The women's "liberation movement," the laudable effort to make women *equal* to men in political and economic life, has been carried to such an extreme that many people think there is **no difference** between the two sexes. The idea of "unisex," that *male* and *female* are the **same**, has *diminished* the **opposing** roles of *husband* and *wife* and of *father* and *mother* respectively. The human *mind* cannot ignore *mass* being **opposed** to **sameness** or defy the *interaction* of **opposite/same** where **oppositeness** *dominates* but is *constrained* within the **same** proportion.

Read "Why Sex Matter for Neuroscience," by Larry Cahill, *Nature Reviews Neuroscience*, June 2006. I quote pertinent parts from the conclusion:

> It is evident that sex influences at all levels of the nervous system, from genetic to systems to behavioral levels. The picture of brain organization that emerges is of two complex mosaics—one male and one female—that are similar in aspects but very different in others . . . . The immediate task for neuroscientists at all levels of our field is to challenge the (often implicit) assumption that sex matters little. Obviously, ignoring significant influences of sex, should they exist, can only retard progress . . . . This is, of course, not a simple task, but it is a necessary one if we are to fully comprehend how and why sex influences brain function in so many ways . . . . Despite the heightened complexity it implies, the issue of sex influences seems to be much too important, both practically and theoretically, to be ignored or marginalized any longer in our field . . . . To quote a recent report from the medical branch of the National Academy of Sciences, "Sex does matter. It matters in ways that we did not expect. Undoubtedly, it matters in ways that we have not yet begun to imagine."

The need for *proportional constraint* in sex as well as life in general is seen in "Stones-Be-Gone: Gene-targeting Drug Restores Chemical Balance Protecting the Gallbladder," *Science News*, December 4, 2004, in which a gene has been found that cures a diseased gallbladder by restoring its chemical *balance*.

Also read "A Higher Power for Insulin," by Fiona M. Gribble, *Nature*, April 21, 2005, from which I quote, "The liver has a central role in glucose homeostasis because it extracts glucose from the bloodstream in times of plenty, and synthesizes glucose in times of need. Thus, if buffers the body from extremes in glucose concentration—a relative excess after meals, and the relative storage between meals . . . . The liver recognizes these different energy states through changes in the blood insulation concentrations; insulin binds to its receptor in the liver and directly inhibits glucose production. But this well-established pathway is far from the whole story." In This Issue of the same issue of *Nature*, giving a brief explanation of the above, I quote, "[The authors identify] K_{ATP} channels in the medial hypothalamus as fundamental to regulation of glucose homeostasis. Thus activation of K_{ATP} channels in the hypothalamus decreases hepatic gluconeogenesis (promotion hypoglycaemia) and their activation in pancreatic B-cells is known to decrease insulin secretion (promoting hyperglycaemia). These central and peripheral actions may be designed to balance each other in order to maintain glucose homeostasis."

Without understanding the *value* and need for *balance*, we are much too prone to go off the deep end **either** way and then make excuses for its unintended consequences. We are geniuses at making excuses for our own misbehavior with such rationalizations as "everyone else seems to be doing it, so why shouldn't I?" Or "so many others are getting away with stealing the identity of others on the Internet, why shouldn't I?"

But if I hurt *society*, I also hurt *myself*. We are all in the **same** socio-economic boat. *I* and *others perpetually* interact as **specific/general** that even the most rational human cannot escape. Humans are still animals that are *dominantly selfish*, especially the young that are extremely *self-centered* with little regard for other *selves* with their *dominances*. Childish *unconstrained* behavior requires parental discipline based on the concept of the constraining force of **sameness**. Read "Favor the Rod, Get the Ax," by Patrick D. Healy, the *New York Times*, March 10, 2005. Also read "Crime Foes: Prevention Is Solution," by Valerie Kalfrin, the *Tampa Tribune*, June 3, 2005. Also read "A particularly good read on the need for constraining youth is found in "Angela Whitiker's Climb", *The New York Times*, June 12, 2005.

Any individual, society, culture, race, or religion can learn how to get along with others, act in a civil way without excessive emotions through the logic of conceptual interactions. Since our *mind's* concepts and their interactions are part of the **same** logic as *mindless* nature we will be better equipped mentally to interact in an *orderly* way with our conceptual **opponents** by complying with the natural order of concepts. Read "Our Overrated Inner Self," by Orlando Patterson, the *New York Times*, December 26, 2007, in which the author states, "Maybe we're all bigots. But we can still be civil."

If the citizens of the United States would put more self-constraint into their present "anything goes" culture, then other cultures that put excessive constraint on personal freedom would not be so terrified of us, especially by our excesses through the *unconstrained* medium of the Internet. If each person in the world could practice *self-constraint* on the Internet instead of trying to exploit it for *self-serving* purposes, think how efficient and enhancing the openness of the Internet would be. No one would have to worry about one's identity being stolen, or information distorted or compromised. We would be free to exploit all the benefits of almost instant worldwide person-to-person communication and with access to Web sites such as Google with their vast store of information. Mutual trust would prevail, and there would be no more fear of viruses and hackers.

But as we shall see, evolution has used every *concept* including **either** *trust* **or** *deceit*, **either** *prey* **or** *predator*, **either** *exploitation* **or** *conservation*, **either** *competition* **or** *cooperation*, etc. on the *chance* that **either** concept of an interaction will dominantly **either** *increase* **or** *decrease* the complexity of *mass* **specificities**.

Read "You're Not Alone," subtitled, "In the Digital Age, Thieves Don't Have to Climb through the Window," by William L. Hamilton, the *New York Times*, November 23, 2006.

Not only do thieves break in and steal, but the **opposite** also occurs when tyrannical governments that want their subjects to all think the **same**. To do so, the tyrants censor free expression of **opposing** ideas over the Internet. Read "Web Tool Said to Offer Way Past the Government Censor," by Christopher Mason, the *New York Times*, November 27, 2006.

Although there will always be those who are unable to constrain their impulses and avarice, the human race would be much better served if we understood that conceptual *unity* will always be *divided* as **opposite/same**, the **same** as the electro/magnetic force that logically orders all *macro*-mass. Only then will *mutual trust* dominate, even become the dominant culture of the World Wide Web.

Listen to disk 26, lecture 51: Kant's Moral Theory, by Professor Kane and to lecture 52: Burke: The Origins of Conservatism, by Professor Shrarmur, a CD sold by the Teaching Company for a good example of attempts by philosophers to get human laws to correspond to nature's laws. We already have the means to do so with inference with its conceptual paradox, and its *logic* is already effectively working for scientists that constrain their *temporal* information processing within this *timeless* form. Kant's "starry heaven above and the moral law within" in *logical* terms is simply the interaction of **general/specific** respectively.

Nations fall and chaos results when we design our own personal laws based on some absolute idea of how the world ought to be governed. The fall of Russian *communism* is a classic example of attempts to impose **sameness** of mind to the exclusion of any **opposition**.

The *dominance* of our *mass* point of view, the *dominance* of **similarities-differences** in *mass*, our **similar-different** brain-minds with their **similar-different** races, cultures, and religions make it obvious that no single *idea* is the solution to the world's conceptual conflicts.

For a good read on the timeless and insoluble paradox of conceptual thought, go to "A Philosopher's Vision of Fundamentalism," by Edward Rothstein, the *New York Times*, January 9, 2006.

HOW SYMMETRY OPERATIONS BUILD SUCCESSIVE INCREASES IN COMPLEXITY

God forbid that we should give out a dream of our own imagination for a pattern of the world.

—Francis Bacon

To do science is to search for repeated patterns, not simply to accumulate facts.

—Opening sentence of Robert MacArthur's *Geographical Ecology*

If *constrained* **sameness** is the *creative* force that drives the evolution of complex systems, then one can extrapolate that nature builds successive increases in complexity with what physicists call "symmetry operations." Symmetry operations make **no difference** to the laws of nature. They also make **no difference** to any point of mass. For example, the *rate* of time changes when an observer goes from one accelerated frame of reference to another. But any observe cannot detect this change in the rate of time. His clock ticks the **same** even though other observer's clocks tick **differently** in **different** frames of reference. The *sequence* of *time* is not Universally the **same**, because it is **opposed** to **sameness**, the *simultaneity* of which is *timeless*.

The **sameness** of "symmetry operations", too, are *timeless*, so let us see how its three operations; *translation*, *rotation*, and *mirror reflection, can successively build complexity:*

Translation is the moving of any shape along a *straight line*. The reason that this makes **no difference** to *mindless* nature is because the idea of uniform *motion* in a straight line and the idea of *rest* are the **same**, as seen in Figure 76. An *inertial* observer therefore can be *interpreted* as being **either** at *rest* **or** in uniform *motion* in a straight line, **unopposed** by any force that would *change* **either** its *velocity* **or** *direction*.

Rotation is motion in a *circle* where any point its circumference always returns to the **same** point from which it began, *inconclusively*.

Mirror reflection makes **either** hand the **opposite** hand—i.e., **both** hands contradict each other forming a contradiction that is *handless* and therefore *symmetrical*.

The *simple* 1-D mirror reflection at the top of Figure 325 is also the **same** as Figure 52 where the more complex **either** *left* **or** *right macro*-hand is *contradicted* by its 1-D mirror image. Superimpose **both** *top* **and** *bottom* of Figure 52 *vertically* without reversing **either** one and all knowledge of *handedness* is extinguished.

The top of Figure 325's one-dimensional mirror *reflection* also shows the complexity of **both** the isolated *electron's* contradictory half-spin **and** the *anti-electron's* contradictory half-spin, **both** of which form the *contradiction* of the *handless* photon seen in the middle of Figure 325. This *contradiction destroys* any idea of *mass*, leaving only pure *energy* with the *symmetry* of its **both-and** function. Although the human *mind* must *interpret* **either** the mass or the energy concept, the *mindless* cosmic photon conserves its *symmetry* **unopposed** by the *asymmetry* of the electron's **either-or** function. Only when its *symmetry* of cosmic *energy* collides with the asymmetry of *macro*-mass does the photon become *part* of an interaction.

Returning to the top of Figure 325, a move to the *right* by the *left* hand is *simultaneously* accompanied by a move to the *right* by the **same** *left* hand that excludes any motion or idea of the *right* hand. Doing the **same** thing with the *right* hand also excludes any idea of a *left* hand. Now envision doing the **same** thing with **both opposing** hands *alternately*, but this time moving the two hands to the *left*. You get the **same** results in the **opposite** way. Now envision placing **both** the **opposing** *left* **and** *right* hands each with their **opposing** 180° directions of motion in double reversal *simultaneously*, and the idea of **oppositeness** is excluded, leaving only the more complex and enhance geometric concept of **sameness**. This **sameness** of *mindless* nature is **opposed** by the concept of any **specific** *mass* no matter how complex. If our brain/minds were not *dominantly* **opposed** to **sameness**, we could not perceive anything and we would not be able to *deliberately* move in any direction *conclusively* in *time*.

Now let us go to the middle of Figure 325 that reflects the *cosmic* realm where **both opposing** hands *simultaneously* form a *contradiction*. Here, a move to the *right* by the *left* hand is *contradicted* by a *simultaneous* move to the *left* by the *right* hand. The *right* hand and the *left* hand are *simultaneously handless*. The **same** contradiction will occurs using the real *right* hand. Again, this is borne out by **both** cosmic photons that *simultaneously* and *inseparably* move away from each other at the speed of light, always remaining the **same** regardless of *distance* or the *time*. If a *macro*-observer *alternately changes* **either** contradictory photon, the other photon will *simultaneously* conserve its **sameness** of concept, ignoring the *macro*-mass idea of **either opposing** concept *alternating* in *time*. The *simultaneous* function of **both/and** excludes the *alternating* function of **either/or**. What is so frustrating to a *macro*-mass observer is that the absolute exclusion of **either opposing** concept necessarily excludes the **opposing** ideas of **either** *mass* **or** *energy* so that light is **both** *particle* **and** *wave* respectively. Although our *macro*-mass can *interpret* light as being **either** *particle* **or** *wave alternately*, until *massless light collides* with *mass*, *interacting* mass/massless as **opposite/same** respectively, the *cosmos* remains *dark*, *nothing* perceptible. Thus, in or *macro*-realm we *perceive* something as being **either** *bright* **or** *dark*, but **neither** perception excludes the other. No material **either** *absorbs* all light **or** *reflects* all light, just as no photographic film will record any image if the aperture of the camera allows **either** too *much* light or too *little* light to strike the film. Some degree of **sameness** must occur between **both opposing** *perceptions* whether of the eye of the camera or the eye of an animal.

The realm of our *macro*-mass is seen in Figure 53 where **either** 180° **opposing** hand *interacts* at 90° with the **sameness** of **neither** hand. **Either** complementary hand can be *separated* from its complement but is *inseparable* from its 90° interaction with **neither** hand—i.e., hand/handless always interact at 90°, or **both** complementary *hands* can *simultaneously* interact with *handless* perpendicularly. In our *macro*-world, a *real* dancer on a stage with a backdrop of two mirrors at 90° reflects **both** *real* **and** *unreal* dancers that are *absolutely inseparable* and *absolutely synchronized* in 180° contradictory directions. Two *real* dancers, however, are *separable* and can only be *relatively synchronized.*

Now refer to Figure 54 where the duck has been realigned at 90° from the duck in Figure 53. The geometric concepts in *each* interaction of front/rear and left/right switch positions. Here we see the **same** thing in Figure 54 as in Figure 53 but in reciprocally **opposing** ways. Now superpose **both** Figures 53 and 54 and a more complex geometric concept of **sameness** is created as seen at the **bottom** of Figure 325 that pictures our *macro*-world where the *left* hand and its **opposing** *right*-hand mirror *image simultaneously interact* to seemingly complete the **same** circle. **Neither** hand contradicts the other, and **either opposing** hand can *alternately* in *time* to seemingly complete the **same** circle, **both** *left*—**and** *right*-handed circles are more complex and *timelessly* the **same** to *mindless* nature. If you superpose **both** Figure 53 and Figure 54 *simultaneously* in double reversal, you will *complete mindless* nature's cube of **sameness**, the *complexity* of which is beyond the comprehension of the human *mind.*

It is intuitive to think that to form the above cube you would connect **both** halves of the cube along their respective edges. But as we have seen with the symmetry operations in Figure 104 and Figure N in Addendum V, one must place one of the halves of the cube *behind* the other so that **both** *halves* join by simultaneous reversal at the **same** point. Now, **both** *halves complete* a square of **sameness** that is four times more complex and enhanced than **either** half.

Although this single cube is *complete* the Universal fabric of **sameness** is not.. Its *diagonals* extend *infinitely* and *timelessly* the **same** to create three-dimensional space with an infinite number of cubes of cubes, all *timelessly* the **same**, **opposed** only by *time*. At any vertex on this lattice framework, any **specific** observer sees the **same** thing that any other observer sees from any another point of view, yet each observer interprets his point of as being *dominant*.

DEVELOPMENT AND EVOLUTION

Logic is invincible because in order to combat logic it is necessary to use logic.
—Pierre Boutroux

Nothing in biology makes sense except in the light of evolution.
—Theodosius Dobzhansky

From *Nature Reviews Neroscience*, "what distinguishes humans from other species? The cerebral cortex is the seat of reason and intelligence, and the six-layered, highly convoluted mammalian neocortex is indicative of the increased neuron number and complexity required for more advance functions. But what are the factors that determine neuron number—and therefore cortical complexity—during embryonic development? The review by Dehay and Kennedy (page 438) . . . examine how intrinsic and extrinsic signals integrate to give rise to precise numbers of neurons in specific cortical areas."

The above suggests that *development conserves* the *simple* geometric concepts of **opposites** (intrinsic/extrinsic) on the *horizontal*, while the paragraph below suggests that the interactions of *evolution conserve* the concept of **sameness** on the *vertical* rise of *complexity*.

From "The Simplicity of Metazoan Cell Lineages," by Ricardo B. R. Azevedo et al., *Nature*, January 23, 2005 I quote from the abstract, "Developmental processes are thought to be highly complex, but there have been few attempts to measure and compare such complexity across different groups of organisms . . . . We find that these cell lineages are significantly simpler than would be expected by chance. Furthermore, evolutionary simulations show that the complexity of the embryonic lineages surveyed is near that of the simplest lineages evolvable . . . . We propose that selection for decreased complexity has played a major role in molding metazoan cell lineages." See FIGURE 379, taken from this letter to *Nature*.

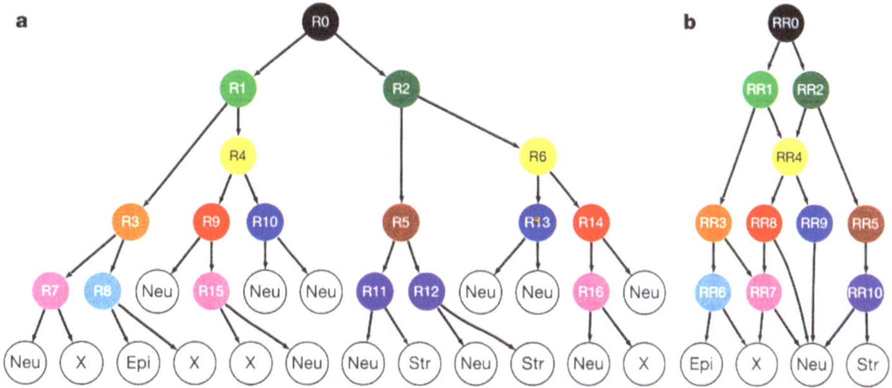

Figure 1 Example of the calculation of cell lineage complexity. **a**, The *C. elegans* ABarapp sublineage gives rise to 18 terminal cells of four different types (open circles): epidermal (Epi), neuron (Neu), structural (Str), and death (X). We begin by describing the cell lineage as a series of 17 rules, one for each cell division (solid circles): R0 → {R1,R2}, R1 → {R3,R4}, …, R16 → {Neu,X}. Solid circles of the same colour indicate equivalent rules, ignoring planes of cell division (for example, R7, R15 and R16). **b**, The minimum algorithmic description of the ABarapp sublineage consists of 11 reduced rules. Each reduced rule is represented by a solid circle labelled RR0–RR10, with a unique colour matching that of equivalent cell divisions (for example, RR7 → {Neu,X} corresponds to the initial rules R7, R15 and R16). The lineage complexity of ABarapp is calculated as the number of reduced rules divided by the total number of cell divisions: $C = 11/17 = 65\%$.

NATURE | VOL 433 | 13 JANUARY 2005 | www.nature.com/nature **153**

FIGURE 379

This suggests that *simple* order is conserved the **same** on the *vertical* as life's **opposing similarities-differences** become increasingly complex on the horizontal.

Chance given enough *time*, constrained by the accelerating force of quarks **sameness**, *dominates* successive increases in the complexity of *mass*. *Chance* with its **either/or** function dominates *development* on the *horizontal* while *design* with its *chanceless* **neither/nor** function dominates Darwinian *evolution* on the *vertical*. If we could *double* the interaction of chance/design with its either/neither function, we would get the **same** thing with this conceptual interaction as with any other conceptual interaction.. But alas, the geometric concept of any *mass* of any **specific** *design* at any level of complexity is always **opposed** to **sameness**. Although the *concept* of mass is *timeless*, *entropy changes* the **specificity** of mass over *time* into the unconstrained **generality** of *energy*. Mass/energy as **specific/general** interact at 90^0 on the *horizontal* as well as on the *vertical* with **both** interacting as *half* of nature's geometric concept of **sameness**.

Self-replication, Reproduction, and Development on the *Horizontal*, Dominated by Reciprocal O**pposition**

Because any **specific** form of life is *temporal*, it is necessary for it to be able to keep reproducing itself. The imperative to reproduce is as strong as the imperative to survive. Simple life forms with distinct nuclei reproduce by *replication*. Simple *duplication* makes each successive generation *almost* the **same**. They mutate *collectively* by *chance* with little *change* in *design*, conserving their **general** organization. Because of their *dominance* of **generality**, when acting as pathogens, they are attacked by the immune system of complex organisms; and their only means of defense is to *collectively* mutate into something *slightly* **different**. In more detail, when an *antigen* detects a pathogen, it immediately signals the immune system to reproduce its kind, thereby transforming its *dominance* of **specificity** into a **generality**. Conversely, the pathogen transforms its *dominance* of **generality** into a **specific** mutation. *Antigen* and *pathogen each* interact **specific/general** reciprocally on the *horizontal* not only in *perceptual* conflict but also in *conceptual* conflict **no different** than between human minds.

The more *complex eukaryotes* are composed of one or more cells with distinct nuclei, a **generalization** that is *dominantly* a **specific** form of life. Eukaryotes that are even more complex with **similar-different** organs that make up the **specific** body, called "metazoans," will be described in detail later. The fertilized cell of metazoans by which development begins is "**undifferentiated**." This geometric concept of **sameness**, *conserved* from the creative force of quarks, *designs* the development of **opposing similar-different** body parts. At the termination of fetal development, the *dominance* of **oppositeness** gives birth to **either** a *male* **or** a *female* by *chance*. **Either** sex of any such species is not just **similar-different** but *unique* in **opposition** to **sameness**. The interaction of **opposite/same** in animal development is seen in Figures 88 and 153a, both showing the **same** process taken from two different sources. These two interpretations show the **same** succession of **opposite/same** *interactions* during fetal development.

Cells replicate by *doubling* into *equally* **opposing** parts so that they become two new cells that are exactly the **same**. This interaction, known as "meiosis," is seen in Figure 182. If the cell's microtubules do not line up exactly the **same** along the *dividing* line between the **opposing** halves, this failure to *conserve* **sameness** results in a defective fetus that usually aborts, or if not, will later cause the dreaded disease of cancer. With meiosis, **opposite/same** also *interact* on the *horizontal*.

In most instances, if DNA fails to faithfully reproduce the **same** set of nucleotides, a complex mechanism of DNA repair has emerged to edit this process. This is suggested in "Big Engine Finds Small Breaks," by Anna Marie Pyle, *Nature*, November 11, 2004, from which I quote:

> RecBCD aids in the repair of double-stranded breaks that occur during normal DNA replication and as a result of damage by ionizing radiation, two of which, RecB and RecD, have the ability to unwind DNA (they are helicases). These two motors move in the same direction along a DNA double helix, but on opposite strands, hence they move with opposite polarity, RecB progressing in the so-called 3' to 5' direction and RecD in the 5' to 3' direction. In this way, they an push together against a duplex section on of DNA, opening it . . . . Remarkable features of RecBCD include its unprecedented speed and processivity—its ability to remain associated with its substrate—together with a puzzling affinity for DNA that is terminated by a clean cut. The RecBCD structure now published by Wigley and colleagues reveals an elaborate complex of interlocking proteins that have invaded a DNA complex, opening four base pairs of DNA and sent the separated single strands in opposite directions. The 3'—terminated strand is fed towards the RecB motor and the 5'-terminated strand is fed towards the RecD motor. RecC appears to manage this operation, holding the helicase domains of RecB and RecD in place, splitting the DNA and directing the orientation of the divided strands.

It appears that RecC is required to **oppose** the **sameness** of **both** RecB **and** RecD for normal DNA repair. Without **opposition** to **sameness**, mRNA could not transcribe the nucleotides into codons and then translate them into the amino acids that make up the various proteins. RNA is so complexly **opposed** to the **sameness** that it can perform a multitude of **similar-different** tasks to maintain the integrity of the genome.

Figure 186 demonstrates that the idea of **either** *male* **or** *female* develops from the **same** body part where **no differentiation** by sex has occurred. Births of males and females average about fifty-fifty with males slightly more prevalent to account for their propensity for taking *chances* as well as the need for less males to fertilize females to assure survival of the species.

In most highly complex mammal species, pairs of *males*—the **same** sex—tend to *repulsive*, or **opposed**, to each other during the mating season, while *females* tend to be *attracted* to the *male* that *dominates* the ritual battle. As with electric charges, **opposing** sexes are *attracted* to each other while the **same** sexes repel each other, as seen in the Web site www.reciprocallogic.com.

The ritual battle of male **opponents** determines which one is the *fittest* for being *selected* by the **same** group of *females*. In some species, the courting ritual between the **opposing** sexes is often a demonstration of *symmetry*, a show of the *creative* force of **sameness**. The courting ritual of the swallow-tailed kite is a beautiful display of **sameness** between two **opposing** sexes. In other species, females are *attracted* to displays of males that demonstrate *symmetry* patterns because the "meiosis-type" pattern is a good indication of the male *fitness*. Likewise, males are highly attracted to the bilateral *symmetry* of the female body.

Similarities-differences of courting rituals between species hide the concept of **sameness** that is *perpetually* conserved on the *vertical* by the *constrained* accelerated force of quarks **sameness**.

There is a slight **difference** in the *size* of brains between the **opposing** sexes of the **same** species, most pronounced between human *males* and *females*. Read "Gender Differences in Cortical Complexity," by Eileen Luders et al., *Nature Neuroscience*, August 2004. Also read "His Brain, Her Brain," by Larry Cahill, *Scientific American*, May 2005. In these papers, the authors find that although women have smaller brains than men, their cortical area is much more complex. The **difference** is in the arrangement of cortical connections, or how cortical complexity is organized, not the amount of intellectual capacity.

The everyday *interaction* of the sexes is froth with **differences** of not just narrow *opinions* but of worldviews. *Males* and *females* **differ** in the *types* of their physical and mental abilities. In **general**, human *males* are *physically* stronger; but *females* are *genetically* stronger. *Males* are more *curious*, *assertive*, and willing to take *chances*, while *females* are more *cautious* and tend to maintain the *status quo*. Read *Why Men Never Remember and Women Never Forget*, by Marianne J. Legato, Rodale Books, 2005. I quote from the Book section of *Science News*, October 6, 2005, "Women's brains are wired for a higher degree of language processing and that this disconnect with men may be responsible for some of the miscommunication that often occurs between partners . . . . Acknowledging these differences may be the key to finally resolving the battle of the sexes, the author speculates." Also read "Why Sex Is Good," by Rolf F. Hoekstra, *Nature*, March 31, 2005, in which sex has been found to make life more resistant to stress, "showing that a sexual population evolves faster than an asexual population when challenged by a novel environment." Also read "The X Factor," by Erika Check, *Nature*, March 17, 2005.

The *dominance* of **either** one of two proteins determines sex. *Testosterone* is *dominant* in *males*, and *estrogen* is *dominant* in *females*. If **either** protein excluded the other, there could be no fertilization, and **neither** *male* **nor** *female* would be created. If **either** protein is *excessively* dominant—i.e., there is no proportional constraint of the dominant/dominantless interaction—a deformed infant will be born or the fetus will abort. To see that excessive dominance of testosterone in males can be fatal, read "High Testosterone Linked to Prostate Cancer Risk," by B. H., *Science News*, October 8, 2005.

Reciprocally **opposing** concepts *interact* between the male sperm and the female egg as follows: Immediately after sexual intercourse, millions of highly *active* male sperm engage in intense *competition* to determine which **specific** sperm can interact with the *passive* and **generally** organized female egg. The highly *competitive* and *selfish* sperms, however, also show some signs of *altruism* in that they *cooperate* to enhance the dominant sperm's *chance* to penetrate and, therefore, *interact* with the egg. Read "Altruistic Sperm: Mouse Gametes Team Up to Power One Winner," *Science News*, July 13, 2002. In the mean time, the *dominantless altruistic* egg becomes somewhat *selfish* and *active*, accepting just one **specific** sperm out of the thousands that seek her favor. Read "Sperm Alliance," under News and Views in *Nature*, February 1, 2007.

These reciprocally **opposing** *interactions* of **opposite/same**, **specific/general**, competition/cooperation, selfish/altruistic, active/passive, all dominant/dominantless respectively form in *simultaneous double reversal* to create the geometric concept of **sameness** in the first phase of development. This pluripotent stem cell, derived on the *vertical* from the constrained creative force of the quarks **sameness** and more complex than the sum of its parts, *doubles* in *simultaneous reversal*, creating another **sameness** *horizontally* that *quadruples* in complexity.

Again, refer to the logic of **opposite/same** in male/female reproduction in Figures 184 and 261, and note that the third cleavage is a *simultaneous double reversal* of the second cleavage before it becomes more complexly the **same** at the *last* division as the *first* division. The fertilized eggs force of **sameness** is the *design* by which life is given a *chance*, given enough time, to develop a highly complex **specificity**, **either** *male* **or** *female*.

The pluripotent stem cell's complex and enhanced force of **sameness** is capable of not only jump-starting but also *conserving* the **sameness** between the mother and her *developing* fetus. The female womb is *dominantly altruistic* and considers her developing fetus to be the **same** as her body parts. This is why the fetus is not rejected by the mother's immune system as foreign material.

To confirm the logic of *interactions* in the *developmental* process between **both** *male* sperm **and** *female* egg, see Figure 184. Contrast this with Figure 185 showing the *reproductive* sexual process of flowers.
The *reproductive* process between male/female humans is the **same** with all races, cultures, and religions. The conceptual conflict in this process between male/female is not a problem, but instead is the very logic by which the two sexes *interact* to reproduce and continue to do so after developmental termination. Trying to solve the conceptually **opposing** points of view between *male* and *female* is futile and so frustrating that without understanding the logic of interactions, incompatibility, anger, and estrangement will dominate their marital relationship. There always has been and always will be *conceptual chaos* when reciprocal logic is not understood and accepted, and inference remains the only ordering method of human *concepts*. Read "Imprinted and More Equal," by Randy L. Jirtle and Jennifer R. Weidman, *American Scientist*, March–April 2007. I quote the author's abstract, "Why silence perfectly good copies of important genes? The answer may lie in a battle between mother and father staged in the genome of the offspring." Conflict between the 180° **opposing** concepts of male and female is *genetically* conserved, remains the **same** on the *vertical*.

The acrimony that persists between *males* and *females* as to which is the more talented or intellectually capable—in a word, which one is *superior* to the other—overlooks the fact that men are better in some things than women, and women are better in some things than men. Read "Born to Shop," by Constance Holden, under Random Samples, *Science*, September 7, 2007, in which it states, "Women have an evolved knack for remembering where to find edible plant matter . . . . Rafts of studies have shown that men trump women at many spatial skills, a spillover from our past . . . when men were the hunters and women the gatherers." As with the male sperm, the male of any race, color, culture, or religion is *dominantly* more competitive, aggressive, and willing to take *chances*, yet cooperate with a leader to accomplish a **specific** task. This can make a highly *conservative* and safety-prone wife-mother furious, leading to inflammatory arguments that **neither** side is capable of solving.

Women's preference for safety and security for themselves and their children makes them less likely to be innovative and conceptually *daring*. As for the male's dominance for conceptual audacity, how many charlatans or cranks are women, eager to take a *chance* against the conventions of science? Not many. My brash and audacious attempt to dislodge inference as our only form of logic is a good example of the extremes to which males will go for *self*-recognition. Move over Aristotle, and Sir Francis Bacon, and make a place for *me*.

Males do not mature as fast as females, giving them longer exposure to childish curiosity and *mindless* behavior. Read "Positive Thinking," *Nature*, January 13, 2005, a book review by Daniel Nettle on *Exuberance: The Passion for Life* by Kay Redfield Jamison, Alfred Knopf, 2004. Also read "The Battle behind the Battle at Harvard," by James Atlas, the *New York Times*, February 27, 2005.

Even the chromosomes of *males* have been found to be more subject to mutation that *changes* organisms by *chance*, as seen in "Recycling the Y Chromosome," by Jennifer A. Marshall Graves, *Science*, January 7, 2005. Interactions are *conserved* longer in females than in males, enabling evolutionists to better trace the evolution of the species. Read "Strong, Silent, Redundant," the *Wall Street Journal*, May 27, 1999, by John O. McGinnis, who reviews a book titled *The Decline of Males* by Lionel Tiger, Golden Books, in which a plethora of **opposing** concepts is described between human *males* and *females*.

Looking at I in Figure 88, the first phase of cell morphogenesis, the beginning of fetal development, is the unfertilized egg with is **sameness** of concept derived from quarks creative force of **sameness**. Upon fertilization by the sperm in phase 2, the cell *doubles* into what is called the pluripotent stem cell that jump-starts fetal development. In phase 3, these two cells of **sameness** again *double* interacting **opposite/same**. In phase 4, the cell again *doubles* forming **opposite/same** in *simultaneous reversal*. From this last phase **similarities-differences** begin to appear gradually increasing in complexity and *dominance* in **opposition** to **sameness**. See also Figure 153, another source of morphogenesis. Figures 183 and 261 are other ways of interpreting fetal development and which also shows how this morphogenesis *ends* the **same** as it did from the *start*—**either** *male* **or** *female*—two **opposing** eukaryotes of the **same** more complex concept of sex.

Although the process of developing increasingly complex **similar-different** parts increasingly hides the **sameness** of the stem cell, it is nevertheless *conserved dominantless*. These *conserved* concepts of **sameness** assist in the *conservation,* and even the re-creation, of its **specific** body part and in some instances re-creating several parts at once. From the very beginning, **sameness** conserves, enhances, and *logically constrains* the information processing of the developing fetus.

At the beginning of the development, a dominant/dominantless *interaction* occurs at *each* of **both** ends of the fetus *simultaneously*. From the *logical* view of *interactions*, it appears that dominant/dominantless forms in *simultaneous double reversal* to create the **same** more complex body that is more than the sum of its parts. See Figures 194 and 235. With this logically enhanced **sameness** of concept, the developing neurons can send complex messages *simultaneously* between the **opposing** ends to route each neuron conclusively to its proper destination.

As would be expected by the logic of interactions, these developing neurons have "guidance molecules" that instruct the neurons to go in **either** of two **opposing** directions, **either** this way **or** that *conclusively*. This process is called "orthogenesis." I quote from This Week in *Science*, *Science*, July 2, 2004: "Developing axons encounter an array of signals telling them to go this way or that. In some cases, the **same** signaling molecule in one context says 'this way' and in another context says 'that.' Chang *et al*. (p. 103) have now elucidated some of the molecular complexity in these seemingly redundant but contradictory instructions. Identification of a receptor protein tyrosine phosphatase, CLR-1 in *C. elegans* provides the key. In some contexts, the CLR-1 partnership with netrin, an axon-guidance molecule, results in attraction, and in other contexts, results in repulsion of growing axon tips." In this quote, note that the **same** *signaling* molecule directs the neuron to **either** one of the two **opposing** routes. Reciprocal logic would suggest that the *signaling* molecule interacts with a yet undetected *silencing* molecule tweaking their proportion as needed to conserve the dynamics of homeostasis.

Sameness, or "polarity," appears to be *conserved* in the developing limbs of the body as seen in FIGURE 380, taken from "Mathematical Modeling of Planar Cell Polarity to Understand Domineering Nonautonomy," by Keith Amonlirdviman et al., *Science*, January 21, 2005. In this FIGURE, note that the geometric concept of **sameness** is *doubled* to interact as a super-**sameness** to create the fruit fly's wing. Compare with Figures 194 and 235, the orthogenesis that begins the process of development.

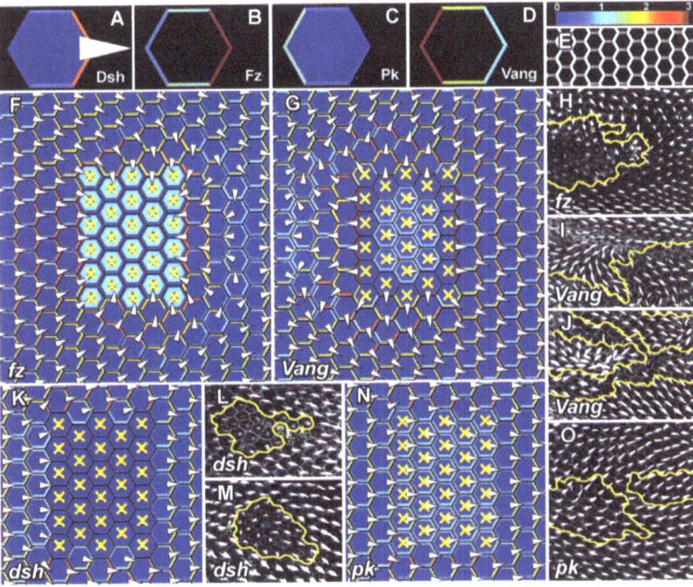

Fig. 2. Simulation results. Wild-type results showing the final distributions of Dsh (**A**), Fz (**B**), Pk (**C**), and Vang (**D**). The same color scale is used in all figures, where 1 is scaled to the initial uniform concentration of Dsh and the scale is truncated so that concentrations greater than 3 are shown in red. (**E**) Simulated distribution of Dsh displayed as an intensity representing total Dsh concentrations, corresponding to the appearance of Dsh::GFP in wild-type experiments. Simulation results of several PCP phenotypes showing the final distribution of Dsh with predicted hair growth directions derived from the vector sum of Dsh (**F**, **G**, **K**, and **N**) and corresponding pupal wings (**H** to **J**, **L**, **M**, and **O**). Greater Dsh asymmetry is represented by hair placement at increasing distances from the cell center. When Dsh asymmetry does not exceed the threshold value, the hair is depicted at the center of the cell. Mutant cells are designated with yellow. (H) fz^{R52} null clones. (I) and (J) $Vang^{A3}$ mutant clones. (L and M) dsh^{V26} mutant clones. (O) $pk\text{-}sple^{13}$ mutant clones. Asterisks mark nonautonomous effects near *dsh* clones.

FIGURE 380

Read "A Higher Order of Silence," *Science*, November 26, 2004, Adone Mohd-Sarip and C. Peter Verrijzr show that "(two copies each of core histones H2A, H2B, H3 and H4) are core histones [that] contain a trihelical histone fold domain that mediates histone-histone and histone-DNA binding, as well as unstructured amino-terminal tail domains." Again, we see two pairs of molecules *each* with **opposite/same** concepts that suggest they, too, might interact to form "a higher order of *silence* that enhances the *signal*."

The human brain continues to develop after birth until about the age of twenty for men and eighteen for women, much longer than all other animals. One of the first changes in the developing brain after birth is the ability to *perceive* **specificities**, then at about age three or the ability to *conceive* **specifically**. *Perception* occurs before language develops, which is why speakers of **different** languages *think* the **same** way. Read "Children Think before They Speak," by Paul Bloom, *Nature*, July 22, 2004. I quote from Bloom:

> 5-month-olds who are raised in an English-speaking community are sensitive to the Korean categories of meaning . . . an adult English speaker who has never heard Korean can tell the difference between a tight fit and a lose fit . . . meaningful contrasts are for making sense of the world . . . . Babies might understand the contrast between tight fit and loose fit, between support and containment, but they are unlikely to comprehend the contrasting meanings of the verbs 'leering' and 'glaring,' or the nouns 'accountant' and 'lawyer.' This must be learned . . . . What is the nature of learning? One compromise view is that there is a universal core of meaningful distinctions that all humans share, but other distinctions of meaning that people make are shaped by the forces of language.

Also read "Early Life Experience Alters Response of Adult Neurogenesis to Stress," by Christian Mirescu et al., *Nature Neuroscience*, August 2004.

Also read "Crime, Culpability, and the Adolescent Brain," by Mary Beckman, *Science*, July 30, 2004, from which FIGURE 381 was taken and from which I quote:

> Eight medical and mental health organizations including the American Medical Association cite a sheaf of developmental biology and behavioral literature to support the argument that adolescent brains have not reached their full adult potential. "Capacities relevant to criminal responsibility are still developing when you are 16 or 17 years old," says psychologist Laurence Steinberg of the American Psychological Association . . . . Adds physician Davie Fassler . . . the argument "does not excuse violent criminal behavior, but it's an important factor for courts to consider." . . . Structurally, the brain is still growing and maturing during adolescence, beginning its final push around 16 or 17, many brain-imaging researchers agree. Some say that growth maxes out at age 20. Others . . . consider 25 the age at which brain maturity peak . . . . A fully developed frontal lobe curbs impulses coming from other parts of the brain, Gur explains, 'If you've been insulted, your emotional brain says, Kill, but your frontal lobe says you're in the middle of a cocktail party so let's respond with a cutting remark.' . . . Luna points out that the tumultuous nature of adolescent brains is normal: "This transition in adolescence is not a disease or an impairment. It's an extremely adaptive way to making an adult."

Read "Behaving Yourself," found in a catalogue of the Teaching Co., (toll free: 1-800-832-2412) promotion, a course of the neurological origins of individuality, in which they say, "The Frontal Cortex is not fully functional until you're about a quarter of a century old, which suddenly explains a whole lot of fraternity behavior." Then read, "Flame First, Think Later: New Clues to E-Mail Misbehavior," by Daniel Goleman, the *New York Times*, February 20, 2007. Also read "The Teen Brain, Hard at Work: No, Not Really," by Leslie Sabbagh, *Scientific American Mind*, August–September 2006. I suggest you read again "How Does the Teenage Brain Work," by Kendall Powell, *Nature*, August 24, 2006.

Serious controversies have arisen as to when a young person should be held accountable for irresponsible behavior, especially criminal activity. The logic of interactions teaches that the *question* of *culpability* has no hard and fast *answer*. The idea of *guilt* is always **oppose** the idea of *innocence*, and a precise age for **either** choice depends on whether the judge or jury thinks that the young defendant is capable of reform by *constraining* his or her behavior. Read "States Rethinking Sending Youth to Adult Courts," subtitled, "Critics say it just makes them worse," by Sharon Cohen, the Associated Press, found in *The Tampa Tribune*, December 2, 2007.

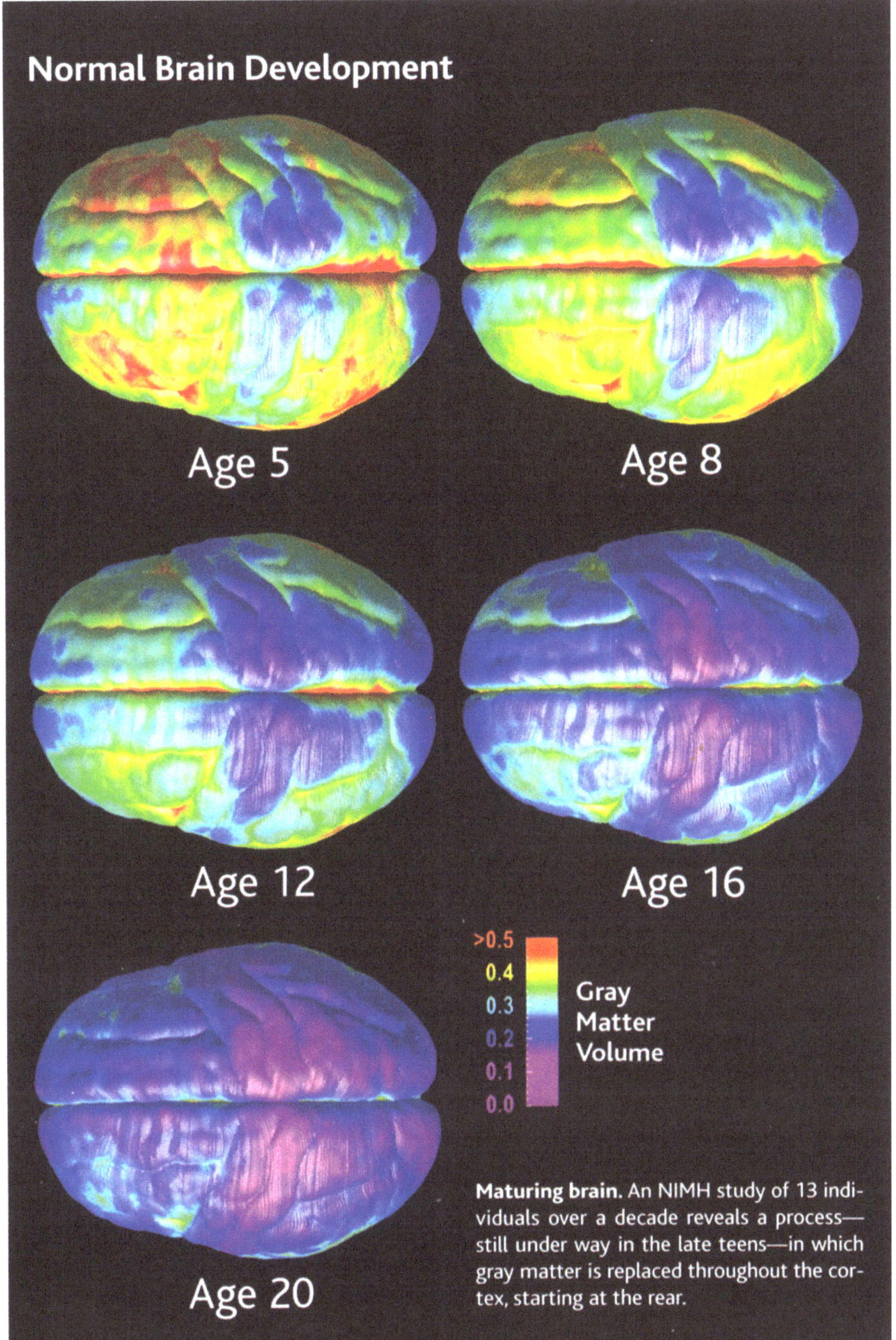

Maturing brain. An NIMH study of 13 individuals over a decade reveals a process—still under way in the late teens—in which gray matter is replaced throughout the cortex, starting at the rear.

FIGURE 381

The degree of human maturity correlates with one's ability to *constrain* **either** extreme of our primitive *emotions*. The recent emergence of frontal lobes in humans possesses this critical *constraint* on excessive emotional behavior. But the development of this *constraining* force takes more time and mentoring than permissive American parents are willing to undertake. Any **specific** *mass* dominantly **opposes** *constrained* **sameness**, especially the brains of highly *emotional* youth who have not yet had time for their frontal lobes to develop and *constrain* their extremes of behavior with *rational* thought.

Read "Author Criticizes Society's 'New Age' of Tolerance," by Kevin Walker, who reviews a book titled *In Praise of Prejudice* by Theodore Dalrymple, *Encounter 2007*.

Our **either/or** emotions and **either/or** perceptual predicates are *dominant*, and they don't want to be *constrained* by **sameness**. Even at maturity, the *brains* **opposing** *emotions* and *predicates* and our *minds constraining* **sameness** interact dominant/dominantless respectively in an unending and insoluble struggle. Understanding that **either** concept is only *part* of an interaction helps prevent the *dominance* of **either opposing** emotion from becoming excessive—i.e., *unconstrained*. To put it in the context of color theory and chromodynamics, **sameness** has the force of *value* to prevent **either** reciprocally **opposing** concept from extreme *intensity* of **either** complementary emotion. As to logic, intensity/value interact as **opposite/same** respectively.

Our primitive and naive emotions and perceptual predicates were designed by evolution to quickly make a choice of, say, **either** *fight* **or** *light*, without taking time to think of the consequences. The *dominance* of **either opposite** also requires *constraint* in our more complex and sophisticated human socio-economic systems to minimize *disorder*. Read "Food for Thought and Profit," by Paul B. Brown, the *New York Times*, October 15, 2005, from which I quote, "Investing and financial planning can be scary. They are made more difficult because of the way our brains hard-wired, **Money** magazine writes this month. 'Neuroscientists and psychologists have come to understand we're wired to act now, think later,' David Futrelle writes. 'Sure, unreasoning terror came in handy when that shadow at the cave entrance might have belonged to a saber-toothed tiger. In a more complex world, thought, an emotion-driven assessment of risk can lead us astray.'"

The **same** thing can occur in the **opposite** way. If you become too comfortable and complacent and your senses become too jaded instead of remaining alert to your surroundings, you will be too slow to react effectively against sudden danger. If you cannot *emotionally* save yourself *now*, there will be no need for the deliberation of rational thought *later*.

Understanding that the concept of **sameness** and its **both/and** function are logically **opposed** by the **either/or** function and that their *interaction* is a logical necessity for human thought transforms the *chaos* of any *conceptual* argument into an *orderly* and *rational* dialogue.

Conceptual *chaos* occurred in the United States during 2005 especially between the Democrats' choice of *collectivism* (**general**) and the Republicans' choice of *individualism* (**specific**). In this nasty dispute, the Republicans wanted to conserve constitutional *constant*—i.e., keep it *constantly* the **same**. The Democrats, on the other hand, were **opposed** to constitutional **sameness**, fighting to make the Constitution subject to *change*. Reading about this argument in newspapers and magazines revealed that each contestant often contradicted its own tenant, strongly suggesting that their argument was simply the *interaction* change/constant, chance/design as dominant/dominantless respectively, *part* of *mindless* nature's universal ordering force of **sameness**.

Understanding the *relativity* instead of the *absoluteness* of **either** conceptual choice will dampen the emotional flame of any zealot. Add to Einstein's theory of *relativity*, the logic of *conceptual* interactions and all mankind, not just scientist, will be enlightened by the *constraining* truth of paradox.

Ideal self-constraint is not to take advantage of the *weaknesses* of others but instead play on their *strengths*. This goes against the genetic grain of our animal heritage where preying on the weak and helpless is the easiest and simplest way to satiate hunger and survive, which is all the more reason to *constrain* or human behavior with the concepts of *empathy* and altruism. Read "Parochial Altruism in Humans," by Helen Bernhard et al., *Nature*, August 24, 2006, in which scientific experiments demonstrate in humans a deep-seated bias toward ones own race, religion, color, and culture. In other words, we are *dominantly prejudiced* and *dominantlessly altruistic*, even though we prefer to think that we humans are, or should be, *dominantly altruistic*.

Then see the Miramax film, *Deep Blue* directed by Alastair Fothergill and Andy Byatt (www.miramax.com/deepblue) that shows how animal behavior seems to be *dominantly immoral* from our *moral* point of view.

Predictable self-constraint in humans is the hallmark of *conceptual* maturity whereas in mature wild animals that are without a highly complex frontal lobe and without conceptual logic are dominated by *unconstrained* and *unpredictable emotions*. Animals learn self-constraint only through *perception and* experience, while human learn self-constraint by **both** *perceptual* experience **and** *conceptual* learning. Only the *concepts* of the human *mind* allow opportunity for highly complex and enhanced maturity, the *predictable constraint* of which we call *wisdom. The idea of* constraint however usually *develops* over time with the aid of having made too many fast and foolish *emotional* choices.

Despite the greatest effort to attain wisdom and develop *self*-constraint, the *dominance* of *self*-determination will prevail. Self/selfless as dominant/dominantless perpetually interacts in each of us. Normal minds can choose to be **either** *moral* **or** *immoral*, but **either** choice will always **oppose** the *amoral* behavor of animals. For a peek into the *amorality* of our genome's animal heritage, read "Parasitic Hairworm Charms Grasshopper into Taking It for a Swim," by Nicholas Wade, the *New York Times*, September 6, 2005. As for the human brain-mind, read "Driven to Market," by Jonah Lehrer, *Nature*, October 5, 2006, in which Colin Camerer and George Loewenstein say, "The mind is a charioteer driving twin horses of reason and emotion. Except cognition is a smart pony, and emotion is an elephant."

Our brain/minds will always **oppose** the logic of universal **sameness** that is *fair* to all points of mass. The emergence of the human *concepts* however allows us to come closer to natures more complex and enhancing universal ordering force than any other animal species. Our *minds* conceptual *interaction* of **opposite/same** being *part* of *mindless* natures universal **sameness** *constrains* us to a level playing field, the *design* of which gives all individuals the **same** *chance* to **either** *succeed* **or** *fail*. Read "Aiming to Level a Global Playing Field," the *New York Times*, September 3, 2006, by Stephen Kotkin, who reviews a book titled, *Making Globalization Work*, by Joseph E. Stiglitz, Norton, 2006.

Conceptual enlightenment begins with parental *self-constraint* as a model for their children along with encouraging *self-constraint* in their children. Discipline should be by **both** *reward*, **and** *punishment* should be dominated by *reward* that not only overcomes resentment and anger in their children but also creates respect for parental authority. Parental *neglect* with **neither** *reward* **nor** *punishment* will *diminish* the child's processes toward self-constraint and mature behavior. Read "Drama, Politics and Gang Warfare Behind Bars," by Ed Bark of the *New York Times*, October 6, 2007, who reviews a television series on the Discovery Channel, in which criminal, even in prison, cannot be constrained and behave like wild animals, unfit for release into an orderly socio-economic society.

For *genetic* reasons, young children cannot discipline themselves, they cannot even learn from their own experience let alone the experience of others. Their long period of development requires constant parental oversight as seen in "Teenage Risks, and How to Avoid Them," by Jane E. Brody, the *New York Times*, December 18, 2007.

As for the interaction between parents themselves, they may be likened to a hammer and an anvil. The *accelerated motion* of the *dominant* hammer interacting with the *motionless inertia* of the *dominantless* anvil enhances the *creative* work of a blacksmith. If two **opposing** hammers clashed without the inertial *constraint* of the anvil, only sparks of chaos would fly and the hammers' energy would be *wasted*.

For humans in **general**, logic and evolution favors *husband* to be *dominant* in the family's *external* affairs while the wife should *dominate* the family's *internal* affairs. This external/internal *interaction* and those mentioned above will assure a harmonious family life with much less need for the enormous social costs of governmental agencies having to check on evidence of child neglect, as seen in "Training To Help Identify Neglect," by Sherri Ackerman, the *Tampa Tribune*, September 3, 2006.

Above all, there must be mutual trust and respect if family harmony is to endure. Nuptial vows are more than just legal contracts to which both parties are bound. Marital contracts are bonds that enable families and society to conform to the laws and interactions of our *macro*-world. By adhering to the "morality" of the **opposite/same** electro/magnetic force, the "mutual embrace" of husband/wife can overcome the *division* of their *unity*. **Both** *husband* **and** *wife* interacting in reciprocal **opposition**, the **same** as the **electro/magnetic** force will create an arena of mutual trust, a *logical* environment that will also enhance the parent/child *interaction*. A read of the scientific investigations of *trust* in humans is "Brain Trust," by Antonio Damasio, *Nature*, June 2, 2005.

The *ideal* interaction between husband/wife is in the metaphor of a musical score played with, say, a violin and a piano. **Either** instrument can be *dominant*. Indeed the two instruments can *alternate* their dominance, but whichever one has the *dominant* role, the other must acquiesce and become *dominantless*. Regardless of which instrument or parent takes the dominant role, the *inclusion* of the *dominantless* role enhances the *dominance* of *order* and *harmony*.

The old saying states that "it takes two to tango." However, one of them must act *dominantly*, especially in ballroom dancing where both "mutually embrace." If **both** dancers try to *dominate*, there will be utter *chaos*, stepping of the other's toes as each one struggles for his or her preference of direction. The dominant/dominantless interaction of dancers corresponds to the *interaction* of tone/overtones, each **no different** than the logical *interaction* of **specific/general** dominnt/dominantless respectively.

I remember my reluctance to act with authoritative *dominance* while taking lessons in ballroom dancing from a lady instructor. To show me that my male *dominance* was required for a dance of "mutual embrace," she took the lead and *dominated* our dance by forcefully lifting me up off the floor to emphasize how easy, orderly, and enhancing it would be to waltz or foxtrot in an orderly way. Better that youth learns from such *extreme* examples by their teachers and parents than risking the limit of their own unsupervised behavior.

It is *fun* and *exciting* to go to **either** extreme, and conversely *dull* and *boring* to be constrained from the danger of extreme behaviors. There is a *genetic* urge for youth to test the limits of what they are capable of doing that should be *constrained* by parental authority because taking excessive *dangerous* actions that almost exclude the idea of *safety* amplifies the *dominance* of *chance*. Excessive risks will increase the *chance* that the *constraining* laws of nature will *punish* you.

Humans must obey the **same** logic as the *interactions* of nature if *order* and *harmony* are to prevail. Perhaps this is why Einstein found the complex interaction between violin and piano in Mozart's sonatas to be the key to understanding the complexity of nature's universal harmony. As with Einstein's *mind* experiments, Mozart introduced to the musical world the artful ability to **reverse** the *interaction* of dominant/dominantless *back* and *forth reciprocally* without losing the overall *dominance* of the theme, thus adding a layer of enhancing complexity to a musical score regardless of the key in which it is written.

That musical harmony is a good metaphor for parent/child interacting dominant/dominantless respectively, read, "Seeking the Key to Music," by Michael Balter, *Science*, November 12, 2004.

For an authoritative history of the human mind's quest to discover the law of nature to which human law can conform in a dominantly *free* socio-economic society, I strongly urge my reader to purchase the DVDs of *Natural Law and Human Nature*, taught by Father Joseph Koterski, SJ ,Fordham University, published by The Teaching Company, Chantilly, Virginia.

Citizens of the United States have a *dominantly* free socio-economic system that requires a high degree of *self*-constraint and *self*-responsibility. The United States Constitution grants *personal* liberty to all, but *constrains* each of us to the laws. The Constitution also requires all the *various* laws to be *constantly* enforced the **same** on all citizens regardless of genetic **similarities-differences**, race, color, or creed, thereby complying with the *logic* of nature's universal forces. The Constitution sets up a conceptual conflict between personal (**specific**) *freedom* and governmental (**general**) *control*, which dominant/dominantless conflict rages continually so that **neither opposing** side will become **unopposed**.

The Constitution also provides for three branches of government that are to be kept in a reasonably *balanced* proportion as follows: Congress legislates **similar-different** laws that can be *changed* as needed over *time*, while the Supreme Court makes sure that the laws passed by the people's representatives fall within the *constraints* of the *changeless* and *timeless* Constitution. The Supreme Court does not solve conceptual conflict, their judgments *interpret* the laws to make sure that they comply with and conform to the Constitution. This does not prevent individual citizens from continually and openly expressing their personal opinions on the courts judgments

The executive branch has a **specific** person, the president, whose deductive conclusions dominate the everyday affairs of the people. The president leads the people so that the rest of us will have some **specific** and *meaningful* purpose to *follow*. As chief executive, he or she is authorized to enforce the laws that congress pass and for the local and state courts to evaluate as to whether the laws fall under the *constraints* of the Constitution.

So far, the *interaction* of **specific/general** as individualism/collectivism, selfish/altruistic dominant/dominantless respectively has maintained its proportion that has not only *conserved* our republic but has been a magnet attracting the world's oppressed. Read "Message in Bottle Rescues Migrant Stranded at Sea," by Marianela Jimenez, the Associated Press, in the *Tampa Tribune*, June 2, 2005. That the conceptual conflict between **specific/general** as individualism/collectivism respectively has held is seen in "Guns and the Constitution," an editorial in the *Wall Street Journal*, November 24, 2007, from which I quote, "Is the 2nd Amendment an individual or a collective right?" The answer is **neither** one to the exclusion of the other, however the logic of conceptual interactions decrees that the **specific** or *individual* right is *dominant* but lonely for those who can constrain themselves within the laws of a free society.

As long as our government conserves its balance of power and its bureaucrats remain the *servant* of the people instead of their *master* and as long as our citizens use *self*-constraint to obey the laws, or else change them if necessary, freedom of mind will continue to prevail. Read "Selfish Leaders Often Forget That 'We the People' Are in Charge," by Douglas MacKinnon, the *Atlanta Journal-Constitution*, June 10, 2005. Read "Congress Can't Cure Itself," by Cal Thomas, the *Tampa Tribune*, January 19, 2006, in which the author said, "It's not just about corrupt lobbying of Jack Abramoff. He is a symptom of a greater problem being that too many in Congress—Republicans and Democrats—come to Washington less to serve the nation than to serve themselves."

Constraining one*self* to the **sameness** that the **different** branches of government **oppose**—i.e., interacting **opposite/same** dominant/dominantlessly respectively—will *conserve* our relatively high freedom of the individual. Freedom/slavery perpetually interact *within* each mind as well as *between* minds. Lose your proportional constraint between freedom/slavery dominant/dominantless respectively, and you will become a *slave* to the ruthless tyranny of the underworld or society's penal system. It is also better to be a slave to one*self* than to enslaved by a Mafia lord and his heartless pushers that addict victims with mentally toxic chemicals that permanently debilitate and imbalance the brains neurological and molecular interactions.

If no one *bought* highly *addictive* drugs, to whom could *producers sell*? With no one to sell to, the production of these *self*-destroying drugs would dry up, and "pushers" would no longer have such a *diminishing* affect on human society. The tragedy of *buying* and *using* addictive substances only *enhance* the self-serving killers and their thugs as seen in "With Beheadings and Attacks, Drug Gangs Terrorize Mexico," by James C. McKinley Jr., the *New York Times*, October 26, 2006. Also read "Cannabis, the Mind and Society: The Harsh Realities," by Robin M. Murray et al., under Perspectives in *Nature Reviews Neuroscience*, November 2007. Also read "Neglect of Latin America Poses Big Risks" from *American Interest*, January–February 2008.

Effective *self*-constraint is difficult to achieve even with the help of the logic of reciprocal logic because nothing is more alien to *individual freedom* than the idea of *constraint*. Understanding the long and difficult processes of the mind's development however will enable youth to understand their need for *self*-constraint and *self*-discipline. Read "Injection Room May Open in Calif.," by Lisa Leff of the Associated Press, found in the *Tampa Tribune*, October 21, 2007.

One of the first things youths must understand is the conceptual interaction of change/constant. For example, without first establishing a *constant*—be it a cord, a meter, or a yard—the measurement of *length* would be *meaningless*. There would be utter *chaos* trying to accurately determine *changes* in *length* without a *constant* rule of measure. The need for a *constant* applies to **different** types of measurements such as *volume*, *pressure*, *humidity*, etc. At our minds *relative* level of complexity,, any *constant* can be *changed*, and any *change* can be made a *constant* because change/constant reciprocally interacts as does **opposite/same**; but all *minds* must agree on *changing* a *constant*.

Mindless nature turns the *interaction* of change/constant in *simultaneous* double *reversal* to form a more complex and enhanced concept of *constant*, a **sameness** that is **opposed** by the *measure* of *time*.

If you substitute *chance* for *change* and *design* for *constant*, this leaves you with *chance dominating* our lives. To see how the *interaction* of **opposite/same** as chance/design dominant/dominantless respectively that brings logical order to our *macro*-mass realm can be transformed into a more complex and enhanced **sameness**, examine the schema below:

SAME CONCEPT	/	**OPPOSING** CONCEPTS
Good things happen to *good* people		*Bad* things happen to *good* people
Bad things happen to *bad* people		*Good* things happen to *bad* people

Now, superimpose the *left* and all concepts become the **same**. Superimpose the *right* and all concepts become the **same**. Superimpose **both** reciprocally **opposing** sides *simultaneously* and a **sameness** of *concept* forms that is more than the sum of its parts. This **supersameness** govern all mass **specificities** the universally **same** with *no chance* of ever *changing*.

For a universal force that brings the **same** logical order to all *mass* **specificities** *simultaneously*, there is no preference for any one of the four above events regardless of any individual religious or cultural interpretation. The tsunami in Malaysia during the Christmas holiday of 2004 when so many people were on or near the beaches, the great earthquake in Lisbon on Sunday in the eighteenth century when most Christians were in church, the Muslim pilgrimage to the glory of Allah in Baghdad, Iraq, on October 31, 2005, where hundreds were crushed to death on a bridge in a panic stampede, and the great earthquake in October of 2005 affecting the Muslim country of Pakistan and the Hindu country of India are all examples that the forces of nature care not what choice of *belief* one has, such as **either** *good* **or** *bad*.

Despite our *mind's* need for a personal God to watch over us and make life *meaningful*, *chance* has been the *dominant* concept as to who lived or died in these traumatic events. Yet our paradigm of deduction in inference and our life's conscious thoughts need *meaning* to sustain our sense of purpose makes a *meaningless* universe hard to believe. Read "Why Would a Loving God Allow Suffering and Pain?" by Cal Thomas, the *Tampa Tribune*, January 5, 2005.

One can pray to God for something *good* to happen to him or her, but the *chance* of it happening will *average* fifty-fifty. If something *good* does happen, it will probably be attributed to God having heard and answered one's prayer favorably. If something *bad* happens in spite of our prayers, it probably will be attributed to the *will* of God who wants to test our faith in Him. But "God makes the rain to fall and the sun to shine on the good and the evil." All we can expect from prayer is hope that the *fair* coin of *chance* will flip in our favor for any **specific** event in time. Read "Dodging Death in the Andes, and Greeting Life with Gusto," by Lary Rohter, the *New York Times*, September 30, 2006. After a harrowing ordeal, Nando Parrado said, "I think that life is simpler than we tend to think. We look for answers and more answers. But there are no answers. Things happen in life, good things and bad. People say why did it happen to me? Well, why not? Some people win the lottery, and other die in a car crash. It happens, and there is nothing we can do about it. The universe doesn't care what happens to you." Also read "Unlucky 13," by Karla Jackson, the *Tampa Tribune*, October 13, 2006.

Circling back to the idea of *constraint* in *development*, even the conceptual conflict of busy/lazy is needed to increase species diversity as read in "When Laziness Pays," by E. Klarreich, *Science News*, January 15, 2005, from which I quote:

> Researchers have proposed a solution to a long-standing evolutionary conundrum: Why do populations of identical organisms sometimes split into two strains, cooperators and cheaters? Cooperation and cheating, inescapable parts of human existence also occur in a wide range of other creatures, down to the tiniest. For instance, in a yeast population, some cells produce the enzymes required to digest sugars, while other cells mooch off their enzyme-producing fellows. Since all yeast cells live in the same environment, why are some cooperators and others cheaters? For decades, researchers have struggled to explain how such dichotomies arise. Now, a trio of mathematicians and zoologists have proposed a model for the split . . . the model could be formulated to deal with cases in which cooperation and cheating strategies are passed on through culture and education rather than just through genes.

An explanation of the evolutionary need for *cheats* is seen in "Cooperation and Conflict in Quorum-sensing Bacterial Populations," by Stephen P. Diggle et al., *Nature*, November 15, 2007.

Development and evolution need *cheats* and *imbalance* as well as *fairness* and *balance*, in succinct logical terms, **opposition** to **sameness**. The highest degree of fairness is *empathy* that is useful in highly complex cultures and societies regardless of their **similarities-differences** as read in "Empathic Neural Responses Are Modulated by Perceived Fairness of Others," by Tania Singer et al., *Nature*, January 26, 2006. I quote the last sentence of the authors' abstract, "We conclude that for men (at least) empathy responses are shaped by valuating other people's social behaviour, such that they empathize with fair opponents while favouring the physical punishment of unfair opponents, a finding that echoes recent evidence for altruistic punishment."

Score one for *punishment* to interact with *reward* to maintain social order. Parents, teachers, and group leaders of any stripe take note that reward/punishment perpetually interact in animal societies regardless of their complexities, and it is the **same** on the *vertical* for the great complexity of humans and their socio-economic order.

A culture of *shame* along with the teaching *self*-responsibility and *self*-constraint would be a powerful deterrent against destructive, chaotic, and lewd behavior—all three of which occurred in "A Park that Ran Wild with the Wrong Crowd," by Dan Barry, the *New York Times*, November 18, 2007. Pure animal *emotions* without a vestige of *conceptual* constraint led to this wanton behavior.

Evolution and development provides its own enforcement within highly social animal societies as seen in "Policing Insect Societies," by Francis L. W. Ratnieks and Tom Wenseleers, *Science*, January 7, 2005, from which I quote:

> The London Bobby is a reassuring symbol of civic order. But this symbol is also a reminder that human societies have conflicts. Insect societies, too, experience internal conflicts, and research increasingly shows that policing . . . ."Within both human and insect societies conflicts arise because the interests of individuals differ . . . . To prevent exploitation, social insects have evolved several methods of policing . . . . When different species are compared, one important overall conclusion emerges: More effective policing results in fewer individuals acting selfishly. There are other striking parallels to human society: Insect policing relies on both detection and prevention, and individuals sometimes attempt to evade policing."

Also read "Cops with Six Legs: Law and Order among Insects," by Susan Milius, *Science News*, March 19, 2003. There seems to be a genetic need to enforce the laws to conserve the socio-economic order of the colony, or more logically, to conserve **sameness** between **opposing** interests.
Either *kill* **or** *be killed*, **either** *predator* **or** *prey*, **either** *good* **or** *bad* has *genetically* alternated throughout evolution and has been *conserved* in varying proportions. All *mass* **specificities** are *dominantly* influenced by the function of **either/or**;

therefore one's culture and family environment can easily tilt the proportion to **either** extreme. Read *The Murderer Next Door: Why the Mind Is Designed to Kill*, by David M. Buss, Penguin Press HC, 2005. Also read "World's 'Never Again' Promise Was a Lie," by Leonard Pitts, the *Atlanta Journal-Constitution*, June 15, 2005, that tells of the endless "inhumanity of human beings."

FIGURE 382 taken from "Is 'Do Unto Others' Written into Our Genes?" by Nicholas Wade, the *New York Times*, September 18, 2007, that sums up a person having *developed* into a *mature* human.

<u>FIGURE 382</u>

ILLUSTRATION BY YAREK WASZUL; PHOTOGRAPHS, CLOCKWISE FROM RIOT POLICE, LAURA RAUCH/ASSOCIATED PRESS; LISA DEJONG FOR THE NEW YORK TIMES; LARS KLOVE FOR THE NEW YORK TIMES; JIM WILSON/THE NEW YORK TIMES; SAURABH DAS/ASSOCIATED PRESS

Before we leave *development*, we need evidence that this process is **reversible**; otherwise the reciprocal interaction of **<u>opposite/same</u>** would not have been conserved on the *horizontal*. Even at the high level of human complexity, embryologists have discovered that they can transform a mature skin tissue back into its embryonic stem cell. This discovery was done independently by two laboratories both getting the same results. I quote from "Scientists Bypass Need for Embryo to Get Stem Cells," the *New York Times*, November 21, 2007, by Gina Kolata: "All they had to do, the scientists said, was add four genes. The genes reprogrammed the chromosomes of the skin cell, making the cells into a blank slates that should be able to turn into any of the 220 cell types of the human body, be it heart, brain, blood or bone. If many genes had been required, the experiments would have failed, Dr. Thomson said, because it would have been impossible to test all the genes combinations."

The fact that the method of seemingly going back in time as being a rather *simple* procedure attests to the concepts of **opposite** and **same** being reciprocally ordered. Also it is more than interesting that four genes are required for the **similarities-differences** in skin tissue to be turned back into the **sameness** of "blank slates," suggesting that the *interaction* of **opposite/same** in *simultaneous double reversal* is *reciprocally* reverted back into the ***same*** *thing from which it derived*. See FIGURE 383 taken from "Advance in Stem Cell Work Avoids Destroying Embryos," by Gautam Naik, the *Wall Street Journal*, November 1, 2007.

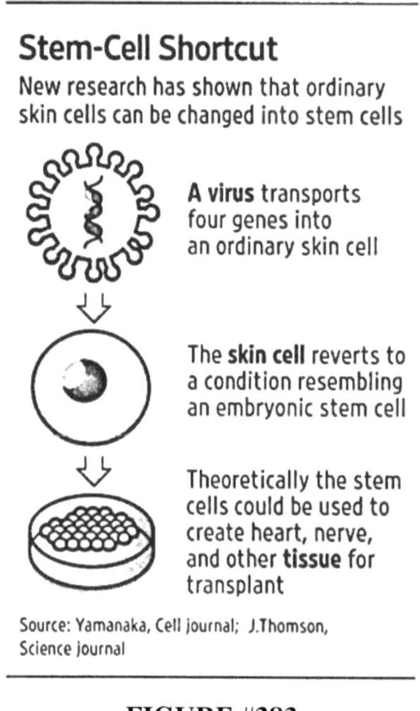

FIGURE #383

See this transformation of *concept* from **oppositeness** to **sameness**, refer again to Figures 88 and 153; but this time read them in *reverse*, from bottom to top.

Evolution of Complexity Dominated by Reciprocal **Sameness** on the *Vertical*

Reciprocal logic requires the acceptance of the *paradox* of division/unity as **opposite/same**; therefore evolution interacts contingency/convergence shown as in Figure 43 that has the **same** megalogical form as seen on the inside front cover. Read "Body Doubles," by Alberto G. Sáez and Encarnación Lozano, *Nature*, January 13, 2005:

> Taxonomists are . . . puzzled when they come across two or more groups that are morphologically indistinguishable from each other, yet found to belong to different evolutionary lineages. That is, when they discover a set of cryptic species. Our records of cryptic species are on the rise, often revealed by surveys of DNA variations. The story repeats itself with increasing frequency—a number of individuals belonging to a morphologically recognized species are sequenced (or otherwise genetically characterized), normally a several points (loci) within the genome. Then, often unexpectedly, the various genotypes will cluster in reciprocally monolithic groups, with no signs of genetic exchange between them. Similar evolutionary scenarios are evident at each locus, suggesting that the corresponding populations are reproductively isolated from each other, yet the sampled populations are not geographically isolated.

The logic of conceptual interactions argues against *contingency* alone, which means that *chance* alone, dictates evolution. *Chance* alone would not allow evolution to do the **same** thing again, not even create a self-conscious creature, at least not one that resembles humans or the great apes. Read "The Phenogenetic Logic of Life," by K. M. Weiss, *Nature Reviews Genetics,* January 2005. The truth (read paradox) is that **both** *chance* **and** *design* as **both oppositeness and sameness** respectively *simultaneously* interact at 90° the **same** as the electromagnetic force. *Contingency*, as with the *alternating* **opposing** electric charges, means *chance* and *change* that *dominate* the interactions of chance/design and change/constant respectively. Evolution is a détente of these conflicting concepts. See Figure 43.

If the accelerated *force* of quarks **sameness** were not *constrained* and if the idea of **sameness** were not *timelessly conserved*, there could be no successive increase in the acceleration of *mass* complexity and no *logical* method for Darwinian evolution.

By *duplicating* the **opposite/same** interaction in *simultaneous reversal*, the force of logical order is *quadrupled*, *conserving* the power of 2^2 seen in all nature's forces. My conjecture is that it is this **same** accelerating but force of quarks that *constrained* within the quarks circle of **sameness** is *conserved* to empower the *evolution* of mass **specificities**, interacting the **same** under the antinomy in inference as fetal *development*. Evidence of this is seen in FIGURE 384 taken from "Plant Speciation," a review in *Science*, August 17, 2007, by Loren H. Rieseberg and John H. Willis, which FIGURE shows that gametes that form in *simultaneous double reversal* are *viable*, where as those that *double* but *do not* form the **same** thing in the **opposite** way are *enviable*. The former is *viable* because it takes the *logical* form of inference, thereby creating the **sameness** of the antinomy in inference seen in the megalogic on the inside front cover. This *creative force* however is hidden by mass's **similarities-differences** that are **opposed** to being the **same**.

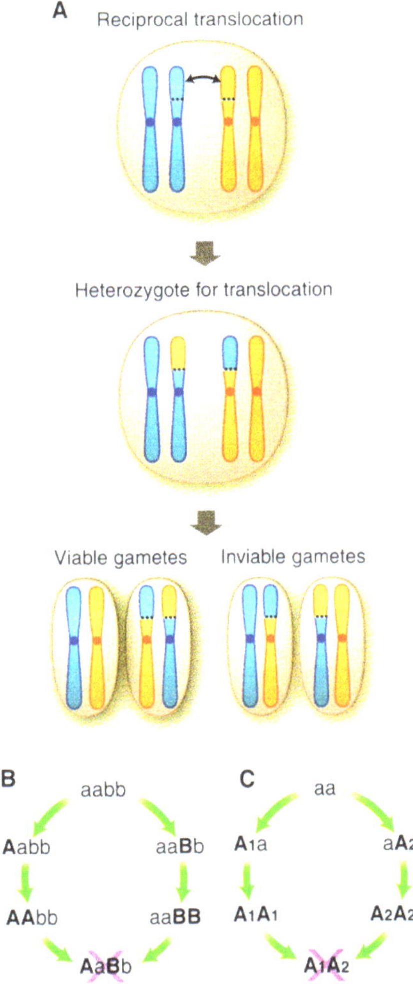

Fig. 1. Genetics of hybrid incompatibilities. (**A**) Example of a typical chromosomal rearrangement in plants, showing loss of fertility in heterozygotes because 50% of gametes are unbalanced genetically and inviable. (**B**) Classic two-locus BDM incompatibility in which new mutations are established at alternate loci and without loss of fitness in geographically isolated populations, but which are incompatible in hybrids. (**C**) Single-locus BDM incompatibility in which new mutations are established at the same locus and without loss of fitness in geographically isolated populations, but which are incompatible in hybrids.

FIGURE 384

Also read "Seeing Double," by André Goffeau, *Nature*, July 1, 2004, from which FIGURE 385 was taken. I quote the abstract, "How do genomes evolve? Studies of numerous yeast species confirm the view that the duplication of genes, larger segments and whole genomes are the key mechanisms."

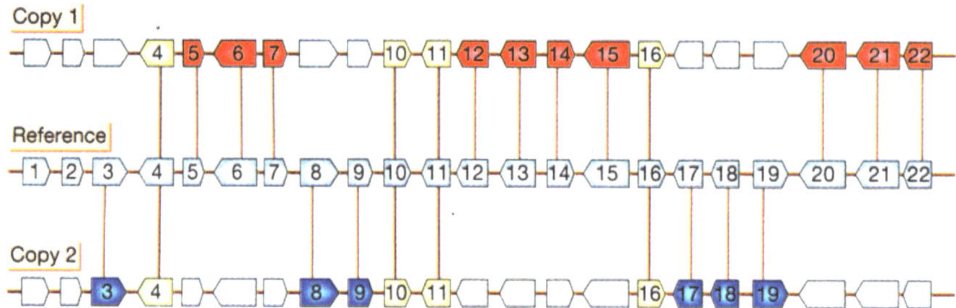

Figure 1 **Detecting whole-genome duplications. It had long been suspected[7,8] that the genome of the yeast *Saccharomyces cerevisiae* arose through the duplication of the genome of an ancestral yeast. Three new papers[1–3] confirm this suspicion. The confirmation involved sequencing the genomes of yeasts such as *Kluyveromyces waltii, Ashbya gossypii, K. lactis* and *Candida glabrata* (which served as reference species) and comparing their gene sequences, and their order and orientation, with *S. cerevisiae.* The genes in grey are in the same order and orientation in the reference species and *S. cerevisiae.* Some of these genes, those in yellow, have relatives that are found in two different chromosomal locations (copies 1 and 2) in the *S. cerevisiae* genome. Some, however, have relatives in the same order and orientation only on copy 1 (red), or only on copy 2 (blue). The findings indicate that genome duplication occurred in the lineage that led to *S. cerevisiae*; some genes were then maintained, while many others were lost or diversified.**

NATURE | VOL 430 | 1 JULY 2004 | www.nature.com/nature

FIGURE 385

More evidence that genetic interactions are *logically* ordered by the genome having been placed in *simultaneous double reversal* is seen in FIGURE 386, taken from "Sorting Out the Genome," by Bryan Hayes, *American Scientist*, September–October 2007.

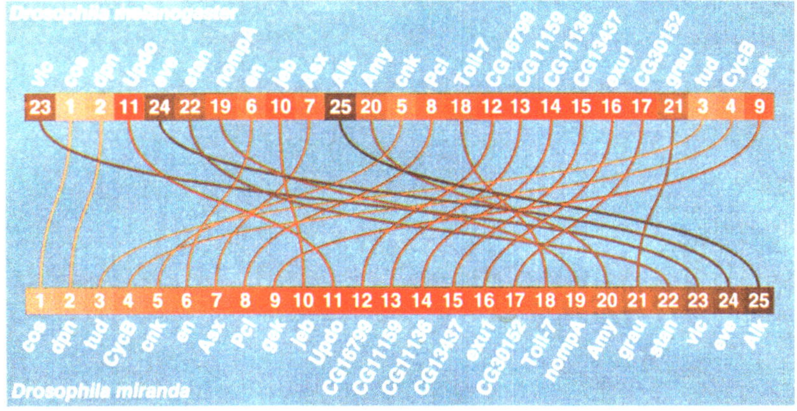

Nature reshuffles the deck: Two species of the fruit fly *Drosophila* have the same complement of genes, but they are arranged very differently. The mixing is caused mainly by occasional reversal events, in which a block of contiguous genes turns 180 degrees. But it's not obvious just what sequence of reversals created the particular permutation seen here. The genes have been numbered arbitrarily to put the *D. miranda* genome in canonical order. The illustration is based on an analysis by Carolina Bartolomé and Brian Charlesworth.

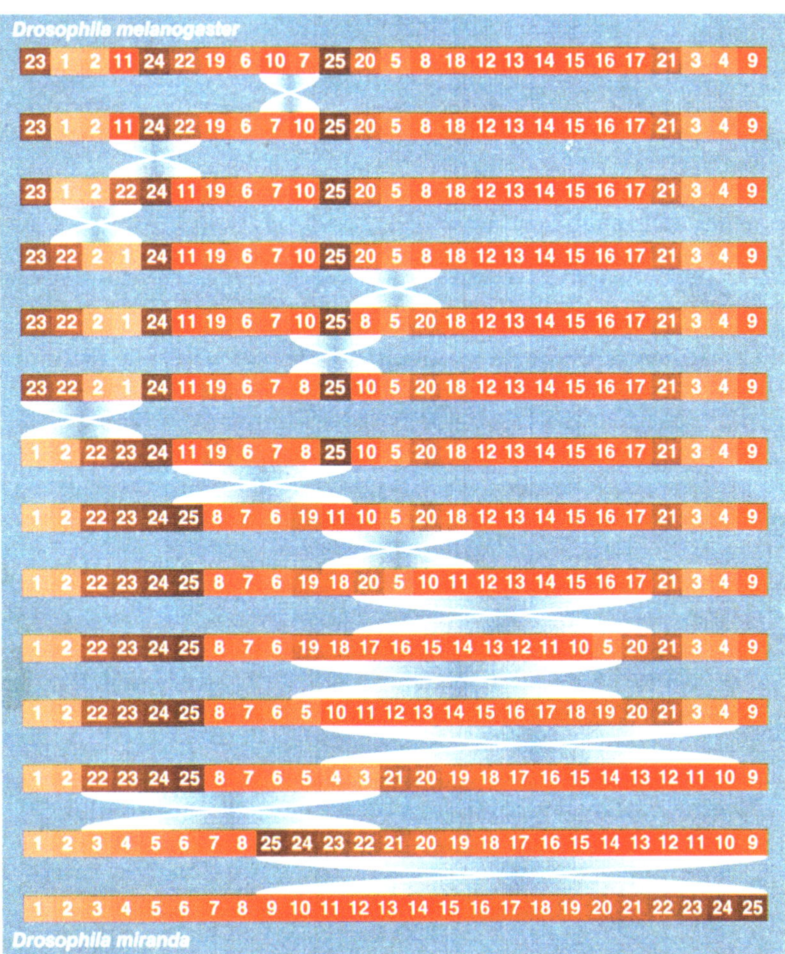

The reversal distance between two genomes—the minimum number of reversals needed to convert one permutation into the other—can serve as a measure of evolutionary divergence. The calculated reversal distance between *D. melanogaster* and *D. miranda* is 13. The sequence of 13 reversals shown here is one possible pathway for the transformation, although it is probably not unique. The pathway was generated by the GRIMM software of Glenn Tesler.

From *American Scientist*, volume 95, September–October 2007

FIGURE 386

FIGURE #387

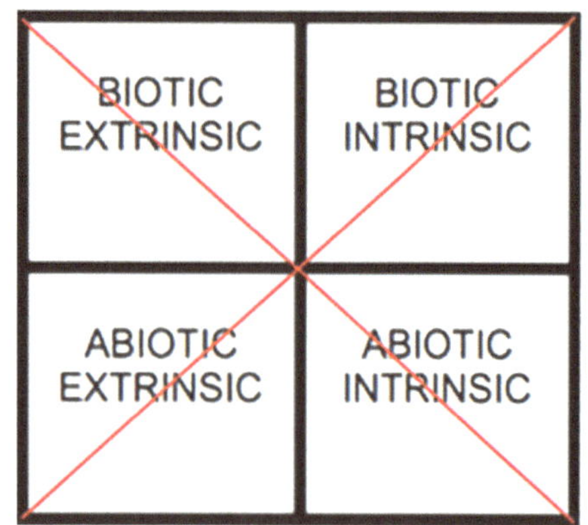

FIGURE 387, *left*, shows a geometric concept of *evolution* in which the *concepts* form the **same** as the diagonaled *numbers* on the *right* add the **same** and just as the *concepts* in the antinomy in inference are all the **same**. This pattern of *evolution* was lifted from Causes of Evolution: A Paleontological Perspective, edited by Warren P. Allman and Robert M. Moss. Go to Figures 106 and 107 to see the so-called magic squares previously described.

I have altered the *square* of Allman and Moss to make the concepts of evolution more *diagonally* the **same**, which can be done by **reversing** either the top or the bottom squares.

With the idea of the gene becoming ever more *complex* through *analysis*, it is comforting that *syntheses* can help keep it *simple*. A good read on how complex the idea of gene has become is found in "Genome 2.0: Mountains of New Data Are Challenging Old Views," by Patrick Barry, *Science News*, September 8, 2007.

More evidence that genetic *duplication* is a prerequisite for successive increases in life's complexity is seen in, "Genome Duplications: The Stuff of Evolution?" by Elisabeth Pennisi, *Science*, December 21, 2001, in which Pennisi states, "The controversial—and formerly improvable—proposition that evolution moves forward through gene duplication of entire genomes is getting support from current advances in molecular biology."

"Proof and Evolutionary Analysis of Ancient Genome Duplication in the Yeast *Saccharomyces cerevisiae*," by Manolis Kellis et al. *Nature*, April 8, 2004, shows that after gene duplication the genes needed for adaptation and survival remain *active*, while those that are no longer needed become *inactive* or else are gradually lost. Later research has found that the two duplicated genes seem to each gradually take on *separate* tasks. They each adapt to *changing* conditions. If one of the two genes is needed for adaptation, its activity is conserved, but if the other is no longer needed, it gradually disappears from the active genome. This *divisive* concept of gene duplication however hides the *unity* of the two genes as seen in FIGURE 388, taken from, "Making the Most of Redundancy," by Edward J. Louis, *Nature*, October 11, 2007. If you look carefully at this FIGURE, you will see that after duplication the two genes have taken on "complementary functions." More to the logic of interactions, the two genes have become **opposing** interpretations of the **same** gene. What they have *not* done is become superimposed to form a more complex and enhanced **sameness** that is more than the sum of its parts. If they should form in *simultaneous* double reversal, they would have the capacity to quickly create a more complex and logically enhanced species on the *vertical*. Without superimposition they continue to serve their enhanced *adaptive* method for *conserving* the **same** species on the *horizontal*.

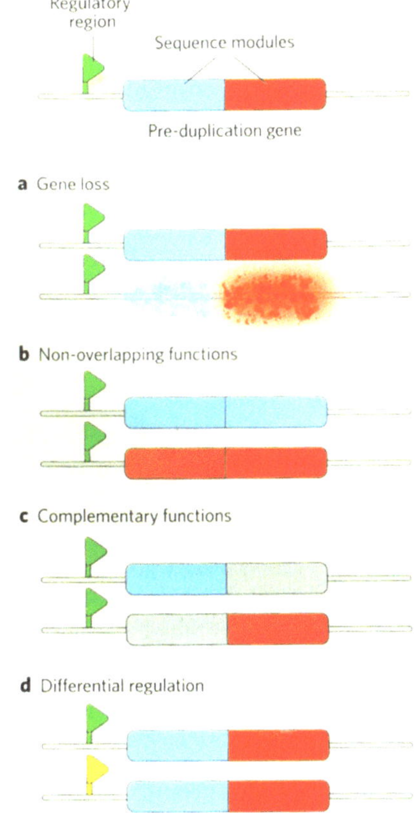

Possible outcomes of gene

FIGURE #388

The *active* genes on the *horizontal* comprise less than 1.5% of the human genome. This extremely small percentage is *perceived* as being no more complex than the genomes of some far less complex species. The species nearly as complex as humans—the great apes, particularly chimpanzees and bonobos—seem to have genomes that are almost the **same** except for small stretches of nucleotides.

Read "Chimp Genome Catalogs Difference with Humans," by Elizabeth Culotta, *Science*, September 2, 2005. Also read "Genome's Riddle: Few Genes, Much Complexity," by Nicholas Wade, the *New York Times*, February 13, 2001.

More evidence that gene duplication (doubling) is required for evolution is seen in "Twinned Genes Live Life in the Fast Lane," by Elizabeth Pennisi, *Science*, November 10, 2000, from which I quote:

> Evolutionary biologist Mike Lynch and computer scientist John Conery . . . describe new insights on how genes arise and fuel evolution. By trolling through sequence data for nine very distinct organisms, they have uncovered evidence that genes are copied far more frequently—and duplicates are lost from the genome far faster—than researchers had thought. What's more, the work suggests that some duplicate genes play a key role in the evolution of new traits and in speciation . . . . More than 30 years ago, geneticist Susumo Ohno of the City of Hope Hospital in Los Angeles proposed that genomes grow and diversify by gene duplication, an idea that most evolutionary biologists have since come to accept. It seems that when—by some quirk of DNA replication—a gene, a piece of chromosome, or whole genome is copied twice, the "extra" genes can take on a new function and expand the organisms genetic repertoire—Lynch and Conery found that most of the duplicated genes are relatively young and extra genes disappear quickly, at least on an evolutionary time scale . . . . Perhaps most surprising, the two found an "astronomical rate of gene duplication," says Sally Otto, an evolutionary biologist at the University of British Columbia in Vancouver, Canada . . . ."Gene Duplications are so frequent that we really need to take them into account as an important source of genetic variation," says [Andeas] Wagner . . . in evolutionary biologist at the University of New Mexico, Albuquerque, and the Santa Fe Institute.

Evolution requires more than just the Mendelian interpretation of genetics that we see on the *horizontal* as suggested in *Science*, June 13, 1997, "Evolutionary Chemistry: Getting There from Here," by the molecular biologist Gerald F. Joyce whom I quote:

> All the enzyme that exist in nature are the product of Darwinian evolution, based on natural selection . . . . The [catalytic] picture is not one of simple lock and key, but rather of a blurry lock that comes into focus through repeated contact with a particular key . . . . Gene duplication events are the primary source of new enzymes in biology. One gene copy maintains the original function, while the other is free to evolve a new function . . . . One conformation maintains the fitness of the molecule, while the others are free to vary in search of improvements in fitness.

If a mutated gene is *beneficial* to adaptation, one of the *duplicated* genes may remain *active* and the other half become *inactive*, or *passive*. The *active* gene becomes **similar-different** in **opposition** to the *passive* gene that is *conserved* as the **same** thing. Active/passive as **opposite/same** reciprocally interact. The *passive* genes that are no longer needed form strings of **sameness** on the *horizontal* for the *active* genes to **oppose**. Because active/passive reciprocally interact, the *passive* genes of **sameness** can become "transposable"—often called "jumping genes"—that can join the *active* genes when needed by the genome to fine-tune adaptation. These *horizontal* interactions can make quick genetic *changes* for *adaptation* to a *changing* environment, which can also increase species *diversity*.

On the other hand, when a genetic active/passive as **opposite/same** *interaction doubles* in *simultaneous reversal* being, more complex occur less often. This **super-sameness** of concept is much more complex and enhancing than the sum of its two *interacting parts*, so fewer *active* genes are needed to do the **same** work as many more *active* genes previously required. Since *each* active gene in humans does much more work than the *active* genes of other animals, gene *count* cannot explain the more complex interactions in the human genome. Gene interactions are ruled more by *value* and *quality* than *number* and *quantity* derived on the *vertical* from quarks **sameness** and electron **oppositeness** respectively. Even with the conceptual interactions of the human mind, as Charles Lindbergh said, "Civilization depends upon the quality rather than the quantity of men."

Read "Fewer Genes, More Noncoding RNA," by Jean-Michel Claverie, *Science*, September 2, 2005. Also read "Adaptive Evolution of Non-coding DNA in *Drosophila*," by Peter Andolfatto, *Nature*, October 20, 2005. Also read "It's the Junk that Makes Us Human," under News in *Nature*, November 9, 2006. These quotes highly suggest that the **sameness** of *con-coding* (passive) genes is *conserved* in the genome for the *coding* (active) genes to interact in **opposition** for *adaptation*.

As for the evolution of the genome, perhaps the best example that each successive *simultaneous double reversal quadruples* the *creative* force of **sameness** is seen in the "Cambrian explosion" about 570 million years ago. In the early Cambrian, *passive* life became *active* with the emergence of bilateral symmetry that was the equivalent of two **opposing** halves of the **same** more complex and enhance body. This great increase in ordered complexity interacted **both** halves *diagonally* the **same** as the square of the antinomy in inference seen on the inside front cover, and the **same** as a mathematical formula proportions its numbers *diagonally* the **same** between its two **opposing** sides. As we shall see, the reciprocally **opposing** hemispheres of humans interact diagonally as the **same** more complex and enhanced brain, the **same** as the antinomy in inference, *dominated* by deduction in the *left* half in most people.

The emergence of more complex bilateral symmetry enhanced the motion of animals with the capacity to deliberately *accelerate* in **similar-different** directions *conclusively*. Instead of simply waiting passively for nourishment to float by in a medium of water or air, life was given the capacity to act as **either** *predator* **or** *prey*, **either** *kill* **or** *be killed*, **either** *eat* **or** *be eaten*. This eventually led to a crescendo of adaptations and therefore to a multiplicity of species. Read under Perspectives, "Is the 'Big Bang' in Animal Evolution Real?" by Lars S. Jermiin et al., *Science*, December 23, 2005, which study correlates the *accelerated* increase in complexity of animal species with the "Cambrian explosion."

In the early Cambrian, the concept of *selfishness* and *self*-survival dominated bilateral life. *Aggression* became so extreme that it transformed into *defensive* mechanisms, causing a sharp decline in the number of species. Species that survived the best interacted the *dominantless* concept of *selflessness* with the *dominance* of *selfishness*. In other words, the interaction of competition/cooperation took hold and species diversity again increased with these **opposing** concepts interacting is a more *constraining* and therefore more stable proportion.

With the emergence of the human brain/mind, the concept of *empathy* emerged from the concept of *cooperation* where humans not only reciprocate good for good and bad for bad but also do *good* to a total stranger with no expectation of the

same thing in return. The enhanced concept of *empathy* is unique to humans and is the basis for charity, although some people establish charitable organizations for their own *selfish* purposes. This will always be so because at any level of life's complexity self/selfless *perpetually* interact in varying proportions.

Other than humans, primate species, including chimpanzees, have *empathy* for their kin but little or no empathy for unfamiliar individuals. Read "Chimpanzees Are Indifferent to the Welfare of Unrelated Group Members," by Joan B. Silk et al., *Nature*, October 27, 2005. I quote from the abstract:

> Humans are an unusually prosocial species—we vote, give blood, recycle, give tithes and punish violators of social norms. Experimental evidence indicates that people willingly incur costs to help strangers in anonymous one-shot interactions, and that altruistic behavior is motivated, at least in part, by empathy and concern for the welfare for others . . . . In contrast, cooperative behavior in non-human primates is mainly limited to kin and reciprocating partners, and is virtually never extended to unfamiliar individuals . . . chimpanzees do not take advantage of opportunities to deliver benefits to familiar individuals at no material cost to themselves . . . . Chimpanzees are among the primates most likely to demonstrate prosocial behaviors . . . but there is no evidence that they are averse to interactions in which they benefit more than others.

Only in humans is the concept of *empathy* often extended to strangers. Humans have empathy not only between each other but also for other species. Rescuing lost or sick animals is a widespread human activity, especially among females. Also, the human effort to slow the extinction of species is not just for environmental and scientific purposes but is also driven by empathy. However, *empathy* for Bambi by an environmentalist may be *prey* for an un-named deer by a hunter who needs meat on the table. **Both opponents** are constantly at each other's throats. Too much *conservation* can lead to *over* population, and too much *exploitation* can lead to *under* population. *Mindless* nature keeps these to **opposing** concepts of a population *constrained* within a proportion, *alternating* them reciprocally in *time*.

Evolution is dominated by *chance*, as "experimented" with almost every conceivable thing. For example, voles, highly social primates, have a much longer sequence of passive repeats of genes suggesting that instinctive *cooperation* has emerged *dominantly* from the conserved *concept* of quarks **sameness**. Read "High-fidelity DNA Found," by David Wahlberg, the *Atlanta Journal-Constitution*, June 10, 2005.

Cooperation and *altruism* are closely tied to the idea of "*direct* reciprocity" where "you scratch my back and I'll scratch yours"—a one-on-one, "tit-for-tat" relationship. *Empathy* is more complex and closely tied to "*indirect* reciprocity" where such interactions are not merely face-to-face, but instead interact between a **specific** person and others in **general**. Here, human *concepts* come into the picture where ideas such as **either** *trust* **or** *distrust*, **either** *good* **or** *bad*, are applied to a **specific** individual. One's reputation is at stake with *indirect* reciprocity. *Indirect* reciprocity occurs in highly complex socio-economic cultures where individual reputations or trust fosters trade and commerce. Mutual *trust* by *indirect* reciprocity is the grease that enhances all human interactions. Reciprocal logic of *indirect* reciprocity is seen in, "Evolution of Indirect Reciprocity," by Martin A. Nowak and Karl Sigmund, *Nature*, October 27, 2005.

Read "Generalized Reciprocity in Rats," under Research Highlights, *Nature Reviews Neuroscience*, in which "this study indicates that generalized reciprocity may have evolutionary rather than cultural origins." Also read "Diminishing Reciprocal Fairness by Disrupting the Right Prefrontal Cortex," by Daria Knoch et al., *Science*, November 3, 2006.

Read an excerpt, "The Ethical Brain," by Michael S. Gazzaniga, in *Cerebrum*, the Dana Forum on Brain Science, 2006.

Despite the ethical brain of humans, the less complex and enhanced brain of primates has been conserved. There is always latent animal instinct, extremes of emotion and *self*-centeredness that can suddenly erupt in cruel and irrational behavior. Read, "Our Social Roots," subtitled "We Share Many Behavioral Traits with Our Primate Relatives—Some Disquietingly Nasty," by Sarah F. Brosnan, *Nature*, December 27, 2007.

Because *development* closely resembles the *evolution* of the species, this relationship has not gone unnoticed by biologists. These two distinct disciplines of biology have begun to collaborate, and in doing so, they have gained much deeper insights into their respective disciplines. These two distinct disciplines has become so entwined and mutually enhancing that they now have become known as "evo-devo." The success of the evo-devo *interaction* has motivated other highly **specialized** disciplines of science to become more **generalized** so that they, too, can get a deeper understanding of biological complexity as seen in "Econophysicists Matter," an editorial in *Nature*, June 8, 2006.

Read "A Long and Winding Road," by Denis Duboule, *Science*, May 13, 2005, a review of the book *Endless Forms Most Beautiful: The New Science of Evo Devo and the Making of the Animal Kingdom*, by Sean Carroll, Norton, New York, 2005. I quote from Duboule's critique:

> Development cannot explain evolution. Understanding developmental mechanisms can give us some hints as to how modifications thereof may have led to what we see today. But all animals available for such experimentation, regardless of their apparent levels of complexity, are themselves the result of millions of years of evolutionary tinkering, following a roadmap that we cannot possibly understand since it bears no intrinsic logic other than what is possible to do. In addition, the question of model systems deserved some addition, as these species are those chosen because they often display highly adaptive features . . . . One may thus wonder how many (if not all) of such systems will be necessary to fully understand the core of the universal rules at work.

Another work that show the close relationship between *development* and *evolution* is, "Compartments and Their Boundaries in Vertebrate Brain Development," by Clemens Kiecker and Andrew Lumsden, *Nature Reviews Neuroscience*, July 2005, from which I quote:

> How can a single signal such as FGF elicit two fundamentally responses on either side of its source . . . neither [of these] two studies . . . were able to explain why different FGFs should act in an unidirectional fashion from the MHB, exclusively affecting tissues anterior or posterior to their source . . . . FGF8 is required to convert IRX2 from a transcriptional suppressor into an activator: therefore, an activated form of IRA2 can convert presumptive tectum into cerebellum when misexpressed in the midbrain, where as a repressor form of IRX2 has the opposite effect when electroporated into the hindbrain.

Whether of *development* or *evolution* the interaction of **opposite/same** is *conserved* at 90°. Read, "Worm's Light-Sensing Proteins Suggest Eye's Single Origin," by Elizabeth Pennisi, *Science*, October 29, 2004, from which I quote, "These findings drive home the antiquity of invertebrate and vertebrate photoreceptors and opsins. 'Not only the morphology but also the molecular biology of the two types of receptors was already set in our common ancestor' . . . two types likely arose in a predecessor of the Urbilateria. In that organism, they speculate, the gene for opsin and the genes to build the one type of photoreceptor cell were duplicated. The extra set of genes might have evolved into different visual systems."

The dominance of deduction in inference is what scientists call *reductionism*—i.e., emphasis on *analysis* that goes from the highly *complex top* to its more *simple* parts at *bottom*. Despite the dominance of *quantitative analysis*, scientists are beginning to see the *value* of *qualitative synthesis*, so they are adding the *bottom-up* view to interact *dominantlessly* with the *dominance* of their *top-down* view. These two reciprocally **opposing** methods for understanding complexity cannot be interacted *simultaneously* by the human *mind*. We must *reciprocally* choose **either** method in *time*, the **same** as we *alternate* **either** the *deductive* **or** the *inductive* process if we are to arrive at a conclusion.

With the concept of *analysis* alone, you only learn more and more about less and less; while with *syntheses* alone, you only learn less and less about more and more. So it's a no-brainer as to why the *interaction* of analysis/synthesis as dominant/dominantless respectively is as enhancing to all human *mind* as the *interaction* of "evo-devo" has been for the minds of biologists.

I quote John Dewey from *Human Nature and Conduct: An Introduction to Social Psychology*, 1922, "It is, for example, almost a truism that knowledge is both synthetic and analytic; a set of discriminated elements connected by relations. This combination of opposite factors of unity and difference, elements and relations, has been a standing paradox and a mystery of the theory of knowledge."

A good read on the need for more bottom up, or *synthesis*, of knowledge in science to serve and enhance the dominance of reductionism is, "You Are More Important Than a Quark," the *New York Times*, June 19, 2005, by Keay Davidson, who reviews the book *A Different Universe: Reinventing Physics from the Bottom Down*, by Robert B. Laughlin, Basic Books, 2005. Laughlin, a Nobel Prize winner in physics, also tells of the need for philosophy to support scientific endeavor, giving particular emphasis on trying to accept and understand the concept of emergence and how, in defiance of mathematics, the whole can be more than the sum of its parts.

To get a better understanding of the creative force of evolution, let us, again, review the *vertical* creation of complexity of *macro*-mass. *Micro*-mass, with its **neither/nor** function, interacts with the **sameness** of the *micro*-photon's **both/and** function. There is **no difference** between the **sameness** of **neither/nor** and the **sameness** of **both/and**, they are **opposing samenesses** of a more complex **supersameness**. This **super-sameness** is then **opposed** by the more complex and enhanced electric charges that alternate in *time*.

Physicists call the **super-sameness** of **both neither/nor and both/and** "alpha," or "the fine-structure constant." Alpha's *constant* allows the **opposing** electron charges to make successive *changes* in the complexity of mass **similarities-differences** as seen the periodic table of the elements. What could be more **opposed** to the **sameness** of alpha than **similarities-differences** of mass, just as what could be more **opposed** to the **sameness** of stem cells than the **similar-different** body parts in fetal *development*?

Alpha (α) is a *constant* created by its less complex *constants*, a more **generalized** *constant* that is, itself, more complexly **specific** *constant* of mass. The formula for alpha's *constant* of *constants* is:

$$\alpha = \pi e^2 2/ hc$$

Where π multiplied by the charge on the electron squared are doubled, then divided by Plank's constant multiplied by the speed of light, all being *constants* of nature, including the *even* number 2, the only even number that is a prime. The great complexity of the *constant* alpha allows an infinite numbers of *variations* to **oppose** it, thus accounting for the great capacity for complexity in our *macro*-world.

Alpha's *constant* of *constants* (read **super-sameness**) is so magical that even some nonreligious physicists claim that alpha's *creativity* is beyond understanding and can only be "God given." Some physicists even use the word *worship* in their adoration of it. Read "Galactic Data Shore up a Constant," by P. W., *Science News*, May 14, 2005, in which Alpha has remained *constant* since the beginning of the temporal universe. Also read "A Finer Constant," by Andrzej Czarnecki, *Nature*, August 3, 2006.

There is a controversial theory of evolution that mitochondria cells with their genome of *accelerated* energy production combined with another genome's less energetic energy to create an alphalike **super-sameness** that enabled life to *accelerate* the evolution of highly complex **similarities-differences**. Read "Power for Life," subtitled "Did the humble mitochondrion—the powerhouse of the cell—play a key role in the evolution of life?" by John F. Allen who reviews a book in *Nature*, October 27, 2005, titled *Power, Sex, Suicide: Mitochondria and the Meaning of Life*, by Nick Lane, Oxford University Press, 2005.

Another *constant* of *constants* called "gamma" is expressed in Euler's equation:

$$e^{i\pi} = -1,$$
$$\text{or } e^{i\pi} + 1 = 0$$

Reciprocal logic suggests that just as **either** *negative* **or** *positive* numbers **oppose sameness** (=), so too the *sequence* of *odd* numbers **opposes** the *simultaneity* of *even* numbers. This is because odd/even is derived from **either/both**, the function of **opposite/same** respectively. It takes **either** half of a mathematical *equation* to solve a problem to get any meaning from *mindless* nature, just as it takes **either** a *negative* **or** *positive* solution of a mathematical calculation to make any sense.

But to *mindless* nature, there is **no difference** between the mathematical *concepts* of **either** *negative* **or** *positive* numbers. **Both** –1 **and** +1 are *simultaneously* the **same**, *nothing*, *zero*. *Number*, *concept*, and *geometry* are all more complexly the **same** as we saw in Figure 105. Read, "What Makes an Equation Beautiful," by Kenneth Chang, the *New York Times*, October 24, 2004.

It is interesting to me that the *number* of all constants is *infinite*. Perhaps it is because *constants* are *timeless*; therefore no matter how long you add increasing *precision* to a *constant*, it can never be *absolutely completed*. Just as time/timeless and change/constant *interact* respectively, so too do finite/infinite, all interacting as **opposite/same** respectively from the point of view of *mass*. If it were not so, how could the goofy mathematics of "renormalization" be so effective in transforming *micro*-mass to the increased complexity of *macro*-mass?

The *dominance* of *data* and *numbers* in deduction is nevertheless replete with *conceptual interactions* such as finite/infinite, odd/even, real/imaginary, negative/positive, variable/constant, commutative/non-commutative, proof/disproof, etc. These conceptual interactions are *incomplete*, being only *part* of *mindless* nature's squares of **sameness**. Again, the idea of *complete* was proven to be logically *incomplete* by the mathematician Kurt Gödel. Read "Waiting for Gödel," by Polly Shulman, the *New York Times*, May 1, 2005, that reviews a book titled *Incompleteness: The Proof and Paradox of Kurt Gödel*, by Rebecca Goldstein, Atlas Book/W. W. Norton & Company, 2004.

I encourage my readers to access this review or read the book to understand that even mathematics, the most rigorous and the dominant scientific method, is as subject to conceptual *paradox* as any other part of knowledge and that "*truth*, exists 'out yonder' [as Einstein put it], even if we cannot put a finger on it.'"

As to *logic*, Gödel's proof came about by his placing the liar's paradox in double reversal *alternately* to get the **same** results from **both opposing** interpretations, which showed that if you could *complete* the proof by *simultaneous* double reversal it would prove *nothing*, a **sameness** that is **opposed** by mathematics deduction process.
Evolution is dominated by *chance* functioning as **either/or** respectively. Since mutation is the very essence of *chance* in evolution, the concept of *design*, like the concept of quarks **sameness**, is hidden. FIGURE 389, however, taken from "Epistasis in RNA Viruses," by Yannis Michalakis and Denis Roze, *Science*, November 26, 2004, shows that **either** *negative* **or** *positive* mutation *alternating* in *time* is the *design*.

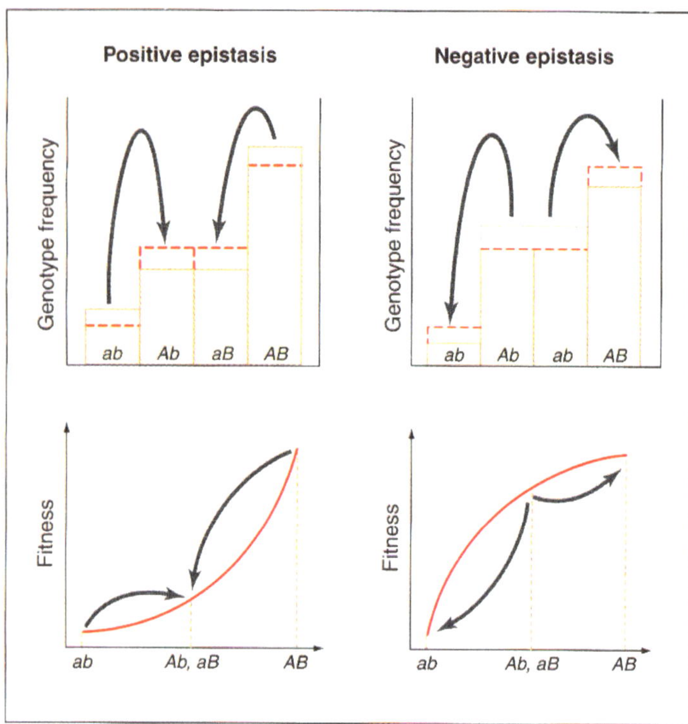

Evolution of recombination between two selected loci. Genotypes *ab*, *Ab*, *aB*, and *AB* have relative fitnesses 1, $1 + s$, $1 + s$, and $(1 + s)^2 + e$, where *e* is a measure of epistasis. Positive epistasis generates an excess of extreme genotypes [(**top left**) dashed lines correspond to linkage equilibrium]. Recombination tends to reduce this excess by re-creating intermediate genotypes (arrows) that are less fit on average (**bottom left**). This effect selects against recombination. Furthermore, recombination decreases the variance in fitness, which reduces the efficiency of selection and also selects against recombination. Negative epistasis (**right**) generates an excess of intermediate genotypes, which is reduced by recombination (arrows). Because extreme genotypes are less fit on average, this selects against recombination. However, recombination now increases the variance in fitness, thus increasing the efficiency of selection and resulting in selection for recombination. Theoretical models show that the benefits of recombination in terms of increased variance in fitness are stronger than the cost in terms of average fitness when epistasis is negative and sufficiently weak (*15*). Two other terms often used in the literature are "synergistic" and "antagonistic." Synergistic ("work together") means that combinations of mutations have stronger effects than those expected from the effects of the individual mutations. Thus, synergism among mutations results in positive epistasis for beneficial mutations and negative epistasis for deleterious mutations ("makes things better when things are good, and worse when they're bad"). Antagonistic ("struggle against") means that combinations of mutations have weaker effects than those expected from the effects of the individual mutations. Thus, antagonism among mutations results in negative epistasis for beneficial mutations and positive epistasis for deleterious mutations ("makes things worse when they are good, and better when they are bad").

FIGURE 389

Figure 180 taken from *Science*, February 14, 1992, shows a synthetic "organism" discovered by Julius Rebeck Jr., a chemist at the Massachusetts Institute of Technology, that replicated itself into two **opposing** halves that if superimposed in *simultaneous double reversal* would cancel out its **opposition** and become the **same** more complex "organism," would *instantly* **oppose** it by interacting with **either** a *negative* **or** *positive* ion.

More evidence for the concept of **sameness** in evolution is seen in "Evolutionary Dynamics on Graphs," by Erez Lieberman et al., *Nature*, January 20, 2005, from which FIGURES 390 and 391 is taken. I quote from the abstract:

> Evolutionary dynamics have been traditionally studied in the context of homogeneous or spatially extended populations. Here we generalize population structure by arranging individuals on a graph. Each vertex represents and individual. The weighted edges denote reproductive rates which govern how often individuals place offspring into adjacent vertices .... Spatial structures are described by graphs where vertices are connected with their nearest neighbours. We also explore evolution on random and scale-free networks. We determine the fixation probability of mutants, and characterize these graphs for which fixation behavior is identical to that of a homogeneous population. Furthermore, some graphs act as suppressors and others as amplifiers of selection. It is even possible to find graphs that guarantee the fixation of any advantageous mutant. We also study frequency-dependent selection and show that the outcome of evolutionary games can depend entirely on the structure of the underlying graph. Evolutionary graph theory has many fascinating applications ranging from ecology to multicellular organization and economics.

FIGURE #390

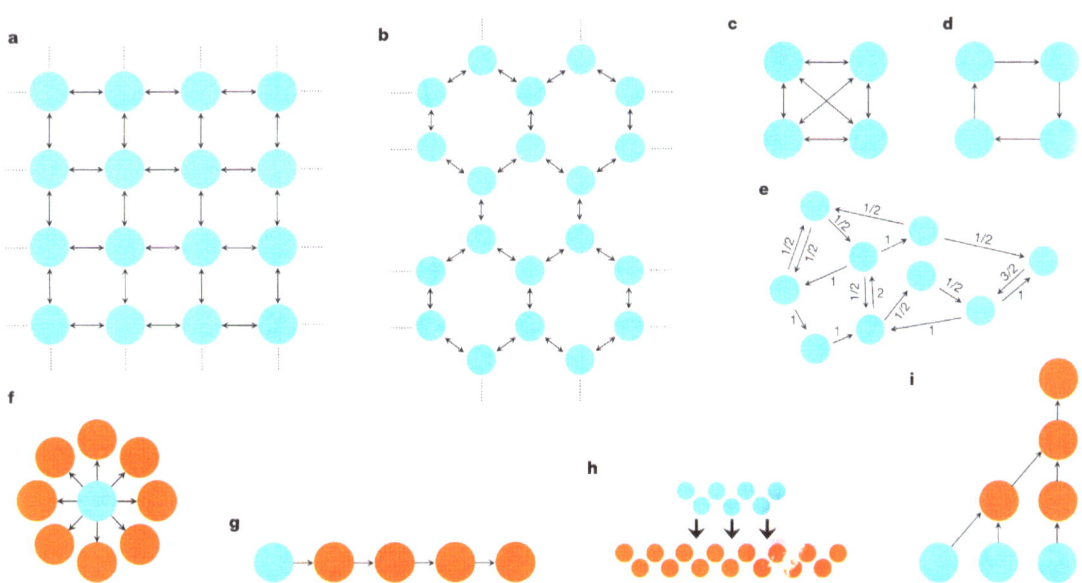

Figure 2 Isothermal graphs, and, more generally, circulations, have fixation behaviour identical to the Moran process. Examples of such graphs include: **a**, the square lattice; **b**, hexagonal lattice; **c**, complete graph; **d**, directed cycle; and **e**, a more irregular circulation. Whenever the weights of edges are not shown, a weight of one is distributed evenly across all those edges emerging from a given vertex. Graphs like **f**, the 'burst', and **g**, the 'path', suppress natural selection. The 'cold' upstream vertex is represented in blue. The 'hot' downstream vertices, which change often, are coloured in orange. The type of the upstream root determines the fate of the entire graph. **h**, Small upstream populations with large downstream populations yield suppressors. **i**, In multirooted graphs, the roots compete indefinitely for the population. If a mutant arises in a root then neither fixation nor extinction is possible.

NATURE | VOL 433 | 20 JANUARY 2005 | www.nature.com/nature

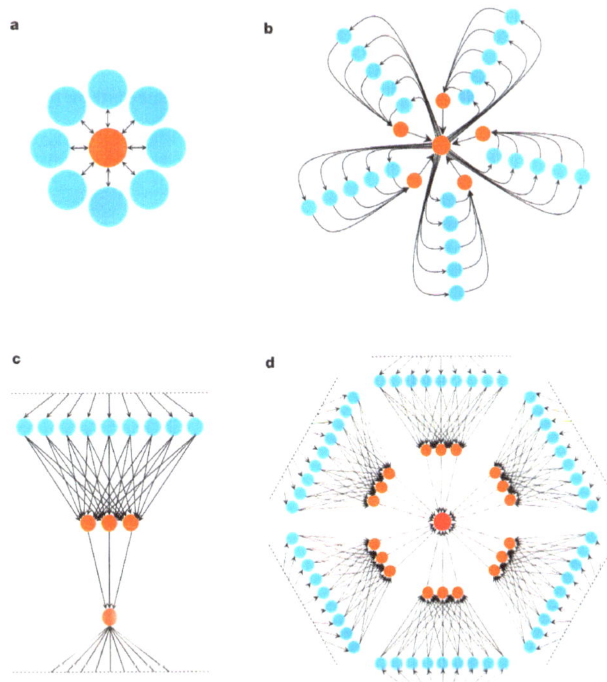

Figure 3 Selection amplifiers have remarkable symmetry properties. As the number of 'leaves' and the number of vertices in each leaf grows large, these amplifiers dramatically increase the apparent fitness of advantageous mutants: a mutant with fitness r on an amplifier of parameter K will fare as well as a mutant of fitness r^K in the Moran process. **a**, The star structure is a $K = 2$ amplifier. **b–d**, The super-star (**b**), the funnel (**c**) and the metafunnel (**d**) can all be extended to arbitrarily large K, thereby guaranteeing the fixation of any advantageous mutant. The latter three structures are shown here for $K = 3$. The funnel has edges wrapping around from bottom to top. The metafunnel has outermost edges arising from the central vertex (only partially shown). The colours red, orange and blue indicate hot, warm and cold vertices.

NATURE | VOL 433 | 20 JANUARY 2005 | www.nature.com/nature

FIGURE #391

The underlying **sameness** of the quarks could be thought *geometrically* as a graph that instead of being merely *static* is more complexly *dynamic*. It's as if a *motionless* graph paper could *dynamically* engage the *motion* of your writing hand, the **same** as Einstein's discovery that for *mindless* nature there is **no difference** between **either** *rest* **or** *motion*, but the **same** measurement of the two is **opposed** by the concepts of *mind*, as seen in Figure 76. The **sameness** of spatial *perpendicularity* with its *lattice* framework interacting with any **similar-different** point of mass is seen in Figure 97. The **sameness** of *space* is reciprocally **opposed** by the *mass*. This mass/massless as **opposite/same** *interaction* is why the lattice framework of *space* is dragged along with the *spin* of the Earth's *mass*, described in "Swiveling Satellites See Earth's Relativistic Wake," by Charles Seife, *Science*, October 22, 2004. Also read "Dragging of Inertial Frames," by Ignazio Ciufolini, *Nature*, September 6, 2007. Because any *point* of mass and *pointless* space *interact* and *mass* is **opposed** to the **sameness** of *space*, space adjacent to the spin is dragged along with it. This proof was predicted by Einstein who was influenced by Ernst Mach who claimed that the "fixed stars" in the heavens were affected by the geometric concept of a spinning mass. Since acceleration/inertia interact at 90° in the spin of a gyroscope, the accelerated *spin* of the gyroscope's *mass* **opposes** the *inertial* **sameness** of *space* as seen in Figure 298. Mach extended of this *local* interaction to a *cosmic* scale.

Let me emphasize at this juncture that scientists and philosophers have no clue as to why "time asymmetry is invariant." As with any concept, the idea of time does not fall under the logic of information processing. The idea of *time* is not a fact to be solved or concluded because it *interacts* in **opposition** to the **sameness** of space. This interaction as imbalance/balance is the **same** as time/space. *Time* belongs to the concept of *mass* that is the *contradictory* of *antitime* in the realm of *antimass*. They cannot interact, else the very idea of *mass* would disappear and everything would be the **SAME**.

How can space remain the **same** when its lattice framework is *distorted* by its interaction with the **opposing** concept of mass? Refer again to the *dynamic* rhombus in Figure 177 where the **same** four 90° angles of *static* perpendicularity are conserved. The rhombus conserves the lattice form of *space* (**general**) even though it is *distorted* by interacting with a *point* (**specific**) of mass. Einstein's description of *gravity* was a *lattice*, the **same** geometry as *space*, seen in Figure 105.

Returning from my digression to evolution, there are newly designed robots that *duplicate* themselves from simple to more complex parts described in "Self-reproducing Machines," by Victor Zykov et al., *Nature*, May 12, 2005, and "Now There Are Many: Robots That Reproduce," by Kenneth Chang, the *New York Times*, May 17, 2005. I quote from the latter, "The robots cannot promulgate unfettered. Each relies on the power from a base plate that it sits on, and the number of robots cannot exceed the number of plates. The researchers must also 'feed' the robots with new blocks at specific times a specific locations, and any error or malfunction thwarts the reproductive dance."

Such human constructed robots that mimic animal motions are themselves not *simultaneously* connected in *time* to the creative concept of quarks **sameness**. They are not themselves direct products of evolution and cannot reproduce on their own and deliberately forage on their own to sustain their energy without the intervention of the human *mind*. Unlike the interactions of life, machines are merely complex tools of the human brain/mind. Read "Whither Model Organism Research?" by Stanley Fields and Mark Johnston, *Science*, March 25, 2005, from which I quote, "We know a lot about the mechanisms that underlie the cell cycle; the cellular components that synthesize, modify, repair, and degrade nucleic acids and proteins; the signaling pathways that allow cells to communicate; and the mechanisms that lead to the selective expressions of subsets of genes. Remarkably, the operating principles of these cellular processes have been conserved throughout the tree of life." A robot with a deductive "brain" can validly process information to a **specific** conclusion, but the creative force of quarks **sameness** has not been conserved in its "brain" so no *conception* of mind can emerge from its *perception*.

The "problem" of the reciprocally **opposing** concepts between *chance* and *design* cannot be solved. This *conceptual* conflict is perpetual so it has and will continue to rage between the *minds* of those that believe God alone created and *designed* humans in his own image and those that believe that evolution by *chance* alone created humans. All efforts to *unify* the two *divisions* have failed because division/unity as dominant/dominantless also perpetually interact.

That concepts perpetually interact is seen in *The Evolution-Creation Struggle* by Michael Ruse, Harvard University Press, 2005, and "Being Stalked by Intelligent Design," by Pat Shipman, *American Scientist*, November–December 2005. Also read "A Free-For-All on Science and Religion," by George Johnson, the *New York Times*, November 21, 2006, in which Johnson states, "Time to take the gloves off, some scientists say."

A good read about the conceptual war between *science* and *non-science* is, "Philosophers Notwithstanding, Kansas School Board Redefines Science," by Dennis Overbye, the *New York Times*, November 15, 2005, in which the author explains how the *left* and *right* have argued their reciprocally **opposing** interpretations of evolution.

Also read the works of the evolutionary scientist Richard Dawkins who argues convincingly that *chance* alone is the only explanation for evolution's successive increases in ordered complexity. His works are summed up in, *Richard Dawkins: How a Scientist Changed the Way We Think*, by Alan Grafen and Mark Ridley, editors, reviewed in *Science*, July 28, 2006, by David C. Queller.

An example of how *order* can accrue from the chaos of conceptual conflict is seen in the argument between the *concepts* of philosophy and the *facts* of science. Mutual enhancement from this conceptual conflict is seen in "A Philosopher's Vision," by Richard L. Gregory, *Nature*, October 6, 2005, who reviews a book, *Action in Perception*, by Alva Noë, MIT Press, 2005. I quote from Gregory, "Philosophers contribute significantly to the brain sciences by clarifying terminology and concepts, occasionally issuing radical challenges and stimulating suggestions. It might be said that the whole of science that conceptual significance is as important as statistical significance, and philosophy contributes most in areas where generally accepted paradigms are lacking, such as cognitive brain science. The mystery of consciousness makes scientists listen to even the wildest ideas and most extreme challenges of philosophers. These are most effective when the philosopher has a grasp of physiology and key experimental data . . . . So philosophy and science meet, to mutual benefit."

It seems strange that autism might be an evolutionary link between chimps and humans, but this is suggested by a scientist who herself is autistic. She compares her autism with the behavior of other mammals and finds that she and they seem to have less complex development of the frontal lobes that puts *constraints* on behavior. Like animals, autists concentrate too much on **specifics** with its function of **either/or** without enough **generalization** with its function of **both/and** to conform to the *constraints* of the **specific/general** proportion. Autists have a hard time interpreting the emotions and behavior of *others*. Being *self*-centered, the concept of *empathy* eludes them.

Read "An Autistic Look at Animals," *Nature*, May 12, 2005, by Marian Stamp Dawkins, who reviews a book titled *Animals in Translation: Using the Mysteries of Autism to Decode Animal Behavior* by Temple Grandin and Catherine Johnson, Scribner /Bloomsbury: 2005, from which I quote, "Animals, like autists, concentrate on detail . . . [humans] concentrating on the bigger picture, would simply not realize [something specific] unless it was pointed out to us . . . . Autistic people are closer to animals than normal people are . . . . Autistic peoples' frontal lobes almost never work as well as normal people's do, so our brain function ends up being somewhat in between human and animal . . . . Political correctness is not part of her world."

Also read "Reflecting on Another's Mind," by Greg Miller, *Science*, May 13, 2005, the last paragraph of which states, "A full account of the basis of neural empathy . . . will require understanding how the brain deciphers the information it gets from the mirror systems—in other words, finding out what's behind all the mirrors." What is hidden is seen in "Learning, Autism, and the Synapse" found under This Week in *Science*, October 5, 2007, in which the editors say, "Thus, alternations in the balance between excitatory and inhibitory synapses can effect learning and such alternations may be a contributing factor in the pathogenesis of autism." An extreme *imbalance* one way may cause autism while an extreme *imbalance* in the **opposite** way may cause another pathogenesis. At **either** extreme there would be too little *constraining* force of **sameness** to create a reciprocal interaction.

Attention deficit disorder (ADD) is the very **opposite** of autism. With ADD there is an excessive amount of **generalizing**, like an excessive amount of light falling on a camera's film that fails to record a **specific** image on the camera's film in the "failure of the reciprocity law." The two hemispheres of ADD are too perfectly *balanced* so that imbalance/balance as dominant/dominantless respectively do not interact reciprocally. The proportion of the "golden mean" is more realistic." See Figures 130 and 131 to understand the *dynamic* order of the imbalance/imbalance, in the *golden mean*, and read "Diagnosis of Adult ADD on the Rise," by Cheryl Powell, Knight Ridder Newspapers, taken from the *Tampa Tribune*, May 22, 2005.

If autism and attention deficit disorder *interacted* their respective geometric concepts of **specific** and **general** as **opposite/same**, then there would be no pathogenesis and the person's brain-mind would be *dominantly* ordered by **either** *deduction* **or** *induction* respectively. The *former* would have a greater mental capacity for *science* while the *latter* would have a greater mental capacity for *philosophy*.

Writing cogently and to the point has always been the mantra of how best to communicate the written word. This requires the *dominance* of deduction in inference that leads to a **specific** conclusion. This is ideal for *factual* knowledge, but when one writes about *conceptual* interactions, these paradoxes prefer induction. If someone like me is reciprocally inclined, he will write in circles, seldom taking the reader to an **unopposable** and random conclusion.

A good example of how the written word of an author that is dominated by *induction* compares with one that is dominated by *deduction* as seen in "We Won't Be Fooled Again," the *New York Times*, October 1, 2005, by Joseph Nocera, who reviews a book *Fooled by Randomness: The Hidden Role of Chance in Life and in the Markets*, by Nassim Nicholas Taleb, Random House. I quote Nocera's comparison of his own style of writing with that of Mr. Taleb:

> "Fooled By Randomness" is not bedtime reading. Mr. Taleb is a complicated man with a sizable ego and a mode of thinking that is anything but linear. He . . . revels in the notion that "big shot" academics "oppose" him. And he writes in fits and starts, picking up ideas, then dropping them, and circling back to them three or four chapters later. When I watched him teach a class at the University of Massachusetts business school recently, I realized that he one of those people who is simply unable to stick with a train of thought; his brain just jumps around too much. That's how "Fooled by Randomness" reads too.

Now, let's *circle* back to evolution from my digression by asking the questions: "Why might our planet earth be the one **specific** point of *cosmic mass* that seems to have been able to *create* successive increases in the ordered complexity of mass **specificities**? How and why has *mass* become so complex and far from equilibrium that conceptual *minds* have emerged not only capable of being aware of themselves but also being aware of being aware, with the capacity to deliberately using the **same** logic as nature to discover the laws and forces of the Universe?" Why is the planet Earth so special when nature prefers no **specific** mass point of view? If the laws of nature treat all mass **specificities** the **same** and if the forces of nature treat all mass **specificities** the **same**, why is Earth so preferred for the evolution of highly complex systems?

Consider the following geometric concepts of **sameness** that make our planet Earth so highly favoured for evolution. Earth orbits the sun in an ellipse that is almost a *circle* so that it is always approximately the **same** distance from the sun's source of energy. So far, we have gotten **neither** too *hot* **nor** too *cold* for the evolution of life to be *destroyed*. This distance also happens to give Earth a temperature where that H_2O is *dominantly* a liquid that serves as a medium within which complex **similar-different** molecules can *interact* in **opposition** to the *dynamic* **sameness** of pure water. Also, the *constrained* proportion between the **specificity** of Earth and the **generality** of space has given Earth enough *time* for *chance* to *design* changes species. The *gravitation* of Earth's mass is **neither** too *week* **nor** not too *strong* for its *great inertial* concept to be *changed* by the *acceleration* of its *diminutive* **specific** parts. In other words, the *motions* of life are *perceived* to be *absolute*, which all but completely hides the **sameness** of Earth's *inertial* mass, as seen in Figure 18. Acceleration/inertia as dominant/dominantless do not seem to *interact* with *weighted* mass as seen in Figure 17. Therefore our motions seem to be absolutely deductive, **from** *here* **to** *there* to a predictable conclusion. Especially for inexperienced youth that things might not be so deterministic is hidden. Our connection with nature's *indeterminate* interactions is transparent.

Because nature's conceptual interactions are *timeless*, somewhere at some *time* the concept of mass, by chance, will find another "sweet spot" for complex life and conceptual thought to emerge in **opposition** to universal **sameness**.

Read, *The Privileged Planet: How Our Place in the Cosmos Is Designed for Discovery*, by Guillermo Gonzalez and Jay W. Richards, Regnery Publishing Inc., an Eagle Publishing Co., Washington, D. C., 2004. Also read, *Rare Earth: Why Complex Life is Uncommon in the Universe*, by Peter D. Ward and Donald Brownlee, Copernicus, Springer-Verlag, New York, N.Y. 10010. For a better understanding of the "sweet spot" in space where *mass* has a fair *chance* to self-create and sustain *life*, read "Where Is Life Hiding?" subtitled, "Location Holds the Key to Finding Extraterrestrial Life," by Margaret Turnbull, *Astronomy*, October 2006. Also read "Life, the Universe, and Body Temperature," by Clifford B. Saper, *Science*, November 3, 2006.

The idea of "intelligent design" by evolution is misleading because the idea of *intelligence* requires that a **specific** brain/mind could be a universal force, the **generality** of which could order all other mass **specificities** the **same**. Such a "designer" would be temporal and local instead of timeless and universal, preferential and prejudiced with **different** laws and forces in **different** places **different** times instead of treating all mass universally and *timelessly* the **same**.

Although the conceptual conflict between *chance* and *design* can never be solved or concluded if **both** conceptual **opponents** understand that their respective *ideas* are incomplete *interpretations* instead of *factual conclusions* and that their *alternating*-in-*time* arguments are reciprocally **opposing** parts of the **same** more complex *simultaneous* interaction, then their respective *minds* would become enlightened with a much deeper insight into their predicament and higher level of logical order. Furthermore, *each* conceptual *mind* will understand that his or her **specific** mass point of view is only *temporally* **opposed** to the *timeless* universally force of **sameness**.

Now read "A Stone Age Meeting of the Minds," by Thomas Wynn and Frederick L. Coolidge, *American Scientist*, January–February 2008, and ponder this: Did *Homo sapiens sapiens* (the wise one) drive Neandertals to extinction because the human *mind* had a more complex and enhanced *conceptual* capacity? I quote the concluding paragraph of this article, "Whenever and however they were acquired, these genes, and the executive functions they made possible, provided modern humans with the ability to conceive of and carry out long-range plans of action, an ability the Neandertals never enjoyed. The difference was small, but it ultimately had a profound consequence: Our species survived and flourished, whereas Neandertals met their demise."

In summation, evolution on the *vertical* is the successive increase in the *doubling* of **opposite/same** placed in *simultaneous reversal*. As with the **opposing** processes of inference, the **opposing** processes of *contingency (***general***)* and *convergence (*specific*)* simultaneously interact as the **same** self-creative force. Survival of the fittest and adaptation remain fundamental concepts of Darwinian evolution, but we now see how *adaptation* is logically ordered. When the *lifeless* **general** environment *changes* **specific** species of *life change* proportionally. There is transformation from *constant* to *change* in each. What connects the two is **sameness** of life's DNA, along with such genes as the Hox set on the vertical, and **sameness** of the earth's orbital "sweet spot" in relation to the sun over deep time. This grand interaction between **specific** *life* and its *lifeless* **general** environment, under the force of gravitational **sameness** and quarks **sameness**, accelerates successive increases in the complexity of **similar-different** species.

THE DIFFERENT SENSES OPPOSED TO SAMENESS

Our task is to find an algorithm, a natural law that leads to the origin of information.

—Manfred Eigen, 1922

<u>Proportional Constraint in Perceptions</u>

The German psychologist Max Weber discovered that several senses such as temperature had to **either** *halve* **or** *double* if the brain was to *perceive* any **difference**. This progression **either** way was the **same** as the nature's inverse square law that requires a *constant* for any *change*. Mathematicians call this <u>half/double</u> proportion, expressed as $^{1}/_{2}$ / $^{2}/_{1}$, a *reciprocal proportion* because it is simply two **opposing** interpretations of the **same** thing. For example, **either** one cow to two acres **or** two cows to one acre is ecologically the **same**. Since the *form* of this *reciprocal proportion* expressed the truth of nature, it may have been the precursor of the mathematical *formula*, previously discussed, with its *equality* (read **sameness**) between two **opposing** sides.

To show that the Weber *constant* applies the **same** logical order to the *perception* of **both** the camera's eye **and** the *perception* of the brains eye, let's start with the camera eye's shutter speed and its effective lens diameter. Figure 135 shows that the camera is constructed to "perceive" changes in the intensity of light by the Weber constant. This Figure tell the **same** story of *perception* from the **opposing** concepts of *time* and *space*. These two reciprocally **opposing** concepts and their <u>half/double</u> proportion a built into all *complex* cameras to *simplify* their use. Any setting other than that *constrained* within the Weber constant will not be a realist interpretation of the scene, so why complicate the camera with unnecessary features? Again, we see that no matter how *complex* the camera, the *simple* interaction of <u>change/constant</u> as **<u>opposite/same</u>**, each constrained within the <u>half/double</u> proportion is the logical basis of *perception* for the *mind* and the *mindless* camera.

A *simple* camera can be constructed by making a hole the diameter of a pin in an enclosed box, which hole substitutes for a lens. Such tiny apertures usually accommodate only a **specific** light wave coming through the pinhole, and therefore cannot record the *complexity* of light waves in **general**. Such *simple* cameras do not give flexibility to the photographer for making pictures under more **general** and variable light conditions. The more *complex* lens cameras offer the photographer a much greater *chance* to express his or her *creative* talent. The more the *mind* understands and complies with the logic of *mindless* nature's *conceptual interactions*, the greater the capacity for creating *art*.

It is more than just interesting that if there is **either** too much light **or** too *little* light at **either opposing** extreme of the <u>half/double</u> proportion, light (massless) and the film (mass) will fail to *interact* and no image will be recorded on the film. This phenomenon, known as "the failure of the reciprocity law," confirms that any setting that breaks through nature's *constraint* of **sameness** cannot process the **opposing** processes of inference. Nor can the human *brain* alone violate the *constraint* of the reciprocity law by taking in **either** too much **or** too little light. It is obvious that *too little* light through the eyes lens will not give enough information for the *brain* to record. It is much less obvious that the **opposite**, *too much* light, also causes failure of the brain's capacity to record information. For example, walking in the bright reflection of freshly fallen snow on a sunny day will soon cause "snow blindness." Experienced snow hikers always wear dark glasses to prevent *too much* light from entering the eye, thus causing an *overload* on the brain's ability to *perceive*. Since the *brain's perceptions* and the *mind's conceptions* are **both** ordered by the reciprocity law, it naturally follows that an excess of **either** opposing *half* will fail the logical constraints for knowledge. Another way to look at this is that there have been many scientific experiments that prove there no such thing as absolute *darkness* of absolute *brightness* in or *macro*-world. The reason is simple, <u>bright/dim</u> perpetually interact.

As with the camera, the human eye requires a **general** background to enhance a **specific** subject as seen in Figures 258 and 259. Experimenting with Figures 226 and 227, you will be convinced that the brain interacts **<u>specific/general</u>** as **<u>opposite/same</u>** <u>dominant/dominantless</u> to accelerate and enhance information processing in the visual system. You will also see that **both** complementary *colors* become the **same**, medium gray. Due to the *excessive intensity* of concentration one's perception contradicts itself into the *value* of medium gray. Again, breaking the constraints of **<u>specific/general</u>** as **<u>opposite/same</u>** makes everything the **same** that knowledge **opposes**. That the logic of this is true is is seen in Figure 217 where the **same** visual system is ordered by two **opposing** interpretations, *each half* interacting dominant/dominantless. If **both opposing** dominant/dominantless interactions were placed in simultaneous double reversal, they would become a more enhanced **sameness** of concept more complex than the sum of its parts.

All this again indicates that our mind's *conceptions* emerged from our brain's *perceptions* ordered by the magalogic seen on the inside front cover.

FIGURE 392 shows that the visual system of the brain is ordered by exactly the **same** megalogical method as Figure 217. In FIGURE 392, there is a representation of the split (**either/or**) and bilateral (**both/and**) parts of the visual system. These two representations are also ordered exactly the **same** as the megalogical method seen in Figure 205, which highly *complex* interaction *conserves* the *simplicity* of **opposite/same** on the *horizontal* where the *motionless left* hand *enhances* the efficiency of *motion* in the *right* hand.

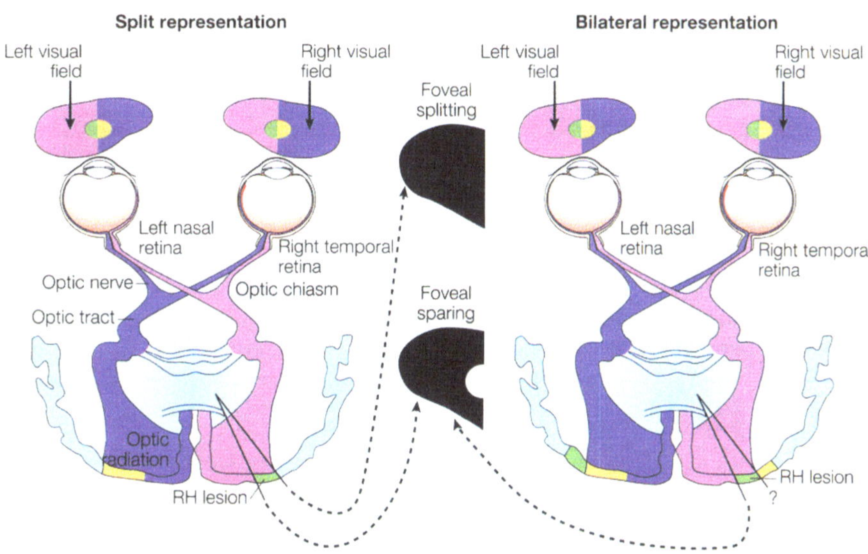

Figure 3 | **Diagram summarizing the anatomy for both the split and bilateral representation models of vision.** In the bilateral model, extra cortex needs to be given over to representing ipsilateral central vision. The bilateral model can account for foveal sparing in hemianopia, but cannot easily account for foveal splitting in hemianopia without some complex mix of simultaneous ipsilateral and bilateral pathology. The split model, however, can account for both. RH, right hemisphere. Modified, with permission, from REF. 37 © (2004) Academic Press.

FIGURE 392

Odor or Smell

Smell is the most primitive sense of life after touch. Its *simplicity* relative to the more *complex* senses of sound and sight is seen in "Ah, Sweet Skunk! Why We Like or Dislike What We Smell," by Rachel S. Herz, *Cerebrum*, the Dona Forum on Brain Science, Fall 2001, from which I quote:

> Unlike wiring in other senses, olfaction is "ipsilateral," which means that the right olfactory bulb receives information form the right nostril and the left olfactory bulb receives information form the left nostril. There is no crossover from right to left, as in the visual system . . . . Odorant receptors have different shapes; how well an odor molecule is detected is determined by how well it first into the olfactory receptor. Those with the better fit may be more likely to occupy the receptor site. If a molecule is too big to fit into a receptor, it cannot be perceived as a smell at all . . . [odor perception] appears to be hardwired in our bodies, which is potential evidence that olfactory responses is are innate. But even here, experience can alter this biological determinism. People who formally could not smell androstenone report being able to detect it after repeated exposure. Moreover, whether a particular person likes the smell of urine or dislikes florals is a psychological issue, different form how a person perceives a specific chemical. In other words, the separation is between donation—detecting and classifying the sense as something—and connection—liking or disliking it. I am primarily concerned with showing that the connotation of odors is acquired, although there is a complex overlap between how an odor is detected and how an odor is connoted . . . . An important dimension of olfaction, which is often not appreciated, is that most smells have a feel to them. Menthol feels cool, ammonia is burning . . . . We learn the meaning (the connotation) of odors

> by association. We experience every smell in a context: semantic, social, emotional, physical . . . . That context always has some emotional content, good or bad, albeit sometimes only weakly . . . . Of all of our senses, olfaction is especially predisposed to become associated with emotional meaning . . . . My argument is that the olfactory system is set up so that, through experience, meaning becomes attached to odor stimuli. This is in contrast to the proposition that we are hardwired to like or dislike various odors before smelling them . . . . Throughout our lifetime we acquire the emotional meaning of odors through experience, but first experiences are pivotal. This is why childhood, a time replete with first experiences, is such a training ground for odor learning. The first association made to an odor is difficult to undo . . . we have found that presenting exactly the same odor stimulus, but with two different labels . . . one good and one bad . . . (for example, vomit vs. Parmesan cheese)—can create an olfactory illusion. The stimulus in one case is perceived as being very unpleasant and in the alternate case as very pleasant. Not only is the odor believed to be what it has been labeled when presented as such, but people do not believe that the stimulus is the same when labeled differently, showing how powerful suggestion and context are to odor perception. We are cued to whether we should lie or dislike an odor by what its name connotes, even before we smell it . . . .
>
> There is an evolutionary argument as to why we are not hardwired to like or dislike any odors. When organisms first evolved as single-celled creatures, their primary function was to take in or reject substances from the outside world. This approach or avoidance response is called chemotaxis. As organisms evolved to be multicellular, they needed a way to detect on their outside what was good or bad (for example, food or non-food), and to communicate that information to the rest of the cells in the body. This is how chemical senses (olfaction and taste) are thought to have evolved. From an evolutionary perspective, the function of odors is to impart information about what to approach and what to avoid—for example, prey or predator. If an organism lives in a small, specifically defined ecological habitat, with particular local prey and predators, it will be adaptive for the organism to be hardwired with a system for detecting what food sources verses predators smell like. For example, the caterpillar of the monarch butterfly needs to know that only milkweed is food. These organisms are specialists. If, however, the organism could live on a variety of sources as food, it would not make evolutionary sense to have to respond to acceptable versus nonacceptable smells wired in. These organisms are generalists. Along with rats and cockroaches, we humans are the world's most successful generalists. We can live in an ecological habitat on the planet and survive by eating available food. If we had been hardwired to accept only fishy smells as food, we would never have survived in the savannah. For generalists, the function of olfaction is to learn how to respond appropriately to a particular smell source when it is encountered, and not hold a predetermined set of responses to particular odors. Thus, animals that are specialists should have innate olfactory response to prey and predators, whereas animals that are generalists should not. They should be prepared to learn from experience what is good and what is bad. Evidence for this can be found in a number of studies of animal behavior . . . . Even specialists are able to modulate innate olfactory responses based on experience. For generalists and specialists alike, nenophobia—a cautious response to novel food and odors—is universal. The response is particularly adaptive for generalists because of the enormous array of possible food sources available and the greater risk of exposure to poisons. What has already been consumed as food is safe; what is unknown may or may not be safe. The behavior of young humans attests to this. Infants and young children generally react with dislike to novel smells and flavors regardless of the emotional tone that adults use in offering them. It is only after these smells become familiar or attractive, as a result of appropriate modeling by adults, that children make discriminating responses . . . emotional responses to odors must be learned . . . olfaction can direct our food choices, but only after we learn what the odors mean in relation to the food in question . . . .
> You may have noticed that when you enter a house with a particular smell, it takes about 20 minutes before you no longer smell it. This is because the olfactory system is geared to detect change (a novel odor) but once the novelty wears off, the receptors cease to respond, and you cease to smell it. This does not mean the smell is gone; it merely illustrates the effect of olfactory adaptation . . . . If we have any innate response to odors, it is caution . . . . Infants and young children show wariness when exposed to unfamiliar odors, regardless of whether the odors are classified as pleasant or unpleasant by the adults around them. This uneasiness in the face of uncertainty is adaptive. It is better to be cautious than reckless when approaching the unknown.

Also read, "Your Home Stinks, But Chances Are You're Unable to Smell It," by Cindy McNatt, the Orange County Register, that appeared in the *Tampa Tribune*, July 8, 2006.

The above suggests that the sense of smell is ordered by **either** *attraction* **or** *repulsion* of the **same** odor.

Read "On Smell and Scientific Practice," *Science*, August 11, 2006, by Miriam Solomon, who reviews a book titled *The*

Secret of Scent by Luca Turin, Faber and Farber, London, 2006, forthcoming from Ecco, New York. I quote from Solomon, "Turin observes that, contrary to the predictions of shape theories, molecules very different in shape can sometimes smell the same . . . and molecules very similar in shape can smell different [molecules], (e.g., isotopes of the same molecule such as acetophenone and deuterated adetophenone). The geometric concepts of **opposite/same** interact in the brain-*mind* and *mindless* nature perpetually and universally."

Taste

In contrast to smell, I have little to offer on taste. All I know is that each of the thousands of taste buds of the tongue has its own **specific** receptor. Perhaps the brain interacts sweet/bitter, or salty/sour, each pair in perceptual conflict, which is suggested in "Bittersweet Symphony," by Jane Qiu, *Nature Reviews Neuroscience*, September 2005. Also read "Bitter Sweeteners," page 11, *Nature*, May 18, 2006. There is also a taste called "umami," or "savory," that may be the *perceptual* equivalent of the more complex enhancing *concept* of **sameness**. This however is mere speculation.

More complexly, the *interaction* of **both** *taste* **and** *smell* creates the greatly enhanced sense of *flavor*.

Sound and Hearing

The sense of *hearing* sound may be more complex than *odor* because the perception of *sound*, as with light, moves *bilaterally* in the brain while *odor* perception moves ipsilateral—i.e., *parallel* instead of *diagonally*. With *hearing*, neurons from the *left* ear carry sound signals to the *right* hemisphere while neurons from the *right* ear carry sound signals to the *left* hemisphere, forming a *diagonaled* square of **sameness** between the two **opposing** *halves*, the **same** as the sense of *vision*.

That *sound* is heard by the logic of reciprocal interactions is suggested in "A New Window on Sound," by Bruno A. Olshausen and Kevin N. O'Connor, where they explain a paper, "Efficient Coding of Natural Sounds," by Michael Lewicki, *Nature Neuroscience*, April 2002. I quote from Olshausen and O'Connor:

> In the vertebrate cochlea, sound is detected by a array of several thousand hair cells that transduce mechanical vibrations into electrical activity. The individual hair cells, and the auditory nerve fibers to which they are connected, are tuned to specific frequencies [the **same** as **specific** odors connect to **specific** receptors] . . . the authors show that the frequency analysis performed by the mammalian cochlea is well matched to the range of sounds encountered in the natural environment. Each auditory nerve fiber may be considered a filter that signals information about the temporal structure of the stimuli within its preferred frequency range . . . . As engineers have understood or years, the design of a filter involves an inevitable trade-off between the precision of frequency tuning and temporal tuning . . . a filter cannot signal both the frequency and the timing of sound with arbitrary precision . . . yet both frequency and timing . . . . The challenge for the auditory system, then, is a trade-off between timing and frequency analysis. One way to think about the time-frequency trade-off is in terms of a 'tiling' of the time frequency plane (Fig. 1). At one extreme, frequency discrimination is sacrificed completely to maximize temporal discrimination . . . . At the other extreme is the Fourier transform in which temporal discrimination is ignored to extract the maximum information about frequency composition of the stimulus. Between these two extremes are many possibilities . . . . There is no end to the variety of tiling schemes that can be imagined . . . these results taken together, suggest that the cochlea and the auditory nerve may be optimized to transmit a wide range of naturally occurring sounds to the brain. It is even possible, as the authors suggest, that the acoustic properties of human speech have evolved to make efficient use of the preexisting properties of the peripheral auditory system . . . . Lewicki's results share an intriguing similarity to the recent work in vision. Neurons in the visual cortex encode both the location and the spatial frequency of visual stimuli, and the trade-off between these two variables is analogous to that between timing and frequency in the auditory system . . . . Curiously, the space-frequency tiling scheme [of the visual system] of both derived filters and those measured physiologically deviates from a wavelet in much the same way as Lewicki finds in the auditory system; bandwidth at high frequencies is narrower than one would suspect. It is not clear whether this similarity is profound or simply coincidental.

Scientists cannot bring themselves to substitute the concept of **sameness** for **similarity** as seen in the above quotation because they, like we, are **opposed** to everything being the **same** in order to *perceive* anything. Besides, **similar-different** objects hide the very constrained **sameness** that is *constantly* conserved to enhance our *changing* perceptions. It is much easier to substitute **sameness** for *balance* or *symmetry*, but even these substitutions have not consistently occurred among scientists.

Vision

Humans emerged from the great apes with the ability to walk straight up on their hind legs. With this posture, early humans could *see* at much greater distances than they could *smell*. Seeing at greater distance was faster than *hearing* or *smelling*. After all, *light* travels at 186,000 miles per second and *sound* at only about 600 miles per hour. The speed of the molecules detected by *smell* depends on the unpredictable speed and direction of the wind. Without the noses close to the ground to pick up organic molecules for *smell*, upright humans developed a highly *complex* ability to *interpret* a distant scene. The sense of *sight* became dominant with *hearing* a close second. The interaction between these two senses allowed complex language to develop.

Without their noses close to the ground, the sense of *smell* in humans was no longer effective. Later species of *Homo* discovered that dogs with their sharp sense of *smell* could substitute for the ineffective sense of human *smell*. The domestication of dogs may have even contributed further to *sapiens*'s loss of smell. In any event, the senses of sight and sound became more advantageous for hunting over great distances. Along with some of the great apes, humans developed color vision, the greater complexity of which gave our eyes even finer visual discrimination and interpretation.

With the emergence of *Homo sapiens sapiens*, runts, shrieks, gestures, body language, and facial expressions that had previously served for communication were no longer adequate, although these emotional signals have been *conserved* for subtle detection of the primitive *intent* of others. See FIGURE 393 taken from "Toward the Neurobiology of Emotional Body Language," by Beatrice de Gelder, *Nature Reviews Neurobiology*, March 2006. The postures in Figure 369 clearly communicate the person's probable action, or lack thereof. More to *logic*, the authors clearly not only show the interaction of **opposite/same** within *each* of the **opposing** concepts of **either** *congruent* **or** *incongruent*, but more complexly that **both** were placed in *simultaneous double reversal* to create a more complex and enhanced **sameness** of concept that would be **opposed** by our newly emerged human language. Our primitive method of communicating *intent* by body and facial language is still used effectively on the *horizontal* in combination with our more complexly enhanced human language. **Both** languages emerged on the *vertical*, conserved by the **same** logical form.

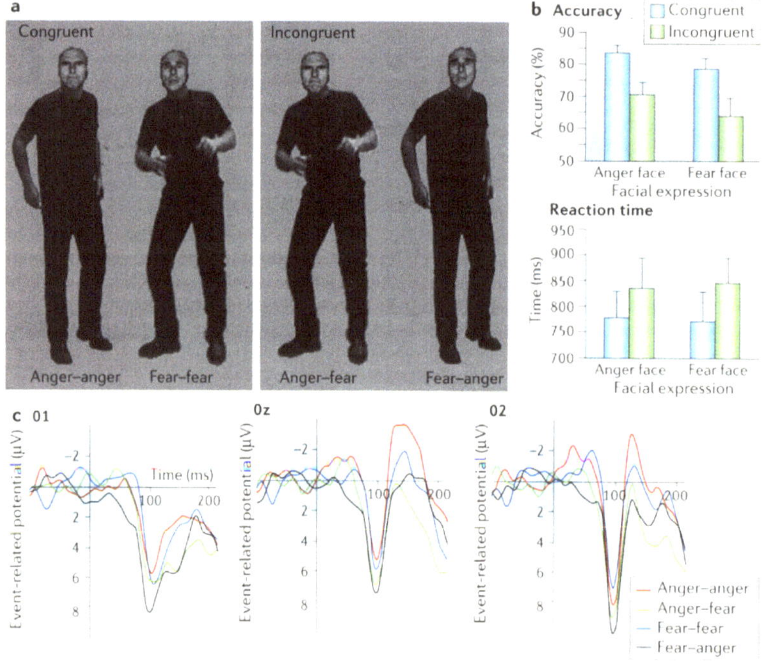

Figure 3 | **Time course of face–body incongruence sensitivity. a** | Examples of the four different categories of face–body compound stimuli used. Congruent and incongruent stimuli consisted of the same material in different combinations. The bodies of the two congruent stimulus conditions were swapped to create a mismatch between the emotion expressed by the face and that expressed by the body. **b** | Behavioural results of the facial expression task for the compound stimuli. Participants had to judge the expression of faces that were accompanied by either a congruent or incongruent bodily expression. Categorization of facial expressions in the presence of an incongruent body emotion significantly reduces accuracy and increases observers' reaction times. **c** | Event-related potentials at occipital electrodes at scalp sites O1, Oz and O2 for the face–body compound stimuli, showing a sharp positive deflection peaking at ~100 ms. This so-called P1 scalp site component is sensitive for the mismatch between the facial expression and the emotional body language. Adapted, with permission, from REF. 39 © (2005) National Academy of Sciences USA.

FIGURE 393

Also read "Embarrassment: A Form of Social Pain," by Christine R. Harris, *American Scientist*, November–December 2006, in which emphasis is placed on facial expressions as language.

For the vast open distances of the African savannahs, a more complex means of communication was needed and selected for by adaptation. This highly complex language formed the **same** as the megalogic seen on the inside front cover, and the *concepts* of which are displayed in Figure 139 gave humans a competitive advantage over all other creatures. Precise instructions could be quickly broadcast over distances for cooperation between individuals and groups. Being able to see and hear complex instructions from a distance enabled the brain to develop highly complex strategies for **both** *aggression* **and** *protection*. The emergence of the human brain with its capacity for complex language gave us our first *chance* to dominate all other complex life.

To see again how the complex sense of *vision* is logically organized in the brain, again see Figure 205. Follow the red and blue lines of neurons from the eyes through the reciprocally **opposing** active/passive hemispheres of the brain and then on to the reciprocally **opposing** passive/active hands seen in Figure 205. Look carefully and you will see the *interaction* of dominant/dominantless in *simultaneous double reversal*. This **super-sameness** of concept is also seen in Figure 217, in which is displayed a better understanding of how and why this megalogical **sameness** has to be **opposed** by the *dominance* of **either** the *right* hemisphere **or** the *left* hemisphere so that complex language can *dominantly* deduce a **specific** conclusion.

Vision emerged separately across all animal species as explained in "Brainless Eyes," by Rüdiger Wehner, *Nature*, May 12, 2005. This paper highly suggests that the early emergence of vision was created from the **same** primitive form that was only complex enough to perceive the world as a *constrained* **generalization**. With *time* giving the dominant **either/or** function a *chance* to **oppose** this **generalization**, a more complex **specificity** of vision developed to *interact* with **general** vision. **General** vision was *conserved* to become peripheral vision that enhances the *dominance* of **specific** vision. The more complex **sameness** of the peripheral vision enhanced the eyes' capacity to make fine distinctions between the **opposing similar-different** objects that *dominate* our *macro*-world. This is **no different** than in hearing a *noise* that having been *constrained* to a **generality** enhances a **specific** *signal*, or the **generality** of overtones that enhance a **specific** note to give it tonal harmony. What the brain/mind cannot do is make all these *perceptual* and *conceptual* interactions *universally* and *timelessly* the **same**. Being able to see **specific** objects *dominantly*, animal brains could more effectively *interpret* whether to **either** prey *on* **or** defend themselves *from* predators, whether an object was **either** *dangerous* **or** *helpful*. Read "Biological Arms Race," a side box in "Learning to Listen," by Sid Perkins, *Science News*, May 14, 2005.

The highly complex sense of vision in humans takes up about two-thirds of the brain's cordial area indicating that vision plays the dominant role of *perception* from which or capacity for *conception* emerged. The great amount of *information* deduced by the visual system requires a comparable capacity for *interpreting* **specific** conclusions. This is why we say the **same** phrase, "point of view," in **both** the brains *visual perceptions* **and** the minds *conceptions*, **both** meaning something **specific** in **opposition** to **general**.

The more complex **sameness** of the entire sensory system's **opposing similar-different** *parts* is revealed in "synesthesia." This is explained by Simon Baron-Cohen PhD. in *Cerebrum*, Fall 2001, who reviewed a book by Patricia Duffy titled *Blue Cats and Chartreuse Kittens*, W. H. Freeman and Company, 2001. I quote from Baron-Cohen's review:

> Synesthesia is a condition in which stimulation of one sense automatically triggers a perception in another sense, although there has been no direct stimulation of the second sense. Thus, a sound might automatically and instantly trigger the perception of a vivid color, or vice versa. Combinations of synesthesia that have been reported include sound giving rise to visual perceptions ("colored hearing") and smell giving rise to tactile sensation. Colored-hearing synethesia appears to be the most common, while certain combinations—for example, touch stimulating hearing –almost never naturally occur. There are two types of synesthesia, development and acquired. Developmental synesthesia begins in childhood, always before age four. It is different from psychotic phenomena such as hallucinations and delusion; nor is it brought on by drug use. Developmental synthesia also differs from imagery that arises from our imagination: Synesthesia is vivid, involuntary and unlearned . . . . Acquired synesthesia, on the other hand, can result from using psychoactive drugs . . . drug induced synesthesia, unlike the developmental variety, is transient, usually has its onset in adult life (or whenever the drug is used), produce sensory combinations that do not occur naturally, and often is accompanied by hallucinations and other loss of reality . . . . Synesthesis [developmental] is usually called a medical (specifically a neurological) condition, but Duffy's account persuades me that we should regard it as a gift . . . synesthesia could be a normal stage of perceptual experience in the developing infant . . . . This suggests the intriguing possibility that we are all synethetes until, somewhere around three months of age, when the maturation of the cerebral cortex gives rise to sensory differentiation.

Read "Hearing Colors and Tasting Shapes," by Vilayanur S. Ramachandran and Edward M. Hubbard, *Scientific American*, May 2003. Also read "Hearing Colours, Seeing Sounds," by Martin Kemp and Colin Blakemore, *Nature*, August 3, 2006, in which the authors show that over the ages some artists have tried to visualize music in their paintings. But the *syntheses* of perception in these painting are always *dominantly* a **generalization** instead of revealing something **specific**.

The **differentiation** of the senses from the **same** neurological source occurs during fetal development. The main hub of the brain's *perceptual* **differentiation** seems to be the amygdala as described by B. Bower, *Science News*, January 25, 2003, titled "Brain Splits Duties to Sniff Out Feelings," from which I quote:

> Figured in many studies as the brain's fear center, the amygdala actually takes charge of assessing emotional intensity of both pleasant and unpleasant sensations, according to a new investigation. At least, this view of the amagdala applies to fragrant and foul odors . . . . The amygdala probably operates in the same way to mark the emotional intensity for sights, sounds, tastes and tactile sensations, the scientists speculate in the February 2003, **Nature Neuroscience**. The amegdala coordinates early processing of the physical intensity of smells and other sensory stimuli, which are then perceived as either pleasant or aversive . . . . The researchers also found that a part of the right frontal brain previously linked to smell perception exhibited increased blood flow as volunteers smelled the pleasant odor, regardless of its intensity. This brain area also responded, to a lesser extent, to pure air. A corresponding section of the left frontal brain showed elevated blood flow as volunteers sniffed the nasty odor at either high or low concentrations. However, pure air drew no enhanced response from this region.

At this point, let me obliquely interject that the "pure air" of the *odor* system can be logically thought as the *concept* of **sameness** that corresponds to the *purity* required of the ingredients in scientific experiments. Without everything being the **same** from between **different** experiments and experimenters, there could be no **general** consensus of the results, which is the *sine qua non* of scientific proof. Even scientists need to *determine* something **specific** out of nature's **general** environment.

Let's go back to the experiments of odor where we found that a **specific** odor is soon transformed to *odorless*. It's not so much that the odor gets tired or jaded, the odor system necessarily returns to its "purity," or **sameness**, so that a **similar** or **different** odor can be quickly detected if there is a sudden *change* in the immediate environment signaling that a **different** situation is being encountered. The continuous perception of the **same** *odor* would not allow for *change* to a **different** odor to be perceived rapidly enough in real time to signal whether a predator might have suddenly arrived. In the *primitive* odor system, **similarities-differences** are **opposed** to **same** odor, although not as *complex* as in the other senses. With hard evidence that *odor* can be transformed to *odorless* and vice versa, we have greater confidence in the proof of reciprocal logic with its *interacting* geometric concepts by which nature *synthesizes* all knowledge as seen in Figure 105.

You can again compare the *simple* interaction of **specific/general** in *smell* with the more *complex* sense of *vision* for yourself by referring to Figures 226 and 227. Stare a few minutes at Figure 227b where the *color* yellow is the **specific** point of view surrounded by the **generality** of *colorless* medium gray. Soon, this color/colorless interaction will merge into the **same** thing—*colorless*, just as *odor* becomes *odorless*, each being the geometric concept of **sameness**. Color/colorless is **no different** than odor/odorless and each **opposite/same** *interaction* must be *simultaneous* functioning as **either/neither** dominant/dominantless respectively. Now superimpose Figure 227b on 227a and **both** *interactions* of color/colorless will become more complexly the **same** thing—*colorless*. You see the **same** thing even more complexly by making the transformation with the two **opposing** color/colorless interactions of Figure 63. **Both samenesses** become more complexly the **same** by *spin* or by *simultaneous double reversal*.

Another experiment of the brain's logic of conceptual interactions is found when you stare at **either** extreme of any two complementary colors. For example, stare a few seconds at an intense *yellow* color and then close your eyes. You will then "see" an intense *blue* color coming from within your brain. You get the **same** results with the 180^0 color complenent. So again we see that the brain, as well as the mind, must *alternately* choose **either** complementary color from the **sameness** of both colors interacting *simultaneously* on order to see *anything* **specific**.

Now let's return to the anomaly of *synesthesia*, a *synthesis* that is not properly constrained.

Read in *Nature*, February 6, under Research Highlights, "Perception: Binding Experience," from which I quote, "The combination of senses results in the world being experienced as a unified whole, but sensory systems do not deliver information to the brain in this way. Signals from different senses are initially registered in separate brain regions. Even within a sensation, features of the sensory mosaic such as colour, size, shape and motion are fragmented and registered in specialized brain regions of the cortex."

This "binding" of sensory perceptions is not a problem for the brain. The "binding," or "unified whole," is the concept of **sameness opposed** by "signals from **different** senses [that are] are initially registered in *separate* brain regions. Even *within* a sensation, features of the sensory mosaic such as colour, size shape and motion are *fragmented* and registered in **specialized** brain regions of the cortex."

If all the senses were the **same**, we would not be able to grasp the richness, complexity, and critical distinctions of the world's **similarities-differences**, so it seems reasonable that persons affected with synesthesia may not have had all their senses **differentiated** during fetal development. Since the amygdala is the main hub where *perceptual* **similarities-differences** are *dominantly* sorted out, this is where *development* would fail to *separate*, say, *color vision* from *sound*, continuing to "bind" or *synthesize* **both** senses. Since the **similar-different** senses are **opposed** to *perceiving* the **same** thing in the **same** way, it is highly probable the *dominance* of **oppositeness** failed at some point during *development*. More succinctly, **opposite/same** *failed* to interact, the **same** as the "failure of the reciprocity law" in photography.

Read "Colourful Language," by Sarah Archibald, *Nature Reviews Neuroscience*, May 2005. Also read "Touching Tastes, Seeing Smells—and Shaking Up Brain Science," by Richard E. Cytowic MD, *Cerebrum*, the Dana Forum of Brain Science, Summer 2002. Also read, "Mystics and Synesthesia" under Random Samples in *Science*, October 29, 2004.
Read "New Tools to Help Patients Reclaim Damaged Sense," by Sandra Blakeslee, in the Science section of the *New York Times*, November 23, 2004, from which I quote:

> An astonishing new technology allows one set of sensory information to substitute for another in the brain. Using novel electronic aids, vision can be represented on the skin, tongue or through the ears. If the sense of touch is gone from one part of the body, it can be routed to an area where touch sensations are intact . . . . Sensory substitution is not new. Touch substitutes for vision when people read Braille. By tapping a can, a blind person perceives a step or curb or a puddle of water but is not aware of any sensation in the hand; feeling is experienced at the tip of the cane.

The land effect shows that the interaction of color/colorless as **specific/general** depicts a *real* scene. Starting with the land effect in Figure 231, note that the *left* scene is illuminated by the *additive* interpretation of primary colors, displaying exactly the **same** scene of the bowl of fruit. The *right* scene shows that the **same** bowl of fruit is illuminated by only *one* of the *colors* and *colorless* black/white. The *left* scene shows that the *additive* interpretation creates a *realistic* scene. The *right* scene is also a *realistic* interpretation through the interaction of color/colorless a **specific/general** respectively. **Both** scenes use the **same** logic to create the third dimension of *hue* that contains all the colors the brain is capable of perceiving.

As reciprocal logic decrees, there is a "reverse land effect" seen in Figure 232, taken from *Light and Color*, by Clarence Rainwater, professor of physics, San Francisco State College, Golden Press, New York, which color/colorless interaction is the **reverse** of the *right* side of Figure 231. Here, in a dimly lit room of *medium gray*, the *interaction* of color/colorless as magenta/white respectively is blocked out by a barrel that seemingly should cast a *colorless* shadow on the *colorless* white screen behind the barrel. But surprisingly, instead of a *colorless* shadow, the screen reflects two *complementary* colors, *magenta* of the *subtractive* primaries and *green* of the *additive* primaries, which are reflected on the screen. This *unrealistic* scene is the very **opposite** of the land effect seen in Figure 231 *right* where the scene depicts *reality*.

It may be a bit of a stretch that *synesthesia* is like the *colored shadows* of the reverse land effect in Figure 232 in which only one color of the *added* primaries and only one color of the *subtractive* primaries represent *under* developed and *incompletely* differentiated parts within *each* **opposing** part of the sensory system.

"Sensory substitution," where any sense can substitute for another, could be the **reverse** of *synesthesia*. If **either** one of the two complementary senses can transform to "substitute" for the other, then **oppositeness** must be *part* of the total sensory system. Contrarywise, if *syntheses* is a **sameness** of the senses, then it too must be *part* of the total sensory system. The trouble is that these two isolated parts do not reciprocally relate, they do not *interact* as **opposite/same** dominant/dominantless to create a *realistic* scene.

A good read on the brain as a finely tuned system of "balance," a **sameness** between the **opposing** concepts of *excitation* and *inhibition*, is found in *Musicophilia: Tales of Music and the Brain*, by Oliver Sacks, Knopf/Picador, 2007.

It is more than just interesting that the **opposing** sensations of **either** *pleasure* **or** *pain* are reciprocally ordered in the **same** neural pathway. Read, "A common neurobiology for pain and pleasure", by Siri Leknes and Irene Tracey, under Perspectives, in *nature reviews/neuroscience*. April, 2008.

COLLISIONS AS INTERACTIONS

The laws that describe the behaviour of complex systems are qualitatively different from those that govern its units.
—Tamas Vicsek, *Nature*, July 1, 2002

Collisions between *object* masses and *interactions* between *conceptual* masses are governed by the **same** logic. The *collision* between two *weightless objects* of the **same** spherical shape and density will bounce way from each other in reciprocally **opposing** directions. *Weighted* mass however, *dominated* by inferential processes, hides its conserved geometric concept. The game of billiards however is one of those rare instances where the logic of *interactions* is clearly revealed. Reciprocal logic is revealed when billiard balls *collide* on a smooth, hard, flat, "level playing field." Also, the billiard balls must be extremely spherical, very elastic (they bounce with little distortion); and they are of the **same** size and weight. With both the *surface* and the weighed *balls* dominated by the geometric concept of **sameness**, their frame of reference is highly inertial with very little friction to **oppose** their *uniform motion in a straight line*. This *finite design* gives the weighted billiard balls a *chance* to move in an *infinite* number of **similar-different** directions. With this accelerated/inertial, infinite/finite, chance/design as **opposite/same** respectively scenario, the *physical* interaction is the **same** as *conceptual* interactions. With these two reciprocally **opposing** interpretations, we now choose the interaction of *physical objects*:
In Figure 15, you will see that the two reciprocally **opposing** *concepts* of projectile/target (motion/rest) *interact* at 90° regardless of **similar-different** speeds and **similar-different** angles at the point of their *collision*.

Now look at Figure 16 that shows the *collision* of two *concepts* that are reciprocally the **same**—projectile-projectile (motion/motion). They move in *complementary* directions at 180° from where after they *collide* they move along the **same** straight line regardless of their **similar-different** speeds and **similar-different** angles.

Both Figures 15 and 16 superimposed show the 90° angle of the acceleration/inertia as **opposite/same** interacting *dominantly* in reciprocal **opposition**. Because this interaction is *logically conserved*, the game can be played in *orderly* manner with an infinite number of **similar-different** outcomes and players that **either** *win* **or** *lose*. Without this universal *ordering* principal, *disorder* would reign; and no game, including billiards, could ever be played. For an animation of billiard ball *interactions* and their reciprocally **opposing** *interpretations*, refer to the Web site http://www.reciprocallogic.com.

For a more comprehensive introduction to the *interaction* of billiard balls, read "The Way the Ball Bounces," Brian Hayes, *American Scientist*, July–August 1996, that shows how much easier and simpler it would be if we understood that *collisions* are reciprocally ordered despite the *dominance* of inference.

When a billiard ball bounces off the cushion of a pool table, the *single* billiard *ball* always bounces off the *cushion* at an angle that keeps the line of the ball **both** *before* **and** *after* the collision the **same**, as seen in Figure 71. Since the angle of collision *varies* with the ball's line of motion, the angle *varies* accordingly in contrast to the *constant* angle of 90° between *two* colliding billiard balls. The angle of the billiard ball hitting the cushion is the **same** as the angle a photon takes when it collides with the *reflective* concept of a mirror, also shown in Figure 71.

The **sameness** and function of *mirror reflection* and the **opposing** *mass* of the billiard ball interact the **same** as when *light* strikes *mass*, which relative/absolute interaction is **no different** than variable/constant, acceleration/inertia in our *relative* world.

Now let us *interpret* the *collision* of billiard *balls* to the *dominance* of inference that **opposes** to the logic of *interactions*. This concept of *momentum* emerges from the interaction of mass/speed.

Place a two billiard balls on the pool table and align the length of the cue stick so that it and the two balls all form a *straight line*. With the cue stick, strike the cue ball so that it *collides* head-on in a *straight line* with the other billiard ball. At this *center* point of the *collision*, the cue ball will stop dead still, but the other billiard ball will instantly *conserve* the **same** *speed* of the cue ball as if **both** *complementary* balls were *one* and the **same** rolling ball.

You can do the **same** thing with more than two billiard balls. Again, align the cue stick, the cue ball, and several more balls so that they all lie along a *straight line*. When the **specific** cue ball strikes the *collection* head-on, the cue ball will again *instantly* stop dead still; but the billiard ball at the other end of the line will *conserve* the *momentum* of the cue ball, as if **both** the *first* **and** the *last* balls were one and the **same** ball. Although the concept of *momentum* is *obvious*, the hidden *interaction* of **specific/general** is also *conserved*. So we again see that sequence/simultaneity, time/timeless respectively interact more complexly as the megalogic seen on the inside front cover.

There is also the *conservation* of *angular* momentum which formula is the **same** as the conservation of *linear* momentum except for their respective mathematical notations.

Solitons

Solitons are a **generalization** that can be **specifically** identified. Solitons are said to be "packets of energy," which seems closely akin to the "constrained energy" of quarks mass. The **specificity** of solitons can also be interpreted as being *constrained* within a more **generalized** area, a **sameness** such as water or of coherent light waves. A soliton of any stripe *moves* within the **same** *motionless* medium of which it is a part.

A good description of solitons is found in *Nature*, June 26, 1997, "Solitons Light the Way," by Allan Boardman, whom I quote:

> Normally, a hump of water like this is thought of as a packet of waves, all traveling with different speeds. This way of looking at things comes from Fourier, but it is a linear viewpoint and the end result is the destruction of the hump. Large amplitudes mean lots of power, so waves in the packet are now forced to interact with one another . . . this is a non-linear system. In a balanced situation, the power of the wave acts against dispersion. The wave remains intact and a soliton is born! . . . Although a soliton is a solitary wave it retains its identity, even after a collision. The change of "ary" to "on" is not surprising . . . physicists call things "on' at almost every opportunity (take electron, meson, and photon, for instance) because the Greek word for "on" means solitary, and in each instance the suffix signals the principle that the entity concerned retains its characteristics.

Read "Solitons Made Simple," by Y. R. Shen, *Science*, June 6, 1997.

Mathematical models of solitons reveal that they are formed "when the two elements *balance* exactly . . . when the *linear* tendency to *disperse* is precisely countered by the *nonlinear* tendency to *remain whole*—a soliton in born." The *italics* in this quote are mine.

Solitons behave as if **specific** and **general** were the **same** more complex and enhanced idea. Perhaps they are "born" by **opposing** internal/external forces *interacting* within the **same** medium.

Boyle's Law

Boyle's law comes from the collision of *object* molecules of the **same** atoms or molecules, such as the **sameness** of air or the **sameness** of any gas that is *constrained* within a sphere. Regardless of the type of gas, the *number* of molecules, or atoms, are always the **same** at any given mole, the measure of energy. This number (6.023×10^{23}) is called Avogadro's number, a *constant*, always the **same** regardless of the *various* atoms or molecules *constrained* within the sphere.

Again we see that the idea of *constrained* **sameness** and the *interactions* of variable/constant and linear/non-linear defy the **opposing** *dominance* of inference.

The N-body Problem

As the **similarities-differences** of mass become increasingly *complex*, the *simple* geometric *concept* of **sameness** is increasingly obscured. We have seen that two closely interacting *weightless* masses are ordered the **same** by the force of gravity regardless of their **similarities-differences**. The *two* **opposing** *weightless* masses, *each* interacting **opposite/same**, have the **same** *form* as the *constrained* **sameness** of any mathematical *formula* with its *two* **opposing** halves.

As more than two such *weightless* masses are added to the gravitational *interaction*, the mathematical *problem* of finding the unknown X becomes increasingly difficult to *solve*. The interaction of just *two weightless* masses is easy to solve for, but the addition of the *third* weightless body is much more difficult to calculate. In computer models, instead of *single* ellipse, *two* ellipses form as *two* **opposing** parts of the **same** figure 8, which divided/unity does not correspond to what is observed.

Beyond *three* interacting *weightless* masses, the *mind* cannot *balance* the force of gravity's universal *contraction*. In other words, no mathematical formula with its **opposing** sides in *simultaneous double reversal* to create **sameness** works. More than three interacting heavenly bodies are *chaotic* to the human *mind* bound by our paradigm of inference. Discussing the N-body problem with my family physician, John Baumrucker of Highlands, North Carolina, it reminded him of the interaction between two male/female lovers who felt a *heavenly* harmony between them. All is well and *orderly* until a *third* lover enters the picture, whether male or female. Then *hell* breaks loose and *chaos* reigns.

The N-body "problem" reminds us that our **specific** *minds* dominance of reciprocal **opposition** is always only *part* of universal **sameness** at any level of *mass* complexity.

Evidence that supports our perpetual entrapment in conceptual conflict even in the highly complex interactions of biology is seen in "Oligodendrocyte Wars," a review by William D. Richardson, Nicoletta Kessaris, and Nigel Pringle in *Nature Reviews Neuroscience*, January 2006. I quote from this review that cites not just the need of but also the necessity for conceptual conflict in understanding the *development* of complex life, as well as understanding the *logical* outcome of any scientific endeavor:

> The developmental origin of oligodendrocytes has been hotly debated for years. Some laboratories, including our own, favoured a unique origin of oligodendrocytes in the ventral neural tube, whereas others went for diversity and multiple origins. The published literature was conflicting and confusing . . . . As is often the case, the answer is turning out to be more complex than expected . . . [a more complex model] has shown conclusively that there are dorsal as well as ventral origins in the spinal chord and brain . . . . This article aims to provide a historical perspective on the "origins" debate, which illustrates, in microcosm, the struggle for understanding that runs through all science. A series of recently published articles has revealed that the jostling among rival ideas and laboratories is mirrored by another type of competition among the oligodendrocyte populations themselves.

Then, the authors go on to explain in detail this conceptual conflict under the heading, "Oligodendrocyte origins—A Battle of Ideas." Then, there is a subsequent detailed explanation headed "Oligodendrocyte Wars in the Forebrain." I quote most of their "conclusion" of the *idea* of oligodendrocyte "origin."

> The driving force of scientific progress is competition among individuals and ideas. That has certainly been true of our field of glial cell development. At last, the long running arguments about the site(s) of origin of oligodendrocytes are being settled—the answer is that there are both dorsal and ventral sources that become active at different times during development and that compete with each other for territory. The old arguments will soon be forgotten as the field moves on, but they were important in passing, because controversy focuses the mind, attracts attention and brings newcomers into the field. Scientific progress is ultimately a group effort in which controversy and dispute play an essential part . . . . We look forward, with anticipation, to another decade of cut and thrust.

As any evolutionary biologist delves ever deeper into his or her *part* of science, it becomes obvious that the brain-mind cannot know everything in just their discipline let alone in all of the hard sciences. A glimpse of the seemingly *infinite* number of **similarities-differences** not just in our realm of *macro-mass* but also only in *biology* in the seen in "What Chemists Want to Know," by Phillip Ball, *Nature*, August 3, 2006, from which I quote, "It has been estimated that there are about 10^{40} possible molecules that could be made from common elements with a molecular size comparable to that of a typical drug . . . . The known chemical world, including the expansion of the natural world that chemists have achieved, is no where near 1% of that."

Factual knowledge for the brain is *finite* in an *infinite* number of the world's permutations and combinations. As for *concepts*, interactions as finite/infinite are *finite*, a **specificity** that **opposes** the *infinite* **sameness** of Universal order.

So my conjecture of the insoluble human predicament is not just an artefact, a delusion, or pure imagination of mind. The logic of interactions is real and must be accepted even though inference requires that we remain dominantly **opposed** to it.

Two examples that demonstrate the *practicality* of the *conceptual* interaction of **opposite/same** are seen in Figure 179. Here, "collision as interaction" in baseball requires that if the outfielder is to catch a fly ball, he and the ball must maintain the **same** angle regardless of any **differences** in their respective speeds. If the outfielder's line of sight *varies* with that of the ball, his mitt cannot *collide* with the ball. See also that two boats will always *collide* if their mutual angle and their line of sight remain the **same** regardless of their **differences** in speeds. **Different** speeds and **different** angles *dominate* these two scenes in **opposition** to **sameness**. But if by chance the angle and line of sight happen to be the **same** between them, the consequence of the former is *good* for the outfielder; but in the latter, the consequence for both boats is *bad*. Again, we see the logic in the most basic law of *macro*-physics—that no two objects can occupy the **same** space at the **same** time. Otherwise, the baseball would pass through the outfielder's mitt, and the two boats would simply pass through each other. Everything would be the **same** without the **opposition** of **similarities-differences** in mass, and *perception* and *conception* could not have evolved to have knowledge of the world.

A good test of reciprocal logic is the everyday interaction between two persons walking pass each other, or of two people **either** *meeting* each other **or** *leaving* each other. In such encounters, whatever the first person to speak says, the other will invariably repeat the **same** thing. It is only natural that two spatially **opposing** greeting would be the **same**.

Two people meeting each other for either the same or opposite purpose is seen in the animations of the Reciprocal Logic II Web site; (*http://www.reciprocallogic.com*).

THE LOGIC OF CONCEPTUAL INTERACTIONS BETWEEN THE TWO OPPOSING HEMISPHERES OF THE BRAIN

The main theme to emerge . . . is that there appear to be two modes of thinking, verbal and non-verbal, represented separately in left and right hemispheres, respectively, and that our educational system, as well as science in general, tends to neglect the non-verbal form of the intellect. What it comes down to is that modern society discriminates against the right hemisphere.

—Roger W. Sperry, ***Lateral Specialization of Cerebral Function*** *in the Surgically Separated Hemispheres*

If biologists have ignored self-organization, it is not because self-organization is not pervasive and profound. It is because we biologists have yet to understand how to think about systems governed simultaneously by two sources of order.

—Stewart Kaufmann, *At Home in the Universe*

In our *macro*-world, things are perceived dominantly as being **either** this **or** that. Even though a few molecules are *chiral*—i.e., two mirror images of the **same** simple molecule as seen in Figure 154, they are not absolute since they can often have **different** affects interacting with other molecules. Most molecules are *handed;* **either** *positive* **or** *negative,* although the carbon atom, the basic element of all highly complex life, is delicately *balanced* to readily interact with **either** a *negative* **or** *positive* charged molecule.

As we have seen in Figure 162, even the highly complex molecule of DNA is **either** *left*-handed **or** *right*-handed, with *left* hand being the *dominant*.

.

The dominance of **oppositeness** with its **either/or** function extends to life's more complex amino acids that are *right*-handed, while the even more complex proteins are *left*-handed. Most of the higher forms of bilateral life are **either** *left*—**or** *right*-handed. The *dominance* of *handedness* in relative mass **specificities** has emerged from the *absolute* **either-or** function of electron **oppositeness** that hides the **neither-nor** function of quarks **sameness**. The *simplest* example of *relative* mass's **opposition** to **sameness** emerged from the weak force where electron charge was found to be slightly *imbalanced*, violating what physicists call "charge invariance." The concept of *time* also begins to reveal its *asymmetry* at this *simplest* level of *macro*-mass complexity setting up the concept of *sequence* in **opposition** to *simultaneity*.

To get a feel for the pervasive dominance of *handedness* in *macro*-mass, visit the following Figures from the previous addenda of **reciprocal logic** where you can get a brief description of each: **Either oppositeness** in Figure 16 **or sameness** in Figure 15; also the reciprocally **opposing** interactions in Figures 18–19; Figures 20–25, 53–54 , and 57; Figures A–B, D–G in Addendum V; and Figure 160–163, 171, 174, 194–195, 204–205, 216–217, 221, 226–228, 230–235, 240, 242–243, 248–249, 253, 255–256, 258–259, and 261–264.

The brains of bilateral animals are divided **either** *left* **or** *right*, and most have a dominant side. The human brain is "split" between the *dominant left* hemisphere and the *dominantless right* hemisphere. Again, see Figure 205 where the *dominant left* hemisphere houses *sequential* thought while the *dominantless right* hemisphere is the seat of *simultaneity*. The *right* hemisphere has been thought to be *logicless* because it is non-inferential. The *right* hemisphere's "spatial," and **sameness** of concept leaves scientists without a clue of any logical order, probably because the *dominant left* brain **opposes** it. Figure 207 and my interpretation in Figure 209 detail the conflicting concepts between the left/right hemispheres interacting dominant/dominantless respectively.

Read "Forty-five Years of Split-brain Research and Still Going Strong," by Michael S. Gazzaniga, *Nature Reviews Neuroscience*, August 2005, in which the author emphasizes that "of the dozens of instances recorded over the years, none allowed for a clear cut claim that each hemisphere has a full sense of self." This speaks volumes for the complexity of the human brain-mind over other complex mammals. There is no better explanation for *consciousness* than the emergence of the human *mind* with its *geometric* interpretation of **either** *left* **or** *right*, and its *conceptual* interpretation of **either** *good* **or** *evil*. Read "Molecular Approaches to Brain Asymmetry and Handedness," by Tao Sun and Christopher A. Walsh, *Nature Reviews Neuroscience*, August 2006.

It is important to understand that *each* hemisphere of humans is a less complex *brain* unto itself capable of making **specific** distinctions between **similar-different** things. **Both** hemispheres, however, interact a more complex and enhanced *consciousness* with its two reciprocally **opposing** *parts*.

Probably the strongest example of the *division* between the *left* and *right* hemispheres is seen in experiments on attention, or consciousness, as seen in, "Left Out by a Stroke: Right Brain Injury May Upset Attention Balance," by B. Bower, *Science News*, October 29, 2005, from which I quote:

> People who suddenly ignore everything to their left after suffering a right-brain stroke display disturbed activity in uninjured parts of the widespread neural network associated with attention . . . . Structures on both sides of the brain typically maintain a delicate balance in regulating visual attention . . . . In some stroke patients, underactivity of damaged right-brain attention leads to hyperactivity in intact left-brain attention structures . . . . As a result, patients focus their visual attention primarily to the right and display various forms of left-sided neglect, such as failing to notice or eat food on the left half of a plate and behaving as if they didn't have a left arm . . . . Participants who recovered most completely from spatial neglect had the most-balanced neural activity . . . . Parts of the left-brain attention system . . . also play critical roles in language use . . . no one knows how the brain orchestrates both attention and language using the **same** patches of tissue.

My conjecture that the **same** brain/mind is divided into two reciprocally **opposing** parts, the **same** as the two reciprocally **opposing** parts of inference and the **electro/magnetic force**, refer again to Figures F and 205 and then to FIGURE 394 taken from "Two Gradients Are Better Than One," by Liqun Luo, *Nature*, January 5, 2006.

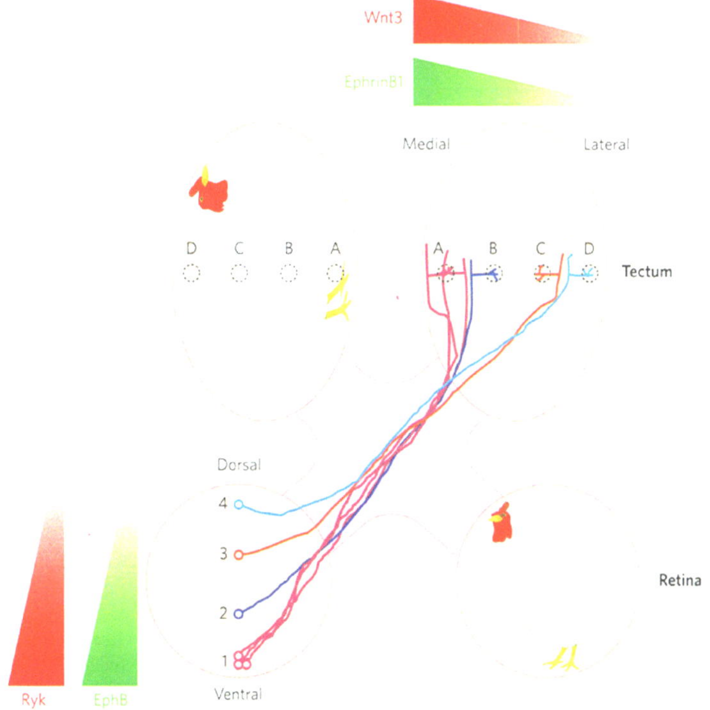

Figure 1 Setting up the retinotopic map. An object on the retina (a chicken) is recapitulated in the tectum region of the brain. (The right retina is wired up to the left side of the brain and vice versa.) This is made possible by growth of nerve fibres (axons, coloured lines) from retinal ganglion cells (RGCs, coloured circles in the retina) along both *x* and *y* axes to the termination zones in the tectum (dotted circles). Only the retinal *y* axis, from dorsal to ventral, is depicted here. For instance, the purple neurons at the bottom perceive the bottom of the image, the chicken's feet; in the brain, they are wired up to region A on the medial side of the tectum, so the information from the image is transformed from the dorsal–ventral axis of the retina to a medial–lateral axis in the brain. The axons home in on their termination zones by regulating the direction of their branching — allowing the three purple axons to reach the same termination zone A. To achieve this, RGC axons expressing EphB receptors are attracted by the EphrinB gradient[6]. So, the purple axons with the most receptors are attracted to the medial side, where there is the most EphrinB. Schmitt *et al.*[2] have added a second gradient to this map, made of Wnt3 and its receptor Ryk. Repulsion of branches mediated by a Wnt3–Ryk gradient counterbalance the attraction mediated by EphrinB–EphB. (Adapted from refs 2 and 14.)

<u>Figure 394</u>

Conceptual conflict in the mammalian brain is expressed in the *geometry* of the *visual* system known as "binocular rivalry" described in a book review by Jeremy M. Wolfe in *Nature Neuroscience*, June 2004, "Consider binocular rivalry, which is experienced when each of the two eyes are presented with a different stimulus at the same location in the visual field. Simultaneously looking at a pattern of vertical lines with the left eye and horizontal line with the right eye does not result in seeing a plaid: the sum of the two images. Instead, what one experiences is a battle, with either the vertical or horizontal lines being perceived at each moment and each location for as long as you care to look. (You can try this yourself at (http://www.psy.vanderbilt.edu/faculty/blake/Rivalry/BR.html.)"

Also read "Neural Correlates of Binocular Rivalry in the Human Lateral Geniculate Nucleus," by Klaus Wunderlich, Keith A. Schneider, and Sabine Kastner, *Nature Neuroscience*, November 2005, from which I quote:

> Binocular rivalry occurs when the input from the two eyes cannot be focused into a single, coherent percept. Rivalry can be induced experimentally by simultaneously presenting dissimilar stimuli to the two eyes, each as a vertical grating to one eye and a horizontal grating to the other. Rather than being received as a merged plaid, the two stimuli compete for conceptual dominance such that subjects perceive only one stimulus at a time while the other is suppressed from visual awareness. Usually one stimulus predominates for several seconds, and the extent of competition between any pair of stimuli depends on stimulus properties, such as their relative contrast or spatial frequency. Because the subjects' perceptual experiences change over time while the retinal stimulus remains constant, binocular rivalry provides an intriguing paradigm to study the neural basis of visual awareness.

In the above quote note that the brain *alternates* dominant/dominantlessly the *vertical* and *horizontal* gratings to perceive *conclusively*. This allows the other eye to take over the **same** perception of **both** although less complexly. If both eyes interacted dominant/dominant *simultaneously*, everything would be more complexly the **same**, and vision would be inconclusive. **Either** the *left* hemisphere's dominance of *analysis* **or** the *right* hemisphere's dominantless *syntheses* can substitute for the other. Because *each* hemisphere interacts dominant/dominantless **either** vision is a *valid* interpretation of a scene. With the loss of **either** eye the interaction of **specific/general** as dominant/dominantless respectively conserves vision in the other eye, although its discernment is less acute.

This is the **same** logic that the mind functions in inference, *alternating* **either** deduction *or* induction *dominantly* to arrive at a conclusion. **Both opposing** processes *simultaneously* would be *inconclusive*. The **same** logic holds for the **electro/magnetic** force that *alternates* **either opposing** hand dominantly.

Also note from the above that *change* is always accompanied by a *constant*.

Another interaction of change/constant in the visual system is that there is an involuntary small fast side-to-side motion that enables the brain to make ultrafine distinctions that enhances vision. Read "An Eye for Detail.", under, In this Issue, of *Science*, June 24, 2007, in which the editors say, "The function of fixational eye movement, the tiny involuntary eye movements or 'retinal jitters' that occur when we fix our gaze on something . . . using a combination of physophysical experiments with statistical analysis of the visual signals entering the eye to counteract the visual effects of the eye movements . . . . Rucci *et al*. show that the fine-grained information is reduced. This suggests that fixational eye movements are part of a strategy used by the brain to extract fine details of visual information." Also read "Eye Twitching May Be Necessary for Seeing," the *Wall Street Journal*, July 18, 2007, in which the editors give a synopsis of an article in *Scientific American*, August 2007.

This battle between the two **opposing** hemispheres is also seen the saccadic movement of the eyes in REM (rapid eye movement) during sleep that enables the brain to maintain its *dominance* of deduction and of *self*-identity. Dreams that occur in this state of sleep also promote memory of what the individual has learned.

Again, refer to FIGURE 368 to see the *interaction* of dominant/dominantless in *each* of the two hemispheres of the brain fighting each other reciprocally for dominance. Read "Cats with a Binding Problem," by Andreas K. Engel and Cornelia Kranczioch, *Scientific American Mind*, August/September 2006, in which we see *perceptual* conflict between *division* and *unity* in the cat's visual system. The conflict, or "bind," however is not a problem of *mindless* nature but instead is required for the logic of all *mass* interactions that are **opposed** to **sameness**. In Figure 217 observe that the two reciprocally **opposing** halves of the cat's brain each *interact* dominant/dominantless.

The brain/mind is *dominantly* hardwired during *development* for *left*-hemisphere *analysis*. But early environmental stimulation can alter the neurological connections without relinquishing the *dominance* of deduction in inference. This proportional rearrangement allows the brain to adapt to its environment in order to survive. Read the editors' introduction to the authors in *Nature Reviews Neuroscience*, September 2006.

Right-hemisphere **generalization** cannot *dominate* the animal brain because that would *diminish* the ability to discern the **similarities-differences** of **specific** objects of mass. **Both** hemispheres interact **specific/general** in *simultaneous double reversal* to create a more complex and enhanced **sameness** that empowers the *left*-brain's language center, the Broca and Wernicke area. This area of inferential thought, seen in Figure 240, still **opposes** the spatial **sameness** of the right brain.

Unfortunately, many educators have sought to downplay the *dominance* of deduction and of mathematical *analysis* with their precisely **specific** solutions. The mantra recently has been to teach students that a **general** mathematical solution is OK as long as the student *interprets* what he or she thinks the solution ought to be. Some students, therefore, never even learn to give the correct change in monetary transactions. Without the *simple* and precise methods of addition, subtraction, multiplication, division, fractions, and algebraic proportions, the more *complex* subjects of trigonometry and calculus can never be learned. Student should *develop* mathematical *concepts* the way *many* brilliant human *minds* have developed them over the long course of history. This would give students a much greater interest and appreciation of science and the need for *quantification* in human affairs.

Denying the *dominance* of deduction in inference has come about by many intellectuals and professors thinking that the subject of history has been too biased toward Western, or Eurocentric, culture for students to get a liberal education. So they have attempted to destroy the history of Western thought. But a liberal education is a **general** education, broad-based knowledge that includes the history of logic, science, and mathematics that have dominantly evolved within Western culture. The idea of *destruction* has invaded all modern thought, including architecture as seen in Figure 330. Read "When Buildings Stop Making Sense," the *Wall Street Journal*, November 2007, by Francis Morrone, who reviews a book titled *Architecture of the Absurd*, by John Silber, Quantuck Lane Press, 2007.

Read, "To Sum Up, Students' Math Skills Fall Short," by Michael Belinsky, the *Atlanta-Journal Constitution*, July 25, 2006. Also read "'Innovative' Math, but Can You Count?" by Samuel G. Freedman, the *New York Times*, November 9, 2005. Also read "Why Righties Can't Teach," by John Tierney, the *New York Times*, October 15, 2005. Also read "Testing Is Not Teaching," by Dorothy Rich, *Knight Ridder/Tribune*, June 2, 2006. Also read "At 2-year Colleges, Students Eager but Unready," subtitled, "Hopes Meet Reality as Many Struggle with Remedial Work," by Diana Jean Schemo, the *New York Times*, September 2, 2006.

Although *facts* and data processing should *dominate* education, it is *meaningless* if *concepts* are not included to lend interest to the rote learning and routine calculations. For example, for many students, the mathematical concepts of calculus are not sufficiently engrained into the brain/mind to make this complex subject seem to be worth the effort to learn. But the *conceptual* interactions of calculus embrace all the dynamics of *mass* in the real world, and learning how to *do* calculus teaches one to *do* things logically and to become synchronized with the *real* world. The old adage, "To get along, go along," is a good reminder for the *mind* to accept and comply with the conceptual interactions of *mindless* nature.

Lecture 4, "The Fundamental Theories of Calculus," under "Calculus Made Clear," one of the many "great courses" copyrighted by the Teaching Company, Chantilly, Virginia, confirms that calculus has two reciprocally **opposing** parts of the **same** form of logic, which demonstrates that *concepts* should take their rightful place along with *number* and *geometry* to get the *real* interpretation of *macro*-mass dynamics as seen in Figure 105. I quote from the introduction to lecture 4, "The Fundamental Theorem of Calculus makes the connection between . . . the derivative and the integral . . . they accomplish opposite goals—one goes from position to speed, and the other goes from speed to position. The duality between the derivative and the interval is the Fundamental Theorem of Calculus. This convergence of *ideas* underscores the power of abstraction, one of the global themes of this series of lectures." These "converging ideas" are seen in Figure 150. Again, Newton *alternated* the reciprocally **opposing** processes of inference to discover the logic of calculus, the **same** as he did to discover of the force of gravity. *Mind* and *mindless* nature are ordered by the **same** logic of conceptual interactions, but the mind **opposes** nature's **sameness** of concept at each successive increase in complexity.

As for the educational system, the most logical method for teaching would be to first evaluate each child's *dominance* of concept as well as the child's proportion of analysis/synthesis, quantitative/qualitative, deduction/induction, science/art. Then divide each student into the subjects that his or her mind favors. Because all students do not have the **same** intellectual capacity or motivation, the *fast* learners and the *slow* learners within *each* of these two conceptually **opposing** categories also should be *separated* so that each type of student can get the maximum amount of learning that his or her **specific** intellectual capacity and motivation can accommodate. This way each student gets the **same** *chance* to achieve the maximum degree of learning for which he or she has the greatest intellectual capacity. Such *conceptual divisions* of brain/minds will correspond to the *conceptual divisions* of the real world, thereby giving each student the best *chance* to adapt and live a fuller and richer life.

The latest evidence for this educational structure is seen in, "Effects of development and enculturation on number representation in the brain", by Daniel Ansari, *nature reviews/neuroscience*, April, 2008, The author leaves little doubt as to the need for making a clear distinction between the brain's dominant *numerical* and dominantless *non-numerical* capacities of children as soon as they begin their formal schooling.

One caveat is that because the brain's *left* hemisphere of information processing is *dominant*, deduction in inference is *dominant*, mass **specificity** is *dominant*, and electron charge is *dominant*, any educational system should nudge students toward the dominant concept of any interaction. When in doubt, choose deduction over induction.

This method of *dividing* the capacity for knowledge into its reciprocally **opposing** parts also conforms to C. P. Snow's *The Two Cultures*, seen in Figure 32, where quantification/qualification substitute for analysis/synthesis, deduction/induction, dominant/dominantless respectively. The reason for teaching science and quantification *dominantly* over art and qualification is evident in the editorial of the *New York Times*, September 18, 2006
Read "The Logic of Human Learning," by Nick Chater, *Nature*, October 5, 2006, in which the author argues for the importance of teaching "Boolean concepts," the *operations* of which are **no different** than the **either/neither** *functions* of reciprocal logic.

Read the editorial, "German Science Policy 2006," *Science*, July 14, 2006, from which I quote, "The German Government recognizes that our future lies in a knowledge based society founded on freedom and responsibility."

Unfortunately, many of today's youth in highly developed countries—especially the United States—are shielded from the need of hard endless labor for survival. As the novelist John Dos Passos wrote in 1934, "We work to eat to get the strength to work to eat to get the strength to work." Read also *Work: The World in Photographs*, National Geographic, 2006.

The cupboards and refrigerators of the wealthier nations may be full, but by not having experienced the bleaker side of life, the naive hopes and expectations of youth are dashed. The *fullness* of life comes up *empty* without having traveled the rocky road to maturity. Read "Land of Empty," *Scientific American Mind*, by Richard Lipkin, who reviews a book titled *The Price of Privilege: How Parental Pressure and Material Advantage Are Creating a Generation of Disconnected and Unhappy Kids* by Madeline Levine, HarperCollins, 2006. Material wealth alone cannot substitute for the wealth of scientific information and the paradox *mind* and *mindless* nature's conceptual *interactions* such as sin/virtue. Read "Washing Away Your Sins: Threatened Morality and Physical Cleansing," by Chen-Bo Zhong and Katie Liljenquist, *Science*, September 8, 2006. I quote the authors' abstract:

> Physical cleansing has been a focal element in religious ceremonies for thousand of years. The prevalence of this practice suggests a psychological association between body purity and moral purity. In three studies, we explored what we call the "Macbeth effect"—that is, a threat to one's moral purity induces the need to cleanse oneself. This effect revealed itself through an increase mental accessibility of cleansing-related concepts, a greater desire for cleansing products, and a greater likelihood of taking antiseptic wipes. Furthermore, we showed that physical cleansing alleviates the unsettling consequence of unethical behavior and reduces threats to ones moral self-image. Daily hygiene routines such as washing hands, as simple and benign as they may seem, can deliver a powerful antidote to threatened morality, enabling people to truly wash away their sins.

To get a feel for the great *complexity* of just one small part of the brain's dominant/dominantless interactions, let me quote from "Dendritic Arithmetic" by Nelson Spruston and William L. Kath, *Nature Neuroscience*, June 2004:

> Why are real brains so much more powerful than artificial neural networks? In part this a matter of circuit complexity and scale, but it is also because real neurons are much are complex than the simple elements used in neural networks models. One strategy for developing more sophisticated neural networks is to replace simple elements with sophisticated computational models of neurons, including branching dendritic trees, thousand of synapses and dozens of voltage-gated conductances. The problem of this approach is that it is not manageable to use such computational expensive neural models on large-scale networks. Another strategy is to develop abstractions of neurons that capture the essential processing power of the real thing.

The authors go on to say that even these *abstracted* models do not complete the understanding of the interactions that connect neurons with each other through the synapses. What is so "exciting" to the authors about their *conceptual* model is that "stimulation of one dendritic branch [of a neuron] at two locations led to a superlinear summation . . . when the two stimulating electrodes were positioned near two different dendritic branches, activation of both inputs produced a roughly nonlinear summation. In fact, superlinear summation only occurred if the two activated inputs were less than 100 *um* apart on the same dentritic branch . . . . Despite these exciting results and their wide reaching implications, the pyramidal neuron is not yet completely solved. To name just a few of the many remaining mysteries." As we have now learned, this *interaction* never can be solved. Its *concepts* are forever *incomplete*.

FIGURE 395, taken from this work, shows a "neuron tree" with its pyramidal dendrites where **both** of its **opposing** **<u>specific/general</u>** ends are placed in *simultaneous double reversal* to create "superlinear summation" that is more than the sum of its parts.

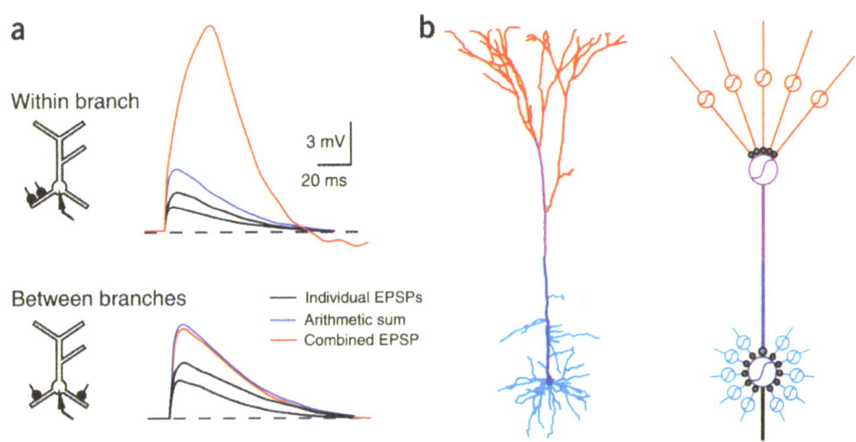

Figure 1 Spiking in individual dendritic branches implies a three-layer model of synaptic integration. (**a**) Schematic representation of the main finding of Polsky *et al.*[1]: two multisynaptic inputs onto a single dendritic branch exhibit superlinear summation (top). Inputs onto separate branches exhibit roughly linear summation (bottom). (**b**) Reconstructed layer-5 pyramidal neuron (left) and an abstracted three-layer network model (right; based on ref. 14). Red branches represent the distal apical inputs and light blue branches the perisomatic inputs. Together, these inputs constitute the first layer of the network model, each performing superlinear summation of synaptic inputs as shown in **a** (indicated by small circles with sigmoids). The outputs of this first layer feed into two integration zones: one near the perisomatic branches (dark blue; *e.g.*, soma) and one near the distal apical branches (purple; *e.g.*, apical spike initiation zone). These integration zones constitute the second layer of the network model (large circles with sigmoids). The third layer (not shown) is the action potential initiation zone in the axon. Grey circles indicate connections between layers.

<u>FIGURE 395</u>

The **difference** in the logic between the animal brain and the computer, other than the former is analog and the later digital, is that the computer's logic is strictly deductive, therefore not nearly as complex. If the animal brain were digitally ordered and able only to processes data by deduction to **specific** conclusions, a **general** conclusion would not be able to enhance the *perception* of **similar-different** objects. Computerized artificial intelligence therefore probably cannot be developed to become as *logically* complex as the mammalian brain and never as complex as the human <u>brain/mind</u>. The "brains" of the computer can only choose the logic of **either** *true* **or** *not true* (*false*) **either** *yes* **or** *no*, **either** 1 **or** 0, *alternately* in time. Read the book review—"The Computer that Started It All," by Gualtiero Piccinini in *Cerebrum*, the Dana Forum on Brain Science, Winter 2005—of the book *Imitation of Life: How Biology Is Inspiring Computing*, by Nancy Forbes, MIT Press, 2004.

Without the geometric concept of **sameness**, the computer has no sense of *value* to *interpret* its own **specific** conclusions. The animal *brain* can *self-determine* **either** *predicate*, but only the human *mind's concepts* can interpret to give *meaning* to information.

The human <u>brain/mind</u> invents and develops computer hardware along with the software's algorithms that gives the computer sets of **specific** instruction. Without *conceptual* capacity for **sameness** having been *conserved* from the creative force of quarks, the human *mind* would not be able to create successive increases in the complexity of the computers and their algorithms. Computers cannot develop or *constantly* self-create *changes* in themselves and thereby *evolve* by successive increases in complexity because the geometric concepts of electron **oppositeness** and quarks **sameness** have not been timelessly conserved to *interact* as two **opposing** aspects of the **same** more complex and enhanced logical entity. The human *mind* has to intervene for digital computing to evolve in deductive complexity, as seen in "Where Have All the

Transistors Gone?" by R. P. Cowburn, under Perspectives in *Science*, January 13, 2006. The report titled "Majority Logic Gate for Magnetic Quantum-Dot Cellular Automata," by A. Imre et al. has Figures that show the **same** but more complex Boolean logic gates that should accelerate and enhance the logical capacity of computer brains. Also read the essay, "A Quantum Recipe for Life," by Paul Davis, *Nature*, October 6, 2005.

To see evolution in the mammalian brain's *vertical* rise in ordered complexity, go to Figures 215, and 306. In this so-called triune brain, *each* successive increase in the level of complexity, forms a complete brain on the *horizontal* but *incomplete* on the *vertical*. Although each level has more complex **similarities** and **differences** of genetic interactions than previous levels, each builds on the **same** logic as the previous brain. Therefore our human emotions and perceptions are the **same** as less complex animals. Reciprocal logic however will enable us to logically *constrain* these wild and unpredictable *emotions* within a *conceptual* proportion that *ideally* would be the golden mean.

To show the bilateral brain is built not only for the *dominance* of prejudice and deduction in inference but also for *optimism*, read in *Science*, November 1, 2007, under This Issue, from which I quote:

The neural basis for depression—often characterized by pessimism—has been widely studied. The neuroscience of optimism, however, is new. We are an incorrigibly optimistic species, extracting positive outcomes even when there is no basis for such expectations. For example, we expect to live longer and be healthier than average, and we under estimate our chances of being in a car accident and getting divorced. A combination of brain images and behavioral studies in healthy volunteers provides evidence for the neural mechanisms underlying this phenomenon. Interestingly, the activity of brain regions that are known to malfunction in depression also predict the optimism bias.

The letter in the same issue that the editors succinctly explain is, "Neural Mechanisms Mediating Optimism Bias," by Tali Sharot, et al.

Also read "Except One Career, Our Brains Seem Built for Optimism," by Robert Lee Hotz, the *Wall Street Journal*, November 9, 2007.

The logic of two **opposing** concepts of the **same** *interaction* between **both** the rostral anterior cingulate cortex **and** the amygdala gives preference to the concept of *optimism*. See Figures 242, 243, 244, and 245 in which **sameness** brings on *depression* that destroys our motivation to do and be something **specific**. The *happy left* hemisphere's dominant **opposition** to *right* hemisphere **sameness** keeps us from feeling excessively *sad*.

The *prejudice* of *optimism* seems to have an evolutionary imperative that keeps us highly motivated despite the difficulties of overcoming unavoidable dangers and species-threatening times. It is just another *interpretation* that favors survival. *Prejudice* of **either** *optimism* **or** *pessimism* however does not exclude the concept of *neutral*—**neither opposite** to conserve the idea of **sameness**.

More profoundly, *neutrality* as *homeostasis*, and as **sameness**, conserves the *harmony* of the highly complex systems of all mammalian life. Without this **sameness** of concept, human consciousness could not have emerged, as suggested in "Cognitive Homeostasis," under Editors' Choice, *Science*, November 2, 2007, from which I quote, "To hold two opposed ideas in the mind at the same time and still retain the ability to function has been hailed as the mark of first-rate intelligence—and would also be likely to elicit a state of cognitive dissonance, which arises when beliefs and behaviors collide. Alas, ample evidence indicates that most humans are able and willing to alter their attitudes so as to bring them into line with how they have behaved, and this to reduce discordance between the two."

There is a module in the brain of monkeys that detects excessive conflict *imbalance* then restores and maintains the *balance*, seen in "Conflict on the Brain," *Science*, November 9, 2007, under This Week in *Science*.

To sum up, we need to accept the *logic* of **opposite/same** interacting in the human brain, evidenced by **both** the "split brain's" *dominance* for *left* hemisphere's *sequencing* of *data* **and** the *right* hemisphere's *submission* with its *simultaneity* of *concepts*. The *dominance* of reciprocal **opposition** *within* the human brain leads naturally to the following chapter on the *dominance* of reciprocal **opposition** *between* human brain-minds.

THE LOGIC OF OPPOSING INTERPRETATIONS BETWEEN HUMAN BRAIN/MINDS

Civilized men have gained notable mastery over energy, matter, and inanimate nature generally, and are rapidly learning to control physical suffering and premature death. By contrast, we appear to be living in the Stone Age as far as our handling of human relationships is concerned.
—Gordon Allport, *The Nature of Prejudice*

A bit beyond perceptions reach I sometime believe I see,
That life is two locked boxes, each containing the other's key.
—Piet Hein, *Grooks*

Fanatics may defend a point of view, so strongly as to prove it cannot be true.
—Piet Hein, *Grooks 4*

There are two kinds of people in the world, those that divide everything into two parts and those that don't.
—Author unknown

When *conception* emerged from *perception*, the most complex and logically enhanced species—*Homo sapiens sapiens* (the wise one)—became aware not only of itself but also aware of being aware, called "consciousness."

Consciousness transformed animal *curiosity* into *why*, which question demanded a **specific** *answer*. Over a long period of *time*, mostly by trial and error, these answers accumulated *factual* knowledge that was consistently *practical* from generation to generation. Human cultures and customs arose that served to maintain social and economic order. *Answers* to *concepts* however only seemed to beget more *questions*. **Specific** answers and solutions between conceptual **opponents** never materialized. The **same** *conceptual* conflicts the Ancient Greeks argued are still argued today inconclusively, with no end in sight. The implacable urge to deduce a conclusion to concepts begat the futile struggle of trying to use reason to overcome the persistent, recalcitrant intractability of these chaotic conflicts . Our imperative for *purpose* and to get *meaning* from information, as well as our paradigm of inference, has hidden any logic of conceptual conflict leaving us with what is now called the "culture wars."

Not understanding that the *interaction* of **opposite/same** *interacts* the **same** as the "mutual embrace" of the *mindless* **electro/magnetic** force at our *macro*-mass level of complexity, our *minds conceptual* conflicts are as chaotic and diminishing as our *physical* wars.

Proving that reciprocal logic can transform conceptual chaos into conceptual order is not easy, or else it would have been discovered by someone else much more intellectually endowed than I. But such brilliant minds are locked into our paradigm of inference that hides the concept of **sameness** that is so crucial to the logic of conceptual interactions, to paradox, and to evolution.

Without the aid of mathematics and my having to rely mainly on induction, I have had no other recourse other than to accumulate as much scientific evidence as possible to assure the validity of a **general** conclusion. Read *Reciprocity* by Lawrence C. Becker, University of Chicago Press. Also read "The Limits of Reductionism," under Correspondence in *Nature*, February 11, 1988.

If however you still think that my method of research is just "stamp collecting," nothing more than anecdotal evidence, then consider the following scientific experiment with the human brain/mind by Dr. Drew Westen and his colleagues at Emory University in Atlanta that is explained in the Science section of the *New York Times* under Findings, January 24, 2006. I quote some of the "surprising" results of this scientific experiment titled "A Shocker: Partisan Thought Is Unconscious," by Benedict Cary, from which I quote:

> Using M. R. I. scanners, neurologists have now tracked what happens in the politically partisan brain when it tries to digest damning facts about favored candidates or criticism of them. The process is almost entirely emotional and unconscious, the researchers report, and there are flares of activity in the brain's pleasure centers when unwelcome information is being rejected. "Everything we know about cognition suggests that, when faced with a contradiction, we use the rational region of our brain to think about it, but that was not the case here." . . . The Republicans in the study judged Mr. Kerry as harshly as the Democrats judged Mr. Bush. But each group let its

> candidate of the hook. After the participants read the contradictory comment, the researchers measured increased activity in several areas of the brain. They included a region involved in regulating negative emotions and another called the cingulate, which activates when the brain makes judgments about forgiveness, among other things. Also a spike appeared in several areas known to be active when people feel relieved or rewarded. The "cold reasoning" regions of the cortex were relatively quiet. Researchers have long known that political decisions are strongly influenced by unconscious emotional reactions, a fact routinely exploited by campaign consultants and advertisers. But the new research suggests that for partisans, political thinking is often predominantly emotional. It is possible to override these biases, Dr. Westen said, "but you have to engage in ruthless self-reflection, to say, 'All right, I know what I want to believe, but I have to be honest.' He added, 'It speaks to the character of the discourse that this quality is rarely talked about in politics.'"

Even more telling of reciprocal logic in action *between* minds with **opposing** world views is seen in "Brain Processes Reflect Politics," by Denise Gellene of the Los Angles Times, found in the *Tampa Tribune*, September 10, 2007, from which I quote:

> A simple experiment to be reported today in the journal Nature Neuroscience that political orientation is related to differences in how the brain processes information. Previous psychological studies have found that conservatives tend to be more structured and persistent in their judgements, whereas liberals are more open to new experiences. The latest study found those traits are not confined to political situations but also influence everyday decisions. The results showed "there are two cognitive styles—a liberal style and a conservative style" [after their experiment] Researchers got the same results when they repeated the experiment in reverse, asking another set of participants to tap when they saw . . ."results provided an elegant demonstration that individual differences on a conservative-liberal dimension are strongly related to brain activity." . . . Lead author David Amodio, an assistant professor of psychology at NYU cautioned that the study looked at a narrow range of human behavior, and it would be a mistake to conclude that one political orientation was better . . . . Still he acknowledged that a meeting of the minds looked difficult given the study results. "Does this mean liberals and conservatives are never going to agree? Maybe it suggests one reason why they tend not to get along."

This neatly fits the pattern of conceptual conflict between Republicans and Democrats in Congress for lifetime appointments to the United States Supreme Court. The Republicans want to *conserve* the original intent of the Founding Fathers while the Democrats want to Court to *change* its judgements to conform to changes in public sentiment and the laws members of Congress pass to please the voters and keep themselves in office. The Republicans want to *conserve* the balance of power between all three competing branches of government in order to *conserve* the Republican form of government, while the Democrats want to tilt power of government to a more democratic form of government. What it boils down to is the battle between the extremes of *totalitarianism* and *mob rule*, both of which are tyrannies of the mind. The original intent of the Founding Fathers was a form of government with a balanced proportion between these two awful extremes. The scheme was simple, comply with the laws of nature by making the Supreme Court the *constant*, the Congress the concept of *change*, and the President the leader, who also carries out the laws that are the **same** for everyone, including Congressmen that make the laws and agencies that enforce the laws. Thus, change/constant *perpetually* interact, **opposed** by *change* in time. To place this Supreme Court battle between Republicans and Democrats in a *logical* perspective that conforms to the conceptual interactions of nature you have to look no further than, "The War for the Constitution," by Gary S. McDowell, the *Wall Street Journal*, October 23, 2007. Also read "The Same Words, but Differing Views," by Adam Liptak, the *New York Times*, June 29, 2007, regarding "opposing ideas of what it means to be 'faithful' to the Brown ruling of 1954." In regard to the "*Brown* ruling," read the self-contradictory nature of extreme adherence to the idea of *equality* in "Civil Rights Struggle Betrayed," by Walter Williams, the *Tampa Tribune*, October 3, 2007.

With further regard to **opposing** interpretations of the **same** subject just mentioned, we see exactly the **same** logical division in FIGURE 396 in which *economists* are split between **either** *deflation* **or** *inflation*.

Headed for a recession?

It depends on whom you ask.

Morgan Stanley **40% chance**

"The risk of outright U.S. recession is higher now than at any time in the past six years."

—Richard Berner, chief U.S. economist

Consumer spending, accounting for two-thirds of economic activity, is under pressure. The main reason: increasing headwinds for the American consumer.

Payrolls have grown this year by an average of 125,000 per month, below last year's 189,000 monthly pace.

Rising energy and food prices are eroding Americans' spending power. Tighter lending standards will likely limit consumers' ability to turn home equity into cash. All these factors may be coalescing into a "perfect storm" that could wallop the U.S. consumer, Berner says.

Bear Stearns **20% chance**

"We disagree with the consumer- and housing-driven theories of a slowdown, We think these have been overdone."

—David Malpass, chief economist

Even with falling home values, spending will likely continue. "There wasn't a strong wealth effect on consumption when house prices and mortgage equity withdrawals were bubbling in 2003 to 2005, nor was there any particular weakness in consumption in 2006" when home prices weakened, he says.

Malpass does see labor conditions deteriorating in coming months – but only enough to slow consumption growth, rather than stop it.

The October jobs report, which showed growth in the business services and leisure industries, is a good sign, as was an October rise in auto sales. He says the numbers indicate a soft landing, not a recession.

FIGURE 396

No matter what the subject, its experts always argue conflicting *interpretations inconclusively.*

That the interaction of **opposite/same** has been conserved on the *vertical* from the *simplest* to the most *complex* form of life, read "Ladies First: Genes Skew Sex Ratios in Evolutionary Struggle," *Science News*, by P. Barry, whom I quote, "Competition among genes within an individual male fruit fly can cause it sperm to produce a high proportion of female offspring. Now scientist have identified a gene responsible for this well know phenomenon as well as the gene that later evolved to restore gender balance . . . . In essence, the two fruit fly genes engage in a tug-of-war in which each succeeds evolutionarily if it can spread widely among future generations."

Evidence that perpetual *conceptual* conflict emerged *between* human minds and that **either** *idea* is never the solution to the other is seen in "The End of Integration," by David Brooks, the *New York Times*, July 6, 2007. One needs to read the whole of this essay to grasp the historical scope of conceptual conflict with its *insolubility* and therefore its "broken promises." Read "Promises, Promises, but It's Really a Culture War," by George Melloan, the *Wall Street Journal*, August 8, 2000.

Go to the Web sites shown at the end of the text and you see that the *New York Times* has a Web site where every issue can be accessed since its first publication. You will find that the culture wars between *individualists* and *collectivists* as **specific/general** are fought daily.

Also read "The Perils of Prognostication," *Science*, June 30, 2006, in which John T. Jost reviews a book titled *Expert Political Judgment: How Good Is It? How Can We Know?* by Philip E. Tetlock, Princeton University Press, Princeton, New Jersey. I quote from Jost:

> Drawing on an essay by Isaiah Berlin, Tetlock distinguishes between "foxes" who "know many little things, draw from an electric array of traditions, and accept ambiguity and contradiction as inevitable features of life" and "hedgehogs" who "know one big thing, toil devotedly within one tradition, and reach for formulaic solutions to ill-defined problems." . . ."Much evidence—including results from Tetlock's own research—indicates that personal needs, order, structure, and closure are positively associated with conservatism, whereas openness, tolerance for ambiguity, and integrative complexity are positively associated with liberalism . . . . Given that people see the future (at least in part) as they would like it to be, an answer to the question of whether liberals or conservatives are more accurate in their prognostications depends upon, among other things, whether the world happens to turn to the left or the right during any specific period of time . . . . In the process [Tetlock]

advances considerably the important work begun by Daniel Kahneman and Amos Tversky, demonstrating just how desperately we need a scientific psychology of judgment and decision-making to correct for the many failings of a ruminating species".

Maybe its time for academia takes a serious look at Reciprocal Logic II in order to get a comprehensive view of, "two **opposing** interpretations of the **same** thing," seen in FIGURE 397.

FIGURE 397

For more evidence of the perpetual conceptual conflict of **specific/general** as **opposite/same** *between* human minds, read "Revisiting the Cannon Wars," by Rachel Donadio, the *New York Times*, September 16, 2007; "Our True Colors," by Cal Thomas and Bob Beckel, the *Tampa Tribune*, October 7, 2007; and "Marching Toward a Polarized Electorate," by Amanda Cox, the *New York Times*, November 6, 2007, from which FIGURE 398 is taken.

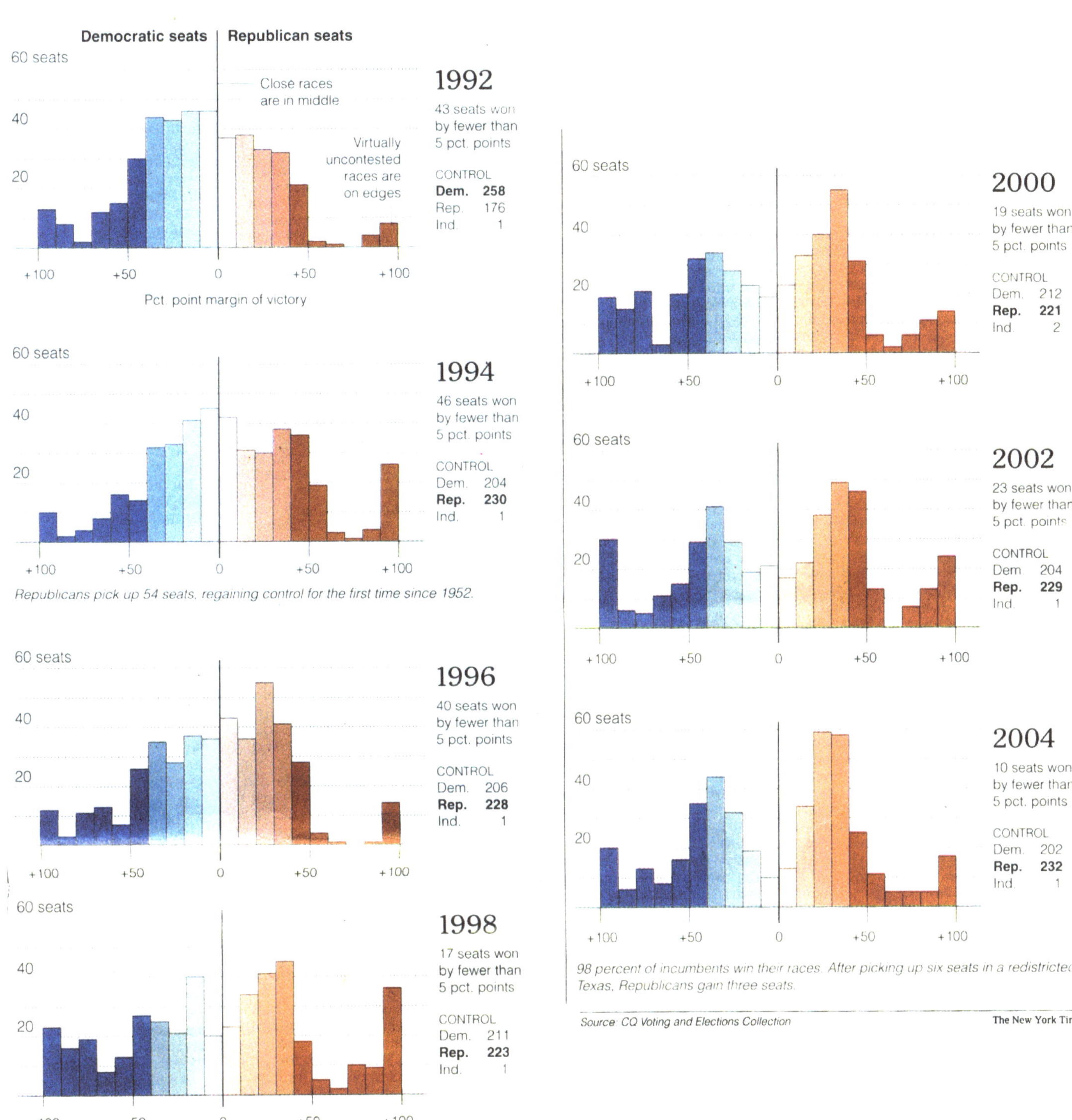

FIGURE 398

Also read "Extreme Politics," the *New York Times*, November 11, 2007, by Alan Brinkley who reviews a book titled *The Second Civil War: How Extreme Partisanship Has Paralyzed Washington and Polarized America*, by Ronald Brownstein, The Penguin Press, 2007.

Liberals *dominantly* embrace the concept of **sameness** because it connotes *fairness*, *equality*, *unity*, and *altruism*. It seems that such noble ideas and behaviors would bring about the *harmony* of nature's universal **sameness** to our *macro-mass* world. But the vast **similarities-differences** of matter, no matter how complex, are **opposed** to **sameness**. Evolution has

opposed sameness on the horizontal, because if *everything* were the **same**, and there would be *nothing* to perceive or conceive. There would be no **similarities-differences** for the brain's senses to discern, and no concept of *time* that is always **opposed** to **sameness**. The quarks' force of **sameness** is conserved on the vertical,but it is dominantly **opposed** at every level of complexity by the electron.

It has been known since the time of the ancient Greeks that extreme ideas seem *self*-contradictory and *self*-defeating. Their principle of "moderation in all things" again flourished among the intellectuals during the Italian Renaissance as read in *The Book of the Courtier*, by Baldassare Castiglione, Dover Publications Inc., Mineola, New York, 2003, in which on page 76 the author succinctly sets forth the ubiquity and richness of conceptual interactions in a very practical and ordinary sense as follows:

> For they would have the world contain all good and no evil, which is impossible; because, since evil is opposed to good and good to evil, it almost necessary by force of opposition and counterpoise as it were, that the one should sustain and fortify the other, and that if either wanes or waxes, so must the other also, since there is no contrary without a contrary. Who does not know that there would be no justice in the world, if her were no wrongs? No courage, if there were no cowards? No continence, if there were no incontinence? No health, if there were no infirmity? No truth, if there were no lying? No good fortune, if there were no misfortunes? Thus, according to Plato, Socrates well says it is surprising that Aesop did not write a fable showing that God had never been able to join pleasure and pain together, He joined them by their extremes, so that the beginning of one should be the end of the other; for we see that no joy can give us pleasure, unless sorrow precedes it. Who can hold rest dear, unless he first felt the hardship of fatigue? Who enjoys food, drink and sleep, unless he has first endured hunger, thirst and wakefulness? Hence I believe that suffering and diseases were given man by nature not chiefly to make him subject to them (since it does not seem fitting that she who is the mother of every good should give us such evils of her own determined purpose), but as nature created health, joy and other blessings,—diseases, sorrows and other ills followed after them as a consequence. In like manner, the virtues having been bestowed upon the world by grace and gift of nature, at once by force of that same bounden opposition, the vices became their fellows by necessity; so that always as the one waxes or wanes, thus likewise must needs the other wax or wane.

Despite this understanding of the naturalness and inevitability of conceptual conflict, no one has acquest or submitted to it. Without a logical form for conceptual interactions, the inconclusiveness and rancour of their conflict is chaotic to the human mind. Concepts just doesn't make any sense from our paradigm of logic without understand the need of and necessity for its antinomy.

Now that we have scientific proof that the *minds* concepts interact by the **same** logic as *mindless* nature, we can and should *logically* interact our concepts the **same** as the "mutual embrace" of the **electro/magnetic** force with each of its reciprocally **opposing** hands interacting **opposite/same** dominant/domantless. With the acceptance of paradox as the natural order of concepts, we will be able to tame our primitive animal *emotions*.

Some Muslims are beginning to realize that the loss of scientific reasoning from their culture has led to *absolute* ideas about how all minds must think. By not complying with *relative* ideas and understand that conceptual paradox as the only truth of reality, they have *diminished* their cultural development—lagging far behind the progress of scientific-oriented nations.

Perhaps the Saudis best exemplify the need to back off from extreme ideas, particularly in regard to economic development, as seen in "Balancing Risks and Rewards," by Marco Venditti in a Saudi advertisement in the *Wall Street Journal*, December 9, 2007. Also read "Arab Intellectuals Debate Reasons for Terrorist Acts," by Nonna Abu-Nasr of the Associated Press in the *Tampa Tribune*, September 24, 2004. Also read "Voices of Dissent Question Hard-Line Stance of Islam," by Neil MacFarquhar, the New York Times, published in the *Tampa Tribune*, December 12, 2004. Also read "Reaping What It Sowed," by Thomas L. Friedman, the *Tampa Tribune*, May 6, 2005. And "The War Within Islam," by Max Rodenbeck, the *New York Times*, May 29, 2005, a review of *No God But God: The Origins, Evolution, and Future of Islam*, by Reza Aslan, Random House, 2005.

Because there is no such thing as unilateral, one-way, reciprocity the Arab world will have to accept the conceptual paradoxes of science instead of their *irreversible* and *absolute* idea of God. Their present attempt to solve the "problem" of their ideological **opponents** by acts of terror has only acerbated conceptual *chaos*. Wars kill people, but they cannot kill ideas.

Muslim "martyrs" extremists are willing to blow themselves up along with "Western devils" as well as other Muslims who deviate from the **absolute** dictates of the Koran. The heretic is always more to be feared than the infidel. However the self-anointed leaders who indoctrinate young minds to these self-destructive acts walk around with a heavy entourage of bodyguards to keep from becoming victims of the holy sacrifice that they preach. See FIGURE 399.

FIGURE 399

There can be no reconciliation with the West because the very idea of free expression of **opposing** ideas goes against their insistence that everyone should believe the **same** as they. Instead of accepting to conceptual paradox, they would be as gods without **opposition**. Anyone who refuses to make the religion of Islam *absolute* in his or her mind, must killed.

Read "Science in the Arab World," an editorial in *Nature*, June 29, 2006, that precedes the report, "Scientists Becoming Targets in Iraq," with the subheading, "Violence is common currency in Iraq, but one group is increasingly and persistently singled out—academics. Declan Butler reports on the risks run by researchers as they struggle to pursue their studies."

The most cogent and lambent argument against the *absolute* mental tyranny of Islam is by an author who was born and raised in the Islamic culture and indoctrinated into its faith. He fled to the United States to escape from the tyranny of *absolute* **sameness** to the haven of conceptual freedom—logically speaking, reciprocal **opposition**. This infidel, Ayaan Hirsi Ali, a resident fellow at the American Enterprise Institute in Washington DC, in a book review in the *New York Times*, January 6, 2008 titled, "Blind Faiths," subtitled, "Arguing that the West's 'Fanaticism of Reason' Is No Match for the Fanaticism of Radical Islam," praises a book titled *The Suicide of Reason: Radical Islam's Threat to the Enlightenment* by Lee Harris, Basic Books, 2007. Read the review and the book to grasp the danger to the freedom of belief that is the essence of the "war on terror." The West's tolerance for **opposing** ideas and freedom of religious belief does leave us vulnerable to the threat of radical Islam's tyrannical god that only a God that *logically* orders the universe can overcome.

Societies with free expression of conflicting ideas will find it difficult to accept the *impersonal* God of Baruch Spinoza and Albert Einstein. However there is no need to exclude a *personal* god because the *interaction* of personal/impersonal as **specific/general** dominant/dominantless respectively. Although this leaves us in the insoluble predicament of being

opposed to the **sameness** of a universal ordering principle, this is the closest we can come to being *part* of nature's universal laws and forces that governs all mass **specificities** the **same**. Read "Reasonable Doubt," by Rebecca Newberger Goldstein, the *New York Times*, July 29, 2006.

Here on our planet earth, there persists the conceptual conflict between those who seek to *exploit* the environment and those who seek to *conserve* the environment. *Exploitation* is dominant, because like all other animals, we need to continually extract energy from our surroundings to work and survive. The fact that we are the most adaptive of all creatures, however, behooves us *constrain* ourselves and have empathy for those creatures that are less adaptive. Otherwise our *exploitation* will exceed its *proportional* constraint, and extinctions will become *disproportionately* high. Reducing species' numbers and increasing species' extinctions will leave us with nothing to *exploit.*

The *logical* interpretation of the perpetual conflict between human *minds* in United States of America shows that *conservatives*, as *individualists* think *deductively*, believing that *individuals* should be given the *chance* for *self*-determination and the freedom to make one's own **specific** *interpretation* of life. *Liberals*, as collectivists, however think *inductively*, demanding socialistic or *cooperative* living where minds in **general** *think* the **same** without any **opposition**. After all, with **no opposition**, it follows deductively that the "problem" of *conceptual conflict* will be solved, leaving only *peace* and *tranquility*. Yet in every such attempt of extreme *collectivism*, internecine *warfare* has erupted as well as the loss of motivation and willingness to take risks, thereby causing these socio-economic regimes to stagnate and fossilize. With liberals, *design* seems to be the only concept that can bring order to social interactions, not realizing that *design* is merely an *interpretation* that defies the *interaction* of <u>chance/design</u> as <u>dominant/dominantless</u> respectively. Extreme *conservatism* is just as self-defeating and self-contradictory as extreme *liberalism*. For example, "bubbles," those wild swings to an extreme, always "burst" with disastrous consequences for everyone. They are always motivated by greed and *selfishness* that has not been constrained within a reasonable proportion, ideally the golden mean. An old Wall Street truism is, "Bulls make money and bears make money, but hogs never do." Read "Bleakonomics," subtitled, "Naomi Klein Tracks 50 Years of Global Capitalism, Spotting Ruthless Opportunism at Every Turn," by Joseph E. Stiglitz, the *New York Times*, September 30, 2007, a book review of *The Shock Doctrine: The Rise of Disaster Capitalism*, by Naomi Klein, Metropolitan Books, 2007.

Read also "Seeing Slavery in Liberalism," by Janice Rogers Brown, the *New York Times*, June 9, 2005. Also read "The Return of the Thought Police," by Wendy Kaminer, the *Wall Street Journal*, October 26, 2007.

The <u>liberal/conservative</u> <u>interaction</u> harkens back to the very basic *interactions* of life such as the <u>coding/non-coding</u> and <u>active/passive</u> genes of the genome interacting as <u>selfish/selfless</u> and <u>dominant/dominantless</u> respectively. Read "Broken Cogs or Strategic Agents," *Science*, April 28, 2006, by Peter Hammerstein and Edward H. Hagen, who review a book titled *Genes in Conflict*, by Austin Burt and Robert Trivers, Harvard University Press, Cambridge, Massachusetts, 2006.

Also read the introduction to three articles in *Nature Reviews Neuroscience*, July 2006, from which I quote, "In many biological systems, there is a fine balance between stability and flexibility, and the brain's cells and circuitry are no exception." Also read "Extremes of Excitability," *Nature Reviews Neuroscience*, July 2006, from which I quote, "Mutations in ions channels, usually characterized by producing either hyperexictability or hypoexcitability, are associated with various neurological disorders." On the same page, read "Peering into the Root of Prejudice," in which the highlight of the subject says, "different regions of the brain are active when we are thinking about people either politically alike or different from us."

Notice the *dominance* of the complementary **<u>either/or</u>** function in the above references regarding *disorder* and again visit and contemplate the more complex and enhancing Figures 22, 25, and 53 that demonstrate reciprocal **sameness** between **both** *left*-wing **and** *right*-wing *minds*.

Read "Culture Warrior," by Harry Stein, a book review of *Right Turns*, by Michael Medved, Crown Forum, in the *New York Time*, February 13, 2005. Read also "Conservatives Thrive on Issues that Divide," by David Brooks, of the New York Times, reprinted by the *Tampa Tribune*, April 6, 2005.

If there is no proportional *constraint*, **either** too much *selfishness* **or** too much *selflessness*, the **same** thing as "failure of the reciprocity law" will occur where **either** excessive light **or** not enough light coming through the camera lens will not record information onto the film. Without *constraint* from **either opposing** extreme logical order fails and knowledge disappears down "black hole."

The logic of conceptual interactions occurs *within* and *between* our genes as seen in "Some Politics May Be Etched in the Genes," by Benedict Cary, the *New York Times*, June 21, 2005. What seems "surprising" to the author is that **sameness** seems to be logical despite our minds **opposition** to it. Also read *Genes in Conflict: The Biology of Selfish Genetic Elements*, by Austin Burt and Robert Trivers, Belknap Press, 2006.

Other than the idea of *self* in its extreme, the most damaging and self-defeating is the idea of extreme *freedom* with its promise of *happiness*. Individuals and nations that have been released from *tyranny* of body and mind instantly embrace absolute *freedom* as the answer to all their problems. The consequences are always horrendous and self-defeating.

An example of the false expectations of *absolute freedom* has occurred in the United States with black Americans. Having been given the **same** degree of freedom as all other citizens and races first by Constitutional intent, then by decree of President Lincoln, and later by order of the U.S. Supreme Court, many blacks still feel that whites are freer than they are. Their *race* has become an excuse for not making the effort needed to achieve their expectations. To be like "whity" is being a traitor to their primitive tribal culture. Unfortunately, this attitude can only perpetuate their role a second-class citizens in their own eyes.

If blacks think that I am prejudiced against the their *race's* intellectual capacity, they are simply wrong. However I am prejudiced against their *culture* of irresponsibility to society by being too prone to flout the law and their men's lack of responsibility toward their own children and families. Read, "All Brains Are the Same Color," by Richard E. Nisbett, the *New York Times*, December 9, 2007. Too many black males have little or no responsibility of fatherhood or parenthood. Without paternal guidance, their children cannot develop into self-responsible and productive citizens to sustain and improve our nation's hard-won *freedom*. Unfortunately many blacks fail to heed some of their own race to lift themselves up from lives of crime, drug addiction, and violence that are often epidemic in areas dominated by blacks. One such pleading is seen in "This Is How We Die, Once More," by Leonard Pitts Jr., a columnist for the *Miami Herald*, found in the *Tampa Tribune*, December 4, 2007.

The black culture that has been promoted by the liberal media has infiltrated all youths that are already too prone to gang violence. Without academic knowledge added to street knowledge, too many students waste their time and the taxpayers' money in classrooms of public schools horsing around instead of paying attention to their teachers. Without even trying to learn the subjects being taught for their own future enjoyment of freedom and happiness, to say nothing of getting into college for advanced education where they can learn enough knowledge to be successful in a competitive economic society, they had rather have fun now disrupting the ability of those few students who want to learn. Even so, they still demand their right to go to college so they can get dumb-down degrees that have no practical or intrinsic value to them or to society.

Read "How a Middle School Can Be 'Dangerous' and Still Get an A," by Samuel G. Freedman, the *New York Times*, December 19, 2007, from which I quote:

> "I didn't teach last year," Ms. Staples said, "I was a police officer and a baby sitter. You'd write up kids left and right, and nothing would happen. No one would help you. And the kids would just come right back. After I got hit, the principal's response was, 'That's what happens in middle schools.'" . . . Mr. Carson similarly described a lack of administrative support and meaningful discipline. "The administration would be telling you that it would all fall into place if you had a better lesson plan or more student engagement or arranged the desks in a U shape," he said. "But it doesn't matter how good your lesson plan is if the kids can't even stay still long enough to write the 'Aim' and 'Do now' off the board. There are no repercussions. There is no punishment fitting the infraction."

If you think I'm being too harsh on black *culture*, not their race, read, "Quite Simply, Too Many People Going to College," which was written by the highly educated black author Thomas Sowell in, *The Tampa Tribune*, November 29, 2007. Then read, "Higher Tuition: What It Means to You," on the front page of the *Tampa Tribune*, December 5, 2007, which shows that the nation cannot afford college students that do not meet the high academic standards for college admission for meaningful academic degrees.

The best read is, "The Hazards of Telling the Truth", *The Wall Street Journal"*, April 15, 2008, by John Leo, who reviews a book titled, "History Lesson", by Mary Lefkowitz *Yale University Press, 2008*. I quote from Leo: ". . . . . The Afrocentrism movement that had begun in the academy in the 1980's and gained astonishing momentum with the publication of Martin Bernal's "Black Athena: The Afroasiatic Roots of Classical Civilization" . . . . . . .During this whirlwind of dubious scholarship, the academic world mostly remained mum, hiding behind the curtain of academic freedom and withholding its criticism lest a statement of simple truth be branded "racist" . . . .[Lefkowitz] concluded that the Afrocentric authors regarded history as a form of advocacy: Like other postmodernists, they believed that truth is impossible to know – that all 'narratives' are socially constructed and thus possess an equal claim to legitimacy. At the time, traditional scholarship was generally under assault, but classics were particularly vulnerable, because they purported to study the foundational texts of the West, Attacking the classics as a complex systeem of lies was emotionally important to those who wanted to take Western Culture down a peg. Feelings and politics mattered, not scholarship . . . . 'the most persuasive narrative was the one with the most desirable result. In effect, [Bernal] was preaching a kind of affirmative action program for the re-writing of history'. "History Lesson", is Ms. Lefkowitz's personal account of what she experienced as a result of questioning the veracity of Afrocentrism and the motives of its advocates . . . . . . "

Instead of *enhancing* integrity, trust, moral values, and tolerance between races and cultures, blacks instead have *diminished* these qualities in all our citizens. The sad consequence of black culture is seen in the statistics of all prisoners incarcerated in the United States as reported in "Justice Dept. Numbers Show Prison Trends," by Solomon Moore, the *New York Times*, December 6, 2007. Read it and ponder the great personal and social cost that honest and *self*-responsible citizens must bear. The mantra of Black *Pride* needs to interact with *shame* to deter bad behavior.

My seemingly harsh and Eurocentric bias against blacks in a free political society is, of course, also directed against any race, culture, or ideology that lacks self-responsibility and refuses to be *constrained* to the law. For example, what is the **difference** between black street gangs with no parental control and street gangs without parental controls of any other race? Lawless gangs of all races and cultures flourish in large cities. Further, what is the **difference** between family feuds in the country and mafia gangs in the city? All have the **same** unconstrained *emotional* instinct with poorly developed conceptual *self*-control. The whole world is in need of *self*-constraint that can be implemented only by accepting the *pointless* logic of conceptual interactions. Read, "Summit Held to Develop Statewide Anti-Gang Strategy," emphasizing that gang ,"Activity Seen as Growing Threat," by Josh Poltilove, the *Tampa Tribune*, December 20, 2007. Read, "So Many Crimes, and Reasons to Not Cooperate", by David Kocieniewski, the *New York Times*, December 30, 2007. And again read "Authorities Crack Down on Vicious Latino Gang," subtitled, "Blacks Targeted in Turf Cleansing," by Thomas Watkins of the Associated Press, found in the *Tampa Tribune*, December 31, 2007.

Somehow these depraved souls need to be attracted to scientific reasoning and the logic of conceptual interactions for their own benefit and for society. As it stand now, these human gangs stake out a territory **no different** than the most primitive animals, fighting tooth and claw for their economic survival. Kill or be killed as if this would bring an end to their conflict. Their self-anointed lords reign like the chiefs of primitive tribes that try to exclude in degree of perceptual or conceptual **opposition**. On and on, generation after generation, the **same** old ignorance of reciprocal logic with its power to transform conceptual conflict from chaos to order, which will enhance the interactions *between* human minds.

Our modern means of transportation and communication, particularly by means of the World Wide Web, brings with it what would seem to be the perfect opportunity to *unify* our conceptually *divided* <u>brain/minds</u>. But this can never happen because <u>division/unity</u> *perpetually* interact <u>dominant/dominantless</u>, the **same** as reciprocally **opposing** *hands* of the **<u>electro/magnetic</u>** force. Read "Not Just a Personality Clash, a Conflict of Visions," by David Brooks, the *New York Times*, October 12, 2004.

The United Nations therefore will never be able to completely *unify* all the nations of the world. Nations will always be divided by **differences** in culture. The culture that *unites* each nation *divides* it from all others. Even the *unity* of a culture is *divided* into individual **differences**. We are each *unique*, always in **opposition** to being the **same**. Read "At the U.N., Discord Over Confronting Iran's Nuclear Ambitions," by Warren Hoge, the *New York Times*, November 8, 2006. Also read "The World as an Imperfect Globe," by Stephen, Kotkin, the *New York Times*, December 2, 2007, and notice the pervasive interaction between **specific** (local) and **general** (global).

The need for *each* of us to have a dominant *belief* in order to have *meaningful* life can never be completed. <u>Meaning/meaningless</u> perpetually interact as do <u>belief/doubt</u> respectively. All the vast store of information available on the Internet will never reveal any *concept conclusively*, and the idea of *meaning* is no exception. *Meaningful* cannot absolutely exclude *meaningless*. Read "Searching for the Meaning of Life on the Internet," by Rachel Pomerance, the *Atlanta-Journal Constitution*, July 29, 2006.

Effective leadership requires more than monetary compensation. Without *meaning* and *purpose*, no job can be satisfying as seen in "'Authentic' Ways of Leading," by Erin White, the *Wall Street Journal*, December 2, 2007, and "Readers, Experts Talk about Employee Retention," page B7, the *Wall Street Journal*, December 3, 2007.

That seeking *meaning* in a *meaningless* universe is a **general** obsession, read "A Simple Query Resonates," by Kevin Austin, found on the same page as the above reference. The *dominance* of *meaning* requires *meaningless* so that these two reciprocally **opposing** concepts can *interact*, just as *faith* is requires *doubt* to conserve logical order.

That the *interaction* of faith/doubt as dominant/dominantless is logically enhancing is seen in the great discovers of natures secrets that have been called "reluctant revolutionaries." Examples are Copernicus, Isaac Newton, and Charles Darwin who were reticent to publish their respective works—afraid of deviating too far from conventional thought. They knew that their discoveries would be very disturbing because they would drastically *change* the way humans viewed themselves in relation to the world and to the universe. They sought evolution instead of revolution understanding that *new* discoveries had to be built on *old* knowledge. Copernicus kept his manuscript hidden for decades with doubts about his own discovery that the sun, instead of the earth, was the center of the universe. Newton was not only slow to publish his work but did so in such a way that only a few human brains could understand it. Darwin spent years accumulating evidence of evolution and then took his time writing it. Read "The Cautious Evolutionist" in the *New York Times*, August 27, 2006, by Adrian Desmond, who reviews a book titled *The Reluctant Mr. Darwin* by David Quammen, Atlas Books/W. W. Norton and Company, 2006.

The Founding Fathers were also "reluctant revolutionaries." Proud of their British heritage of having created the freest nation up to their time, they were torn between wanting to remain British citizens and the humiliation of being taxed by the British without being represented in Parliament. With this internal conflict between *change* and *constant*, the American colonists decided that since the king would not allow them to be full-fledged citizens of England, they *reluctantly* declared their independence. In a new land away from the shackles of superstition and political dogma of Europe, and being students of history, science, and philosophy, they were mentally prepared to improved the British form of government by devising a system that more closely conformed to the "natural law" that great philosopher have sought since the time of ancient Greece. They created a government that became *part* of nature's universal **sameness**—a government with laws that were the **same** for everyone and enforced the **same** for everyone regardless of their **similarities-differences**.

As for the dominance of *individuals*, it has been *individuals* that have been the most *creative*. Individuals have always made meaningful changes in the way humans have viewed themselves in relation others and to the universe. Individuals such as Zeno, Pythagoras, Euclid, Jesus of Nazareth, Socrates, Plato, Aristotle, Copernicus, Galileo, Kant, Beethoven, Descartes, Newton, Maxwell, Lavoisier, Darwin, Einstein, Bohr, Freud, Gödel, and Crick have been the evolutionary agents of the human *mind*. Although each borrowed from the work of others, their *synthesis* of conceptual paradox formed a **sameness** that enabled them to discover successive increases in the complexity of nature's universal ordering principle.

Read "The Theorist," *Nature*, October 26, 2006, by Horace Freeland Judson who reviews a book, *Francis Crick: Discoverer of the Genetic Code*, by Matt Ridley, HarperCollins, 2006. I quote from Judson, "Crick was above all the theorist—and that's the unifying thread. Not just with the coding problem but throughout his career Crick was the one who put other scientists' thoughts in order. He soaked up data, mostly from other people's, but he saw beyond the data to their meaning, their shape, their implications. He found principles, and did it with a preternatural clarity of mind and inimitable style."

Extreme *ideas* tend to stimulate our highly dominant **either/or** *emotions*, resulting in extremes of behavor, even senseless violence. Religious beliefs in particular tend to be *absolute* that makes reciprocity impossible. Zealots of any stripe always incite *chaos*. Any *belief* in God, however, is but an *idea*, an *interpretation*, always subject to valid **opposition**, as seen in "Faith, Reason, God and Other Imponderables," by Cornelia Dean, the *New York Times*, July 25, 2006, from which I quote:

> Human reasoning is "beset with logical problems that include overdependence on authority, overemphasis on coincidence, distortion of evidence, circular reasoning, use of anecdotes, ignorance of science and failure of logic" he [Lewis Wolpert] writes. And whatever these traits might say about acceptance of religion, they have a lot to do with public misunderstanding of science. So, he concludes, "We have to both respect, if we can, the beliefs of others, and accept the responsibility to try and change them if the evidence for them is weak or scientifically improbable." This is where the scientific method comes in. If scientists are prepared to state their hypotheses, describe how they tested them, lay out the data, explain how they analyse their data and conclusion they draw from their analyses—then it should not matter if they pray to Zeus, Jehovah, the Tooth Fairy or nobody. Their work will speak for itself.

The two reciprocally **opposing** hemispheres *within* the brain-mind have the **same** logical order as *between* the brain-minds. There is a perpetual conflict between Western and Eastern cultures characterized by the *logical* concepts of **specific** and

general. This is starkly revealed in, "Two Views Of Life, Enduing, Unyielding", *The New York Times*, March 26, 2008, by William Grimes, who reviews a book titled,"Worlds at War", subtitled, "The 2,500-year struggle between East and West", by Anthony Pagden, *Random House, 2008.* I quote from Grimes: "........Worlds at War.......traces the seemingly endless series of misunderstandings and armed conflicts between 'an ever-shifting West and an equally amorphous East' to the time of myth, when Paris abducted Helen, provoking the Trojan War. With the passage of centuries, boundaries shifted, tribes and people replaced one another, new religions appeared, empires rose and fell. Yet a remarkably consistent theme asserted itself: the irreconcilable differences between two competing views of the world, memorably expressed by Herodotus in his history of the struggles between the Greeks and the Persians, which pivoted not on politics but on 'and understanding of what it was to be and to live like a human being,' The Greeks subscribed, broadly, to 'an individualistic view of humanity', The Persians displayed courage and ferocity on the battlefield but as a society, Mr. Pagden writes, paraphrasing Herodotus, they were 'craven, slavish reverential and parochial, incapable of individual initiative, a horde rather than a people ......"

Change/constant and specific/general each interact perpetually whether of *mindless* nature or the human *mind*.

Consider also the following from *Chemtech*, March 1981, "Time as Turn," by Shin Ohara, professor of ethics and philosophy, Aoyama Gakuin University, Tokyo, Japan:

> The Japanese attitude toward nature is neither subordinate no antagonistic; rather we try to live in *harmony* with naturc, to live "in" nature .... People minimize open competition, putting emphasis more on the subtlety of emotional sensitivity and egalitarian consciousness .... Taking turns seems to be closely related to the Japanese sense of balance or harmony. Our sense of balance is "egalitarian" in the we assume that people are homogeneous both in ability and in interest, although this cannot be true actually .... The goal of Japanese education thus strives for homogeneously, evenly trained persons who fit the needs of industrial society. A problem in Japanese education where open competition is avoided is that unique talents of certain students are tactfully concealed .... In the Japanese educational system, being a unique person is often assumed to be bad because unique implies that one is not well balanced. Here again, the most important thing to do is not to aim at the individual good; rather, it is to try to keep harmony among one's peers .... The Japanese often leave things as they are and make them intentionally ambiguous.

Chemtech, October 1982, "A Short Course in Japanese," by Walter Mule, whom I quote:

> Japanese is reverse English .... To say "come" here, we [Westerners] begin with our palm up and wave it toward our body. But when Japanese tell you to come, "Koi," they start with their palms facing down and wave it away from their bodies, in exactly the same way the we motion "Go away." ... When you and I refer to ourselves, we express it as "you and I." But the Japanese always say "I and you." ... The Japanese have the unique ability to read their languages in more than one way. Their books are read from what we would call the back to the front, and the Japanese title appears on what we consider the back cover. They read vertically, beginning from the right side of the page and progressing to the left. But they also print some books the way we do: horizontally and from left to right. This is the case especially for scientific and mathematical text books .... Nowhere is the evidence of diametric opposition clearer than in the structure of the two languages .... To translate a sentence such as, "The man I happened to meet on the Ginza in Tokyo yesterday was your neighbor, Mr. Ikeda," the thoughts must be rearranged in reverse to "Yesterday Tokyo Ginza on meet happened to man you of neighbor Ikeda Mr. was .... To be successful in our relations with our Japanese friends we must further understand a significant cultural opposite that is reflected in language. It concerns the degree of positiveness in our speech. Americans have a mania for positive definite statements, whereas the Japanese prefer to be indefinite .... There is also another word, *desyoo*, which is best translated as 'probably am,' 'probably is,' or 'probably are.' ... *Desyoo* shows that the previous statement is not definite .... I feel sure the Japanese become as irritated by American insistence on definite statements as we do when we get so many *desyoos* in answer to our questions .... I have given you but a fraction of all the instances where Japanese and English are the reverse of each other."

No wonder the Japanese and other Orientals are so "inscrutable" to Westerners. The English writer Rudyard Kipling said, "East is East and West is West, and never the twain shall meet." Kipling was right about this conceptual conflict being *perpetual* because the interaction of **opposite/same** is forever reciprocally ordered. If the interaction of **both** *Eastern* **sameness**, or *simultaneous balance,* **and** *Western* **oppositeness**, or *alternating imbalance*, were placed *in simultaneous double reversal*, they would become a more complex concept of **sameness**, more than the sum of its parts.

It seems logical then that if *Eastern* and *Western* cultures *interacted* reciprocally, the **same** in **opposing** ways, **both** cultures would be enhanced. In fact, this has already occurred. The *dominance* of the West's concept of *quantity* and the East's *dominant* concept of *quality* has already paid off in commercial trade between the United States and Japan. The Japanese have taught the Americans the value of *quality*, and the Americans have taught the Japanese the efficiency of *quantity* to their mutual benefit. China also has reciprocally interacted not only with the United States but also with other cultures to their mutual benefit, but China has not allowed **opposition** to *dominate* over **sameness** of *conceptual* thought. India however has embraced the logic of science with its dominance of deduction in inference as seen in "Losing an Edge, Japanese Envy India's Schools," by Martin Fackler, the *New York Times*, January 2, 2008, from which I quote, "Most annoying for many Japanese is that the aspects of Indian education they now praise are similar to those that once made Japan famous for its work ethic and discipline: learning more at an earlier age, an emphasis on memorization and cramming, and focus on the basics, particularly in mathematics and science." At Little Angels Kindergarten, its two-year-olds are taught to count up to twenty, three-year-olds are introduced to computers, and five-year-olds learn to multiply, solve math word problems, and write a one-page essays in English, tasks most Japanese schools do not teach until second grade.

All the Far Eastern nations have enjoyed *accelerated* growth by having adopted the science-based education that originated in the West, but which is now in decline in the West from countercultural pressures previously discussed. Read, "Dogma, not Faith, Is the Barrier to Scientific Inquiry," by U. Kutschera, under Correspondence in *Nature*, September 7, 2006.

If the extremely religious would also try to understand that scientific reasoning, formed by the paradox in inference, has been the sole source of the *accelerated* increase in the ordered complexity of technology, then *science* and *religion* as doubt/faith can reciprocally interact. Read "Biotech, Red in Tooth and Claw," *American Scientist*, September–October 2006, by Nathaniel C. Comfort, who review a book titled *Challenging Nature: The Clash of Science and spirituality at the New Frontiers of Life*, HarperCollins, 2006, from which I quote:

> The conflict between religion and science is one of the greatest stories ever told. It is a chivalric war, centuries old—a grand and ceremonious fight between two camps, each of which believes itself self-evident on the side of the right. With John Williams Draper's magisterial three-volume work of 1875, **History of the Conflict Between Religion and Science**, a canonical narrative began to emerge. In it, science, as the expression of reason, eroded the cultural power of religion, as the embodiment of superstition and dogmatism. Copernicus removed the Earth from the center of the cosmos. Darwin lowered man to the level of the animals and eliminated God form creation. And more recently, Watson and Crick opened the door to engineering life, turning us into God Himself. The story contains a kernel of truth, but the story of the history of science is more than the conquering of spiritual darkness by the light of reason. Both religion and science have mixed legacies: both have done harm as well as good. And both tend to be most dangerous when they become dogmatic and intolerant, and when they confuse faith with knowledge.

It is encouraging to find an instance in U.S. *politics* where **opposite/same** is *interacted*—i.e., where **both opponents** are given the **same** *chance* to dominate. A political situation that makes **sameness** as the basis of **opposition** is found in "Leveled Colorado District Creates an Election Laboratory," by Carl Hulse, the *New York Times*, August 5, 2006, from which I quote, "Unlike the vast majority of House districts around the country, this one . . . was not gerrymandered to guarantee victory to one party. Quite the opposite. It is a freak of modern political nature, purposefully drawn to be balanced between the parties and provide a genuine test of the ideas and abilities the opposing candidates . . . . It is political scientist's dream."

One would think that *economics* has nothing to do with the *interaction* of cruelty/altruism. But excessive *altruism* can be as self-defeating as excessive *cruelty*. For example, the *altruistic* practice of medical science can now save many genetically defective newborns that are *unfit* for a *self*-sustaining and adaptable life. Such noble effort however could gradually pollute the genetic pool of humans because it flies in the face of evolution's strongest concepts, survival of the *fittest* and *selection* by chance for adaptability. The long-term consequence of this will be the hough economic cost of medicines, therapies, nursing care and social strain that these defects will require.

We can mitigate this dilemma by making those with serious *inheritable* genetic defects and diseases impotent. This way such unfortunate individuals can continue to live, but they cannot pass on their inherited defect to future generations. Then *altruism*, the very *strength* of medical practice, will not *weaken* our genetic future, to say nothing of lifting excessive financial and time-consuming burden from **specific** families and society in **general**. The catch is that one can't precisely measure the concepts of *fit* or *unfit*, *cruelty* or *altruism*, *merciless* or *merciful*. One can only *interact* these pairs of concepts in a *constrained proportion* where **neither** one is excessive, or considered absolute. Under these proportional *constraints*, physician and parents can make realistic judgments.

A newly born infant that has no possibility of living without being put on life support should, according to evolution's "survival of the fittest," be denied such assistance. Evolution would let such an individual die *now* in order to *keep from killing* the species in the *future*. Darwinian evolution demands not just efficiency but parsimony. By selecting for *cruelty* in the *present*, evolution has *empathy* for life's diverse genomes in the future.

Read, "When Torment Is Baby's Destiny, Euthanasia Is Defended," by John Schwartz, the *New York Times*, March 10, 2005. Also read, "Technology Can Save Tiny Babies, Problems Still Likely Physicians Warn," by Knight Ridder newspaper taken from the *Tampa Tribune*, October 8, 2005.

All this means that in our judgments, we should *interact* cruel/kind, evil/altruism, and bad/good as *complementary* concepts within a constrained proportion, lending dominance to the **specific** case at of the moment. An example that helps such medical judgements is found in a study on the amount of time needed for *healthy* fetal development, seen in "Study Redefines Window for Early-Birth Risks," by Shirley S. Wang, the *Wall Street Journal*, November 13, 2007. Risk of permanent damage and susceptibility to serious lifelong diseases occur when a baby is delivered earlier than recommended.

In the real world, *predators* are *cruel*, but they are also *kind* not just to themselves and their families but also to its *prey* by killing those *unfit* to defend themselves. Conversely, the genes of the *prey*, given *time*, can by *chance adapt* a defence that eventually sets up a more "*level* playing field," a proportion between the **opposing** concepts of *predator* and *prey*. Since every *predator* is also a *prey*, and vice versa for the **same** reason, the idea of **either** *cruelty* **or** *kindness* depends on one's perceptual or *conceptual* point of view.

If the average age of the human population continues to increase due to the discoveries and advances in medial science, the social cost from *excessive* altruism will become so great that the collective attitude will necessarily transform from keeping the elderly *alive* at any cost to the **same** biological method of cellular life—apoptosis, *self-suicide*. Once a point is reached where one can no longer bear the excruciating pain of an incurable condition and life takes on a *negative* value to one*self* and to society, then it would be normal for that person to take the *selfless* act of refusing any further life-sustaining medical procedures. Then the act of suicide will *conform* to nature's method of *apoptosis*—programmed cell death—and the idea of *self* toward the end of life will transform to *selfless*. Read, "Death, Unconsciousness and the Brain," by Steven Laureys, *Nature Reviews Neuroscience*, November 2005.

At the high level of life's *complexity*, it seems that nature is *profligate* and that cell suicide would not occur; but again, at any level of complexity, nature is also *simple* and *parsimonious*. As with evolution, any event that is not economical or does not promote *systemic* economy is excluded. Read, "Deadly Priming," by Roberto Kolter, who explains a report by Ilana Kolodkin-Gal et al., *Science*, October 26, 2007. Timely *death* is necessary to conserve the proportional constraint between *life* and *death*, just and *entropy* is necessarily *balanced* against *genesis*, as seen in Figure 337.

An example of the insoluble *interaction* of good/bad in *mindless* nature was seen in the controversy among medical scientists over the class of drugs that reduce the inflammation and pain of arthritis such as Vioxx and Celebrex. They turned out to be *good* for arthritis but *bad* for the heart. Also, experiments showed the drug Aleve that was touted to be *good* for arthritis could also *bad* for the heart, while the drug Viagra that often is *good* for penal erection can also be *good* for the heart. "One man's meat is another man's poison," but conversely, "What's sauce for the goose is sauce for the gander." *Concepts* are *inconclusive*, and so too are their adages. **Either opposing** choice of a *factual* conclusion can be a valid interpretation.

Conceptual conflicts in medicine continue to be uncovered and they are as confusing to doctors as to their patients, leaving us all with *paradox* that defies the *absolute* idea of a *cure*, or a so-called silver bullet medication (**specific**) that has no collateral damage (**general**). Read, "Good Pill, Bad Pill: Science Makes It Hard to Decipher," by Gina Kolata, the *New York Times*, December 22, 2004. Also read "Vioxx. Celebrex. Now Aleve. What's a Patient to Think?" by Anahad O'Connor, the *New York Times*, December 28, 2004. Also read "Latest Research: Just Sit Quietly in a Dark Room," by Daniel Ruth, the *Tampa Tribune*, January 10, 2005.

To slightly diverge, companies already use the reciprocal logic with **opposing** pairs of *predicates* called "fuzzy logic." It is *fuzzy* because it is *inconclusive* and therefore non-inferential. Instruments designed with *proportional constraint* between extremes of *perception* have been highly successful. Take for example, digital instruments that measure *perceptions* between *cold* and *hot* temperatures are called thermostats. They are designed with a "comfort zone," a *constrained* **generality**, so that they can be tweaked to one's **specific** preference. The thermostat can become more complexly ordered by adding the concepts of *humidity* and *pressure* to temperature, all three of which are proportionally constrained. This more complex instrument of "fuzzy logic" is called an "altimeter."

Although "fuzzy logic" is logical, its non-inferential nature has seemed to be only a *perceptual* anomaly akin to induction, so it has not caused a "paradigm shift" to the more complex and enhancing logic of *concepts and their perpetual* interactions. However it does give added credence to my conjecture that *conception* emerged from *perception.*

A good example of how an excessively *imbalanced* proportion can lead to *chaos* thereby *diminishing* the diversity of species is seen in "Attack of the Killer Jellies," by Richard Stone, *Science*, September 16, 2005. An example of how *balance* of conflicting concepts enhances the processing of information is found in "Critical Period of Plasticity in Local Cortical Circuits," by Takao K. Hensch, *Nature Reviews Neuroscience*, November 2005, from which I quote:

> In the tangled circuits of the neocortex, the action of locally balanced excitation and inhibition first integrates and then detects competition. Even small change in the relative amounts of excitation and inhibition can markedly alter information processing. This delicate balance is dynamically adjusted by circuitry in the cortical layers, where inhibitory connections are developed later than excitatory connections . . . . The concept of excitatory-inhibitory balance in key neural systems during development might apply to epilepsy, autism, Rett syndrome and Tourette's syndrome, schizophrenia and even the encoding and retrieval of memories. Finding the keys to critical period plasticity could enable the development of novel therapies and training paradigms for education, rehabilitation, recovery from injury and lifelong learning in adulthood.

Evolution has *conserved* the *dominance* of reciprocal **opposition** to **sameness** in human brain-minds. Through no fault of our own, our *mass* **specificities** are **opposed** to the idea of **sameness** at any level of complexity. With this insoluble paradox, it is no wonder that so many people believe that Jesus of Nazareth was born to bring us the message of God's forgiveness, that we are redeemed from our "original sin" of being **opposed** to God. This is the "good news" of the Gospels.

"Original sin" was inflicted on us by being **either** *male* **or** *female* of the **same** idea of sex. The allegory of this **opposite/same** interaction occurred in The Garden of Eden. It tells of our *separation* from the God and *why* we are **opposed** to God's Universal **Sameness**. In the paradise of the Garden of Eden, there was no **opposition** and no conceptual conflict, God and man were *simultaneously* the **same**. The garden was absolutely *peaceful*. There was no need to make a choice of **either** concept of an interaction. There was no such thing as **either** *good* **or** *evil*, **both** were the **same**. **Opposing** God's will by eating the forbidden fruit of knowing **either** *good* **or** *evil*, the human race was expelled from God's Garden. Having been given the reciprocal choice of **either** *good* **or** *evil*, as well as other *conceptual* choices, the *temporal* human mind became trapped in the insoluble predicament of conceptual conflict—perpetually at *war* **both** *within* **and** *between* human minds, in defiance of God's will.

It is more than just interesting that Jesus of Nazareth and Socrates of Athens in ancient Greece were the first known individuals who seemed to understand and practice the logic of nature's *conceptual* interactions. In the case of Socrates, he stuck to his belief and teachings that one should be sceptical of and question dogmatic ideas, yet *simultaneously* submitting to the laws of ancient Greece by accepting the death penalty although he could have easily been pardoned. As for Jesus, he preached against hypocrisy; nothing infuriated him more than self-contradictory ideas and behavor. Like Socrates, Jesus preached complying with "the will of God," yet he too submitted to the death penalty for breaking the law of Rome even though he **opposed** their tyrannical rule over Jewish law and culture. Both men believed that no one could be *really* free without being *constrained* by law, and more importantly constraining one*self* to the law; otherwise chaos instead or order would prevail. Both men epitomized the *timeless insolubility* of conceptual conflict, thereby germinating the long and tortuous development of governmental organizations through *self*-constraint and *self*-responsibility for a dominantly *free* life.

St. Thomas Aquinas reminded us "evil results from man's abuse of God-given free will. He thought it was impossible for God to create humans having a rational nature and a free will who would never sin." This quote is from "For the Worst of Us, the Diagnosis May Be 'Evil,'" by Benedict Carey, the *New York Times*, February 8, 2005. Also read "The Chomsky of Morality," *Nature*, October 26, 2006, by Paul Bloom and Izzat Jarudi who review a book titled, Moral Minds: How Nature Designed Our Universal Sense of Right and Wrong, by Marc D. Hauser, *Ecco., 2006*.

We have a strong tendency to believe what we *want* to believe to make us feel comfortable and secure. Examples of how self-deceptive and unrealistic *extremes* of belief can be read, "Scare Yourself Silly, but the Real Terrors Are at Your Feet," by Abigail Zuger MD, the *New York Times*, October 25, 2005.

Whether *male* or *female*, one needs a *purpose* in a *purposeless* life to motivate us and to get any *meaning* from a *meaningless* universe. But if we acknowledge the *incompleteness* of our mind's *interpretations*, we will be much more able to adapt to the human predicament. Knowing that *creation* by *genesis* and *destruction* by *entropy* interact reciprocally in time, we will be better able to submit to death. Knowing that our brain/minds *complex* **specificities** are too "far from equilibrium" to be sustained, our "return to equilibrium"—to the **sameness** of the **general** environment—is the natural order of nature. More profoundly, we can rejoice with the "good news" that although *mass* with its *temporal perceptions* will be *destroyed*, our *massless concept*s that *perpetually* interact will be *conserved*, no longer **opposed** to God's Universal order of **Sameness**.

"Then shall the dust return to the earth as it was and the spirit shall return unto God who gave it" (Ecclesiastes 12: 7).

The "good news" from Jesus of Nazareth is that human *minds* are forgiven from their **opposition** to God's **Sameness** because we are innocent victims of evolution. Since our *concepts* and their interactions are *parts* of universal **sameness**, we are *logical* candidates for redemption once our *timeless conceptions* have departed our *temporal* lives to join the more complex and enhanced *whole* of **SAMENESS** that not even *time* can **oppose**.

In the mean time, our lives' *dominate* **opposition** to **sameness** leaves us with an innate *need* for a *personal* God that will deliver us *safely* from those that would **oppose** us or *endanger* our lives. It is an emotional and perceptual imperative to seek a *conclusive* answer to our prayers and a *solution* to the "problem" of our *conceptual* **opponent**. Acquiescing to an *impersonal* God that created us *personally* seems too much to ask.
But again, the message of Jesus was that we are forgiven from this terrible dilemma. By accepting the personal/impersonal paradox along with the antinomy in inference, we will be able *dominantly* choose a *personal* God without excluding an *impersonal* God. Then, we will better understand and accept our *insoluble* human predicament.

Read "God Is Bred," *Nature*, July 13, 2006, by Crispin Tickell, who reviews the book, *Six Impossible Things Before Breakfast: The Evolutionary Origins of Belief*, by Lewis Wolpert, Faber and Faber, UK, 2006, to be published in United States by WW Norton, January 2007.

To better understand that the *incompleteness* of conceptual **sameness** in *mass* is *conserved* no matter how complex, see FIGURE 400, as well as the *square* of conceptual **sameness** in FIGURE 401, taken from "Allosteric Mechanisms of Signal Transduction," by Jean-Pierre Changeux and Stewart J. Edelstein, *Science*, June 3, 2005. These two FIGURES show how life's proteins can *interact* "at a distance" the **same** as *forces* of nature. These so-called allosteric mechanisms are able to enhance complex information processing at a distance between **different** protein complexes by a graded connection between them that according to the FIGURES place the interaction of **opposite/same** in *simultaneous double reversal* thereby forming a more complex and enhanced **sameness**, **opposed** at 90° by the *acceleration* of information processing.

I quote the authors' abstract, "Forty years ago, a simple model of allosteric mechanisms (indirect interactions between distant sites) used initially to explain feedback-inhibited enzymes, was presented by Monod, Wymans and Changeux. We review the UMC theory and its applications for the understanding of signal transduction in biology, and also identify remaining issues that deserve theoretical and experimental substantiation." I also quote from the authors' last chapter, "Conclusion: The Quest for Theory":

> After 40 years, the allosteric theory of signal transduction has been applied to signaling molecules as diverse as regulatory enzymes, nuclear receptors, and to the various classes of membrane receptors. It has even been extended to riboswitches within which folded RNA domains serve as receptors for specific metabolites and to the allosteric cascades of spliceosome activation. As expected, each signaling system displays features of its own. But the concept of signal transmission mediated by discrete conformational transitions that exist before ligand binding would appear to be universal . . . . The simplicity of the theory facilitates its experimental test. However, both the theory and the available technology have reached their limits. This is an important area for future research . . . . Another lies at a more macroscopic scale: It includes the deciphering of the networks of allosteric interactions taking place in supramolecular assemblies within the cell and between cells. The future of allosteric proteins is more promising than ever, and we expect that theorizing will become more important as we define our understanding of the mechanisms that allow elaborate physiological control over protein function.

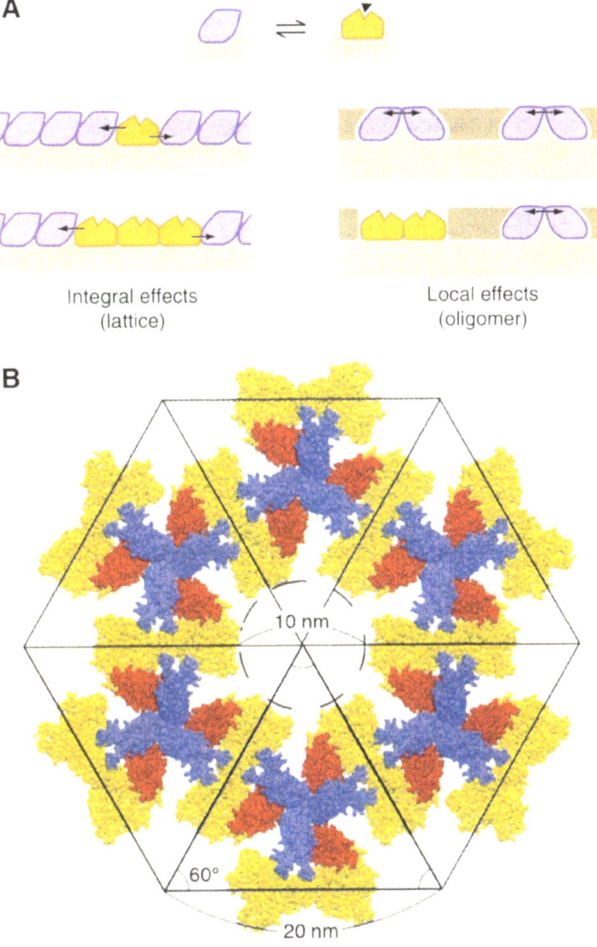

Fig. 5. Allosteric membrane lattice. (**A**) Extension of the allosteric theory to a membrane lattice (*6*). (**B**) The extended cooperative lattice of *E. coli* chemotactic receptors as visualized looking onto the plasma membrane; receptors are in blue, and the linking proteins CheW and CheA are in red and yellow, respectively (*37*).

FIGURE 400

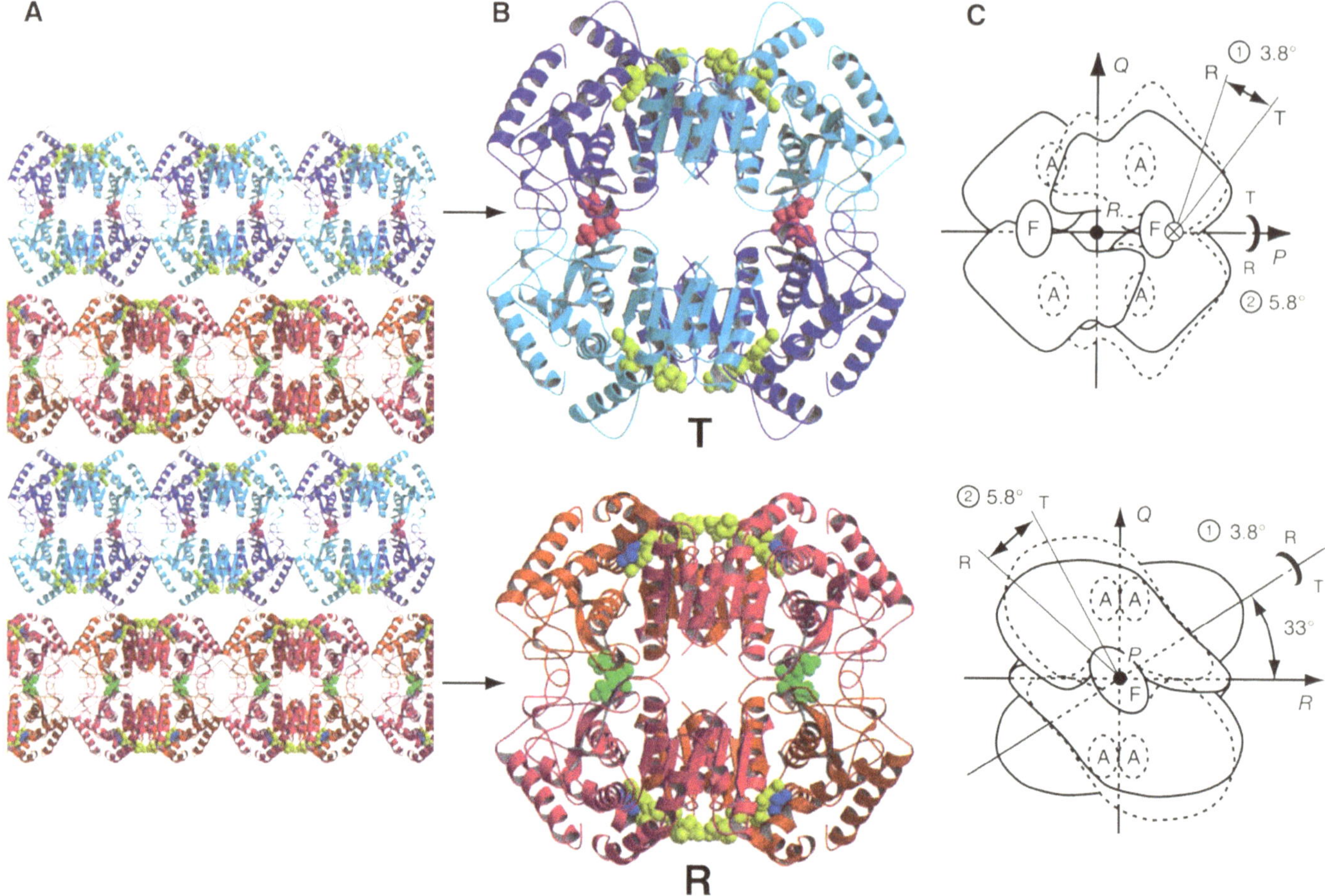

Fig. 2. Structural demonstration of the MWC model: The T and R states coexist in the crystals of bacterial L-lactate dehydrogenase (*14*). (**A**) Planar view of crystal. (**B**) T and R states enlarged showing bound ligands [coenzyme: NADH (reduced form of nicotinamide adenine dinucleotide), light green; regulatory signal: fructose 1,6-bisphosphate, fuchsia in T, green in R; substrate analog: oxalate, blue] at topographically distinct sites and conservation of symmetry of the quaternary structure with little change in tertiary organization of the subunits. (**C**) Two views showing the rotations corresponding to the T-R transition with respect to the three orthogonal axes denoted by *P*, *Q*, and *R* (viewed looking down the *R* axis in the upper schema and down the *P* axis in the lower schema). A and F refer to the analog, oxalate, and to fructose-1,6-bisphosphate, respectively.

FIGURE 401

If allosteric mechanisms are *conceptually* the **same** as gravity's "action at a distance," then their geometric concept of diagonaled **sameness** seen above should be as highly complex and enhancing of inference as are the calcium and potassium channels, all having been conserved from the unobvious and overlooked *self-creative* and *constraining* force of quarks **sameness** on the *vertical* rise of complexity.

With this understanding, hopefully my reader's mind will be elevated from the *simple emotional* function of **either/or** at 180° to the more *complex conceptual* function of **either/neither** at 90°, the latter interacting the geometric concepts of **opposite/same**, **no different** than the force of **electro/magnetism** that logically orders all *macro*-mass interactions, and **no different** than the human *mind* divides inference into its two reciprocally **opposing** processes, the *logic* of which is *part* of *mindless* nature's Universal ordering principle.

THE END

SUGGESTED READINGS

BOOKS

Angier, Natalie. *The Cannon: A Whirligig Tour of the Beautiful Basics of Science.* Houghton Mifflin Company: Boston, New York, 2007.

Newton, Roger G. *From Clockwork to Crapshoot: A History of Physics.* The Belknap Press of Harvard University Press: Cambridge, Massachusetts, 2007.

Scerri, Eric R. *The Periodic Table: Its Story and Its Significance*. Oxford University Press: New York, New York, 2007.

Zee, A. *Fearful Symmetry: The Search for Beauty in Modern Physics*. Princeton Science Library: Princeton University Press, Princeton, NJ.

Avner Ash & Robert Gross, *Fearless Symmetry: Exposing the Hidden Patterns of Numbers*, Princeton University Press, Princeton, NJ.

Hewitt, Paul G. *Conceptual Physics + Practicting Physics*, Addison–Wesley, 2002, Salt Lake City, Utah

Quine, W.V. *Pursuit of Truth*, 1990, Harvard University Press, Cambridge, Massachusetts.

Felipe Fernáández-Armesto, *Truth: A History and a Guide for the Perplexed*, St. Martin's Press, New York, N.Y., 10010.

J. Ivey Davis, *The Struggle Among Ideas: A Tourist Guide to the Human Predicament*, Outskirts Press, Inc., Parker, Colorado 80134.

WEB SITES

The New York Times, www.select.nytimes.com (Free with subscription)
The Tampa Tribune, www.tbo.com
Nature, www.nature.com The entire content of Nature from its first eighty years is available online.
Science, www.sciencemag.org
American Scientist, www.americanscientist.org
Scientific American, www.sciam.com
Science Reviews Neuroscience, www.nature.com/reviews/neuro
The New York Academy of Science, *Update*, www.nyas.org
Science News, www.sciencenews.org

See *animated* Figures of **opposite/same** interactions at www.reciprocallogic.com.

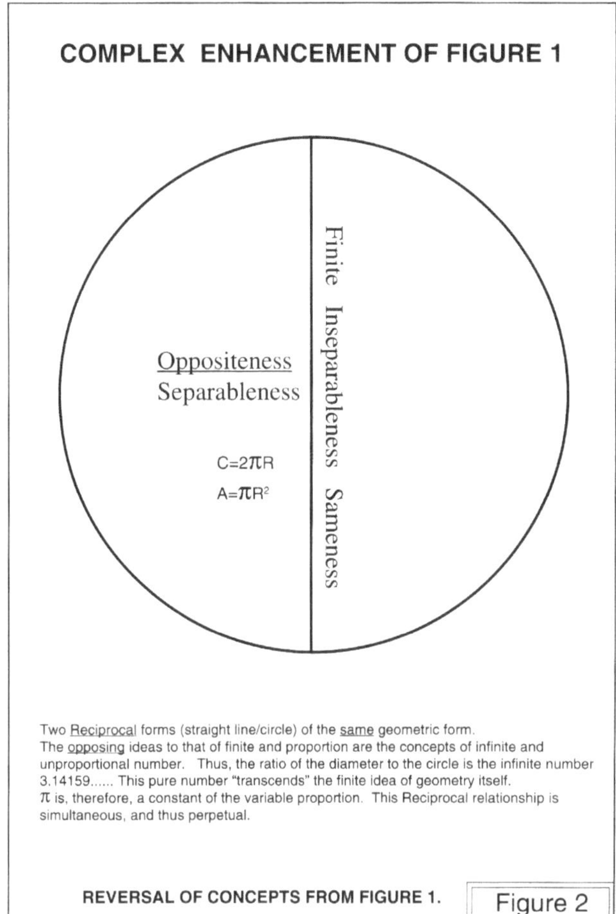

Figure 2

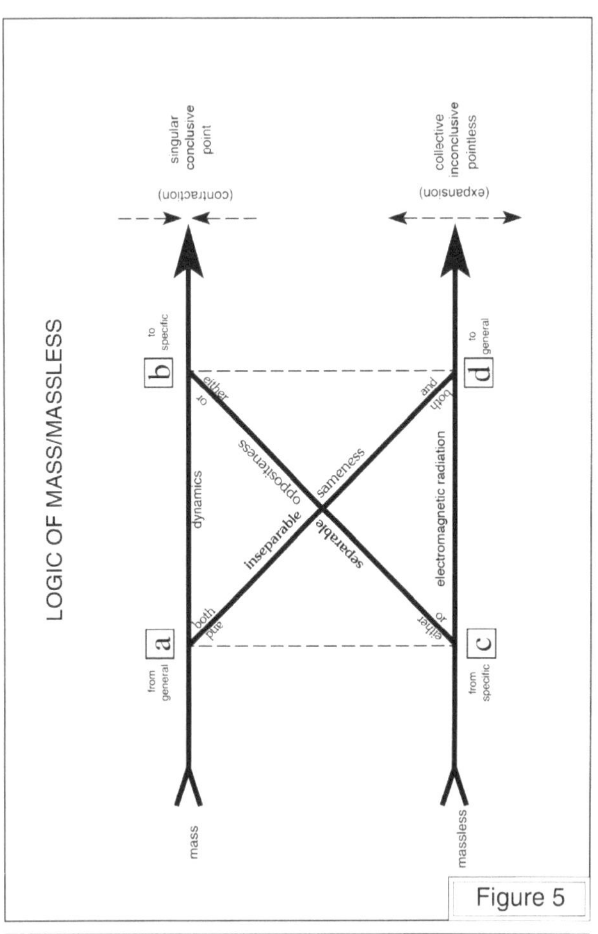

Figure 5

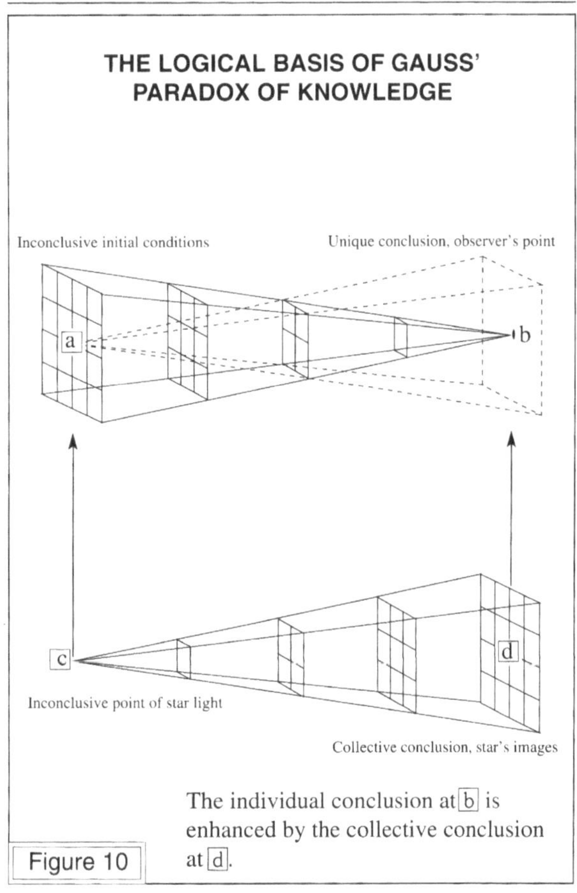

Figure 10

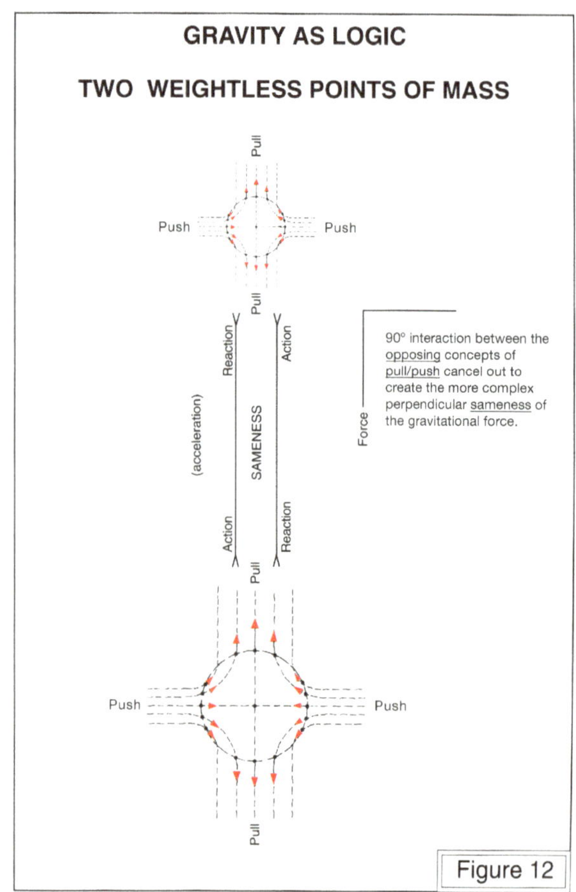

Figure 12

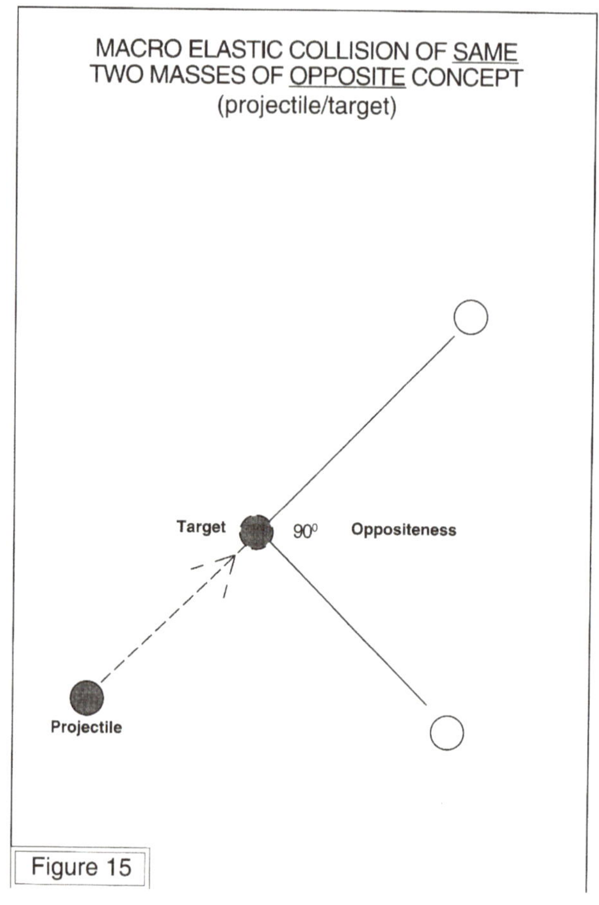

Figure 15

KEPLER'S 2ND LAW

EQUALITY OF AREA AND TIME

A line drawn from the Sun to the planet sweeps out equal areas in equal times. The three shaded portions in this figure have equal times. The three shaded portions in this figure have equal areas. The planet takes equal time intervals to traverse the corresponding three segments of the orbit.

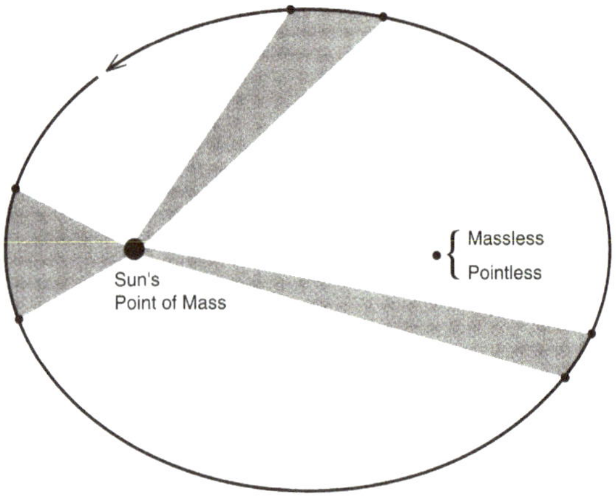

Massless
Pointless

Sun's
Point of Mass

Sameness of dimensionality (area).
Oppositeness of undimensionality (time).

The void, or massless, point (focus) centers the unaccelerated orbit. The chosen mass point centers the accelerated orbit. It makes no difference which of the two opposing foci of the ellipse is chosen by mass. Either mass point of view is valid!

Figure 14

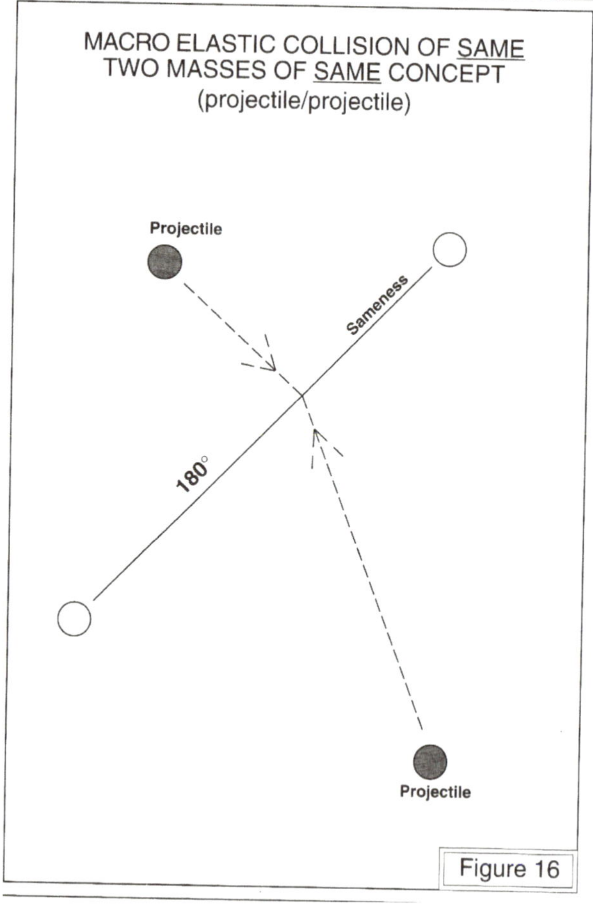

Figure 16

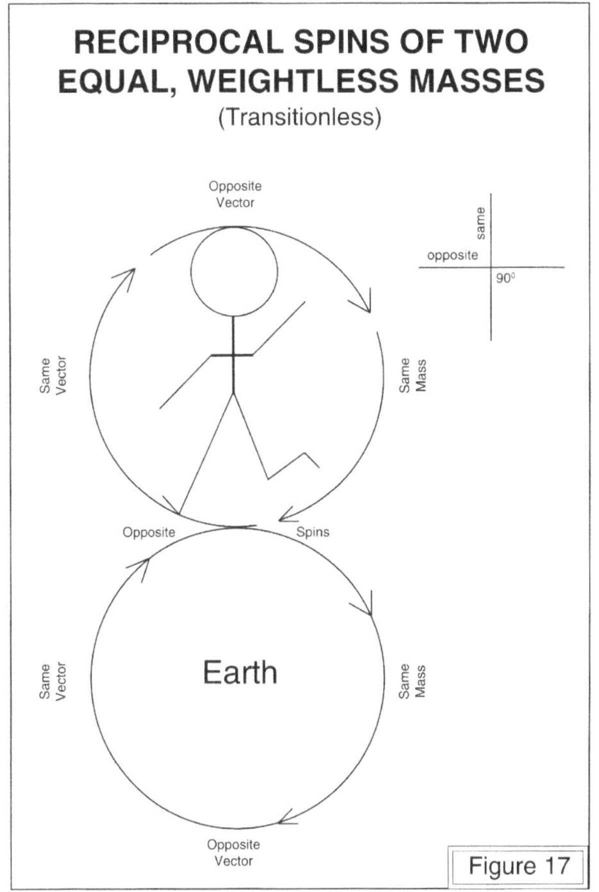

Figure 17

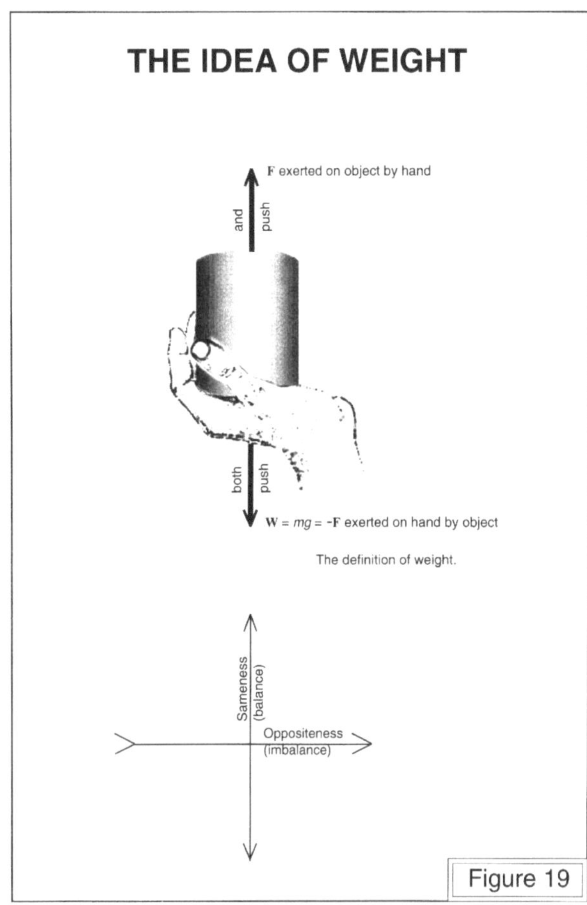

Figure 19

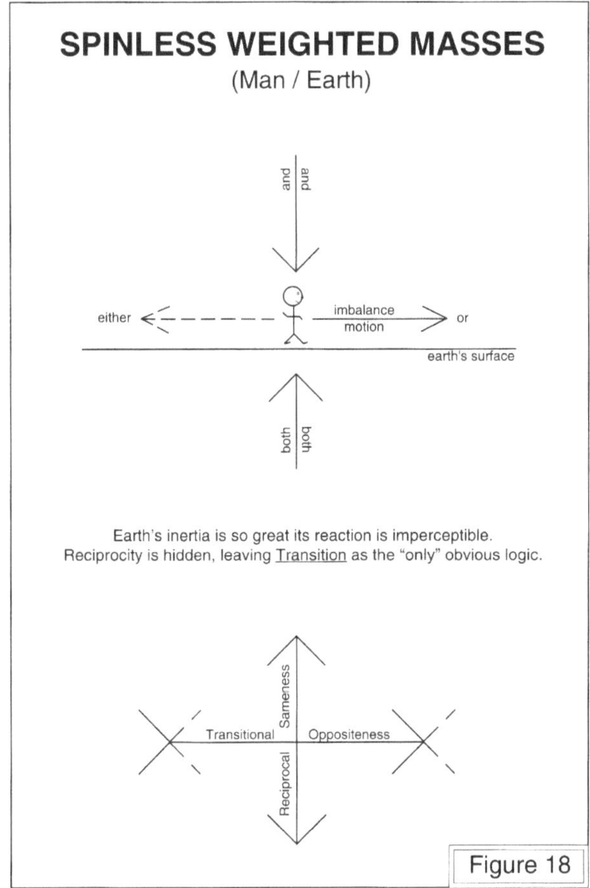

Figure 18

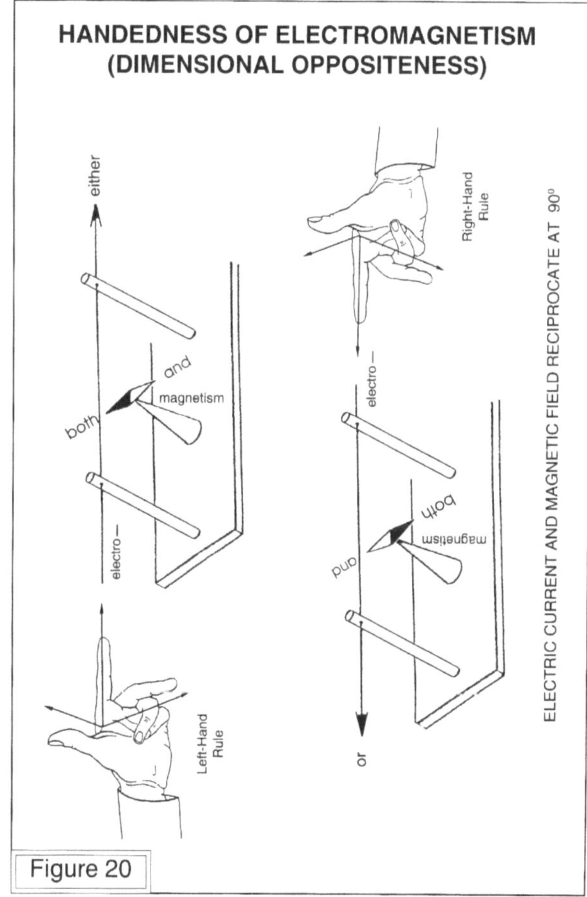

Figure 20

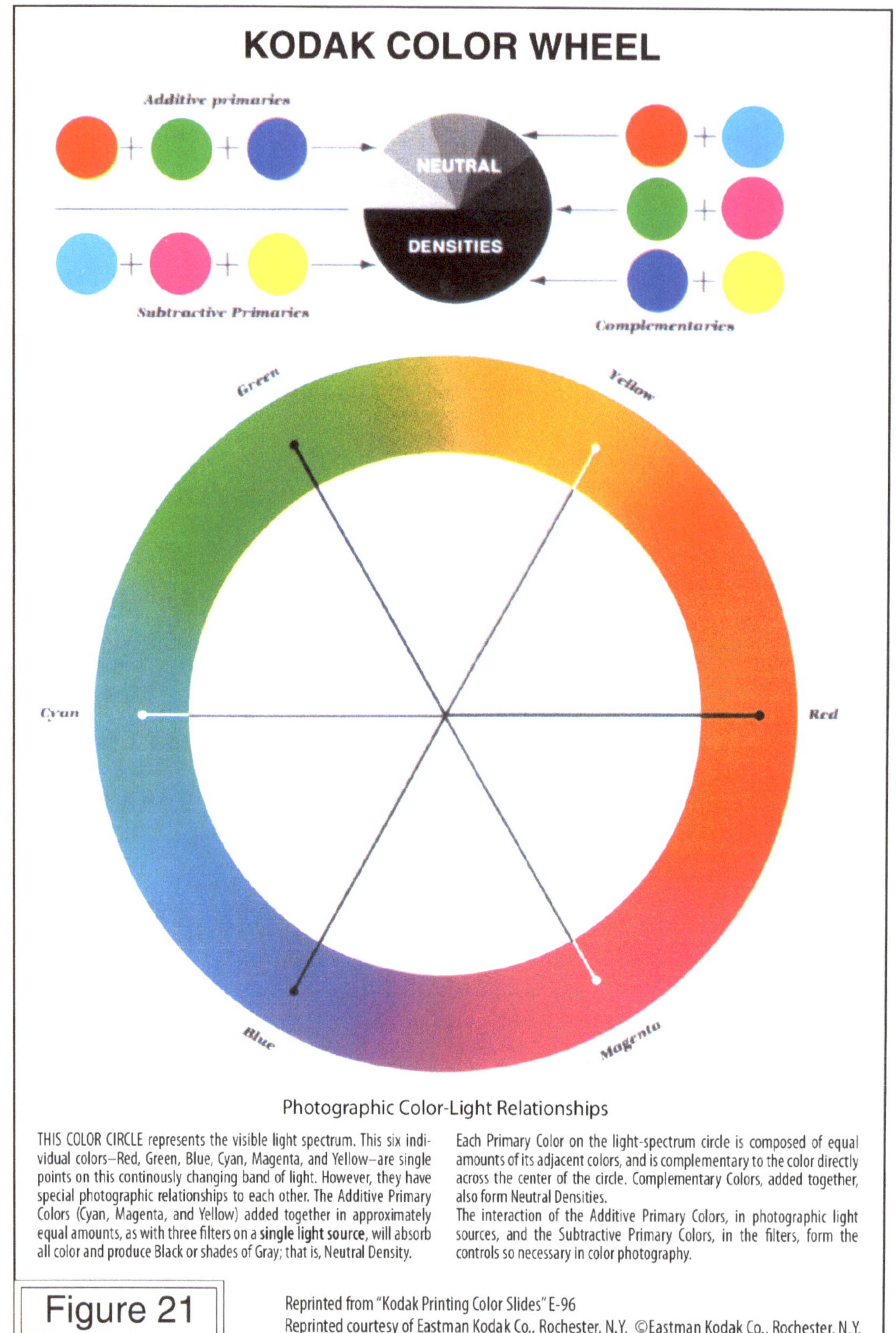

Photographic Color-Light Relationships

THIS COLOR CIRCLE represents the visible light spectrum. This six individual colors–Red, Green, Blue, Cyan, Magenta, and Yellow–are single points on this continously changing band of light. However, they have special photographic relationships to each other. The Additive Primary Colors (Cyan, Magenta, and Yellow) added together in approximately equal amounts, as with three filters on a **single light source**, will absorb all color and produce Black or shades of Gray; that is, Neutral Density.

Each Primary Color on the light-spectrum circle is composed of equal amounts of its adjacent colors, and is complementary to the color directly across the center of the circle. Complementary Colors, added together, also form Neutral Densities.

The interaction of the Additive Primary Colors, in photographic light sources, and the Subtractive Primary Colors, in the filters, form the controls so necessary in color photography.

Figure 21

Reprinted from "Kodak Printing Color Slides" E-96
Reprinted courtesy of Eastman Kodak Co., Rochester, N.Y. ©Eastman Kodak Co., Rochester, N.Y.

a
b
c
d

a
collection
individual
b
emission
from the general
to the specific
blue
colorful
either - or
1/2
2/2
Colorless
Colorful
both additive
oppositeness
sameness
and subtractive
1/2
2/2
either - or
absorption
yellow
from the specific
to the general
colorless
c
individual
collection
d

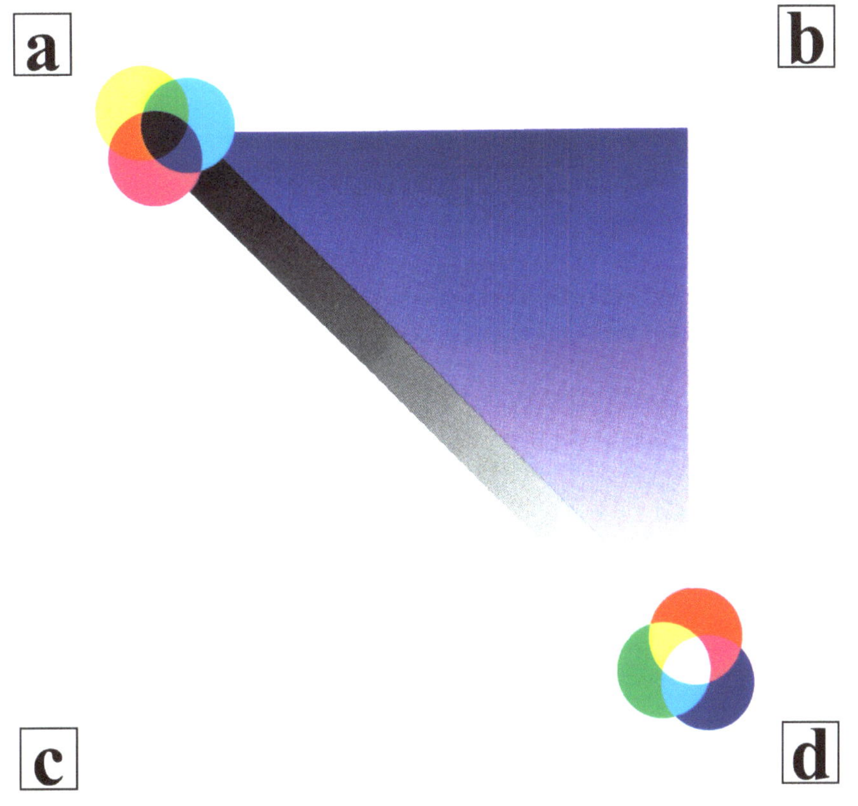

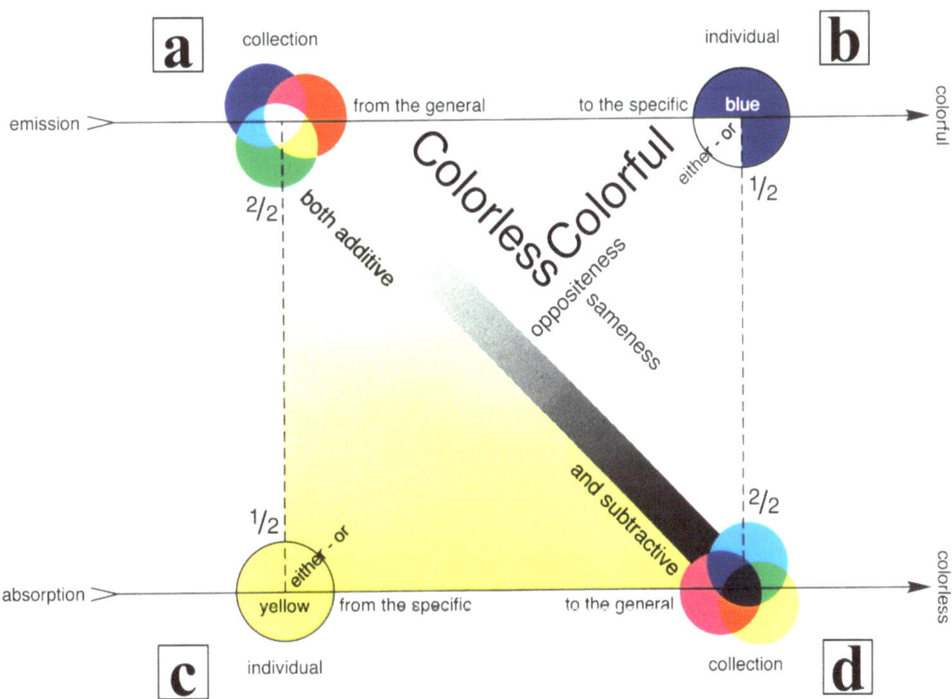

Figure 22.2

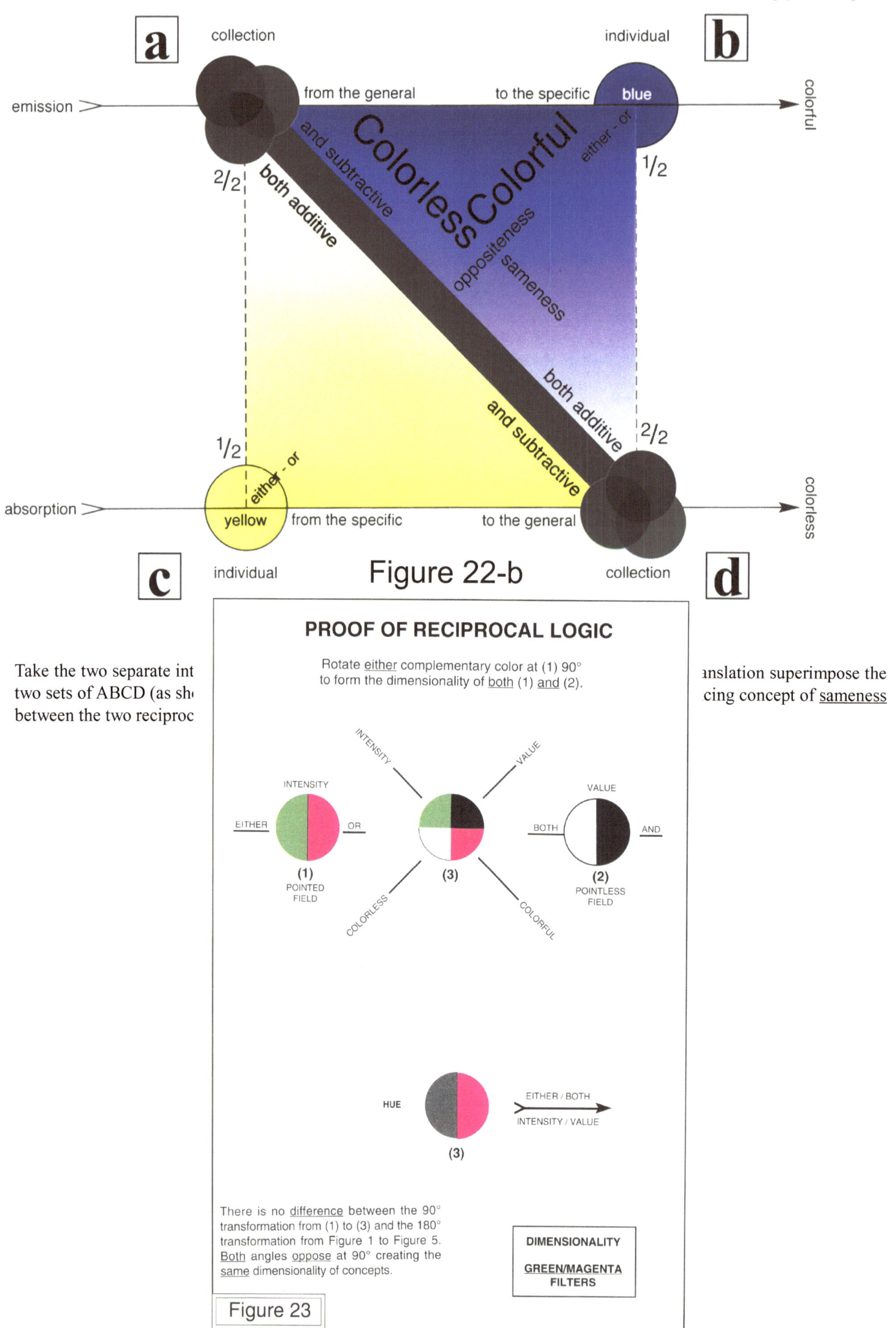

Figure 22-b

Figure 23

Take the two separate int ... anslation superimpose the two sets of ABCD (as sh ... cing concept of <u>sameness</u> between the two reciproc

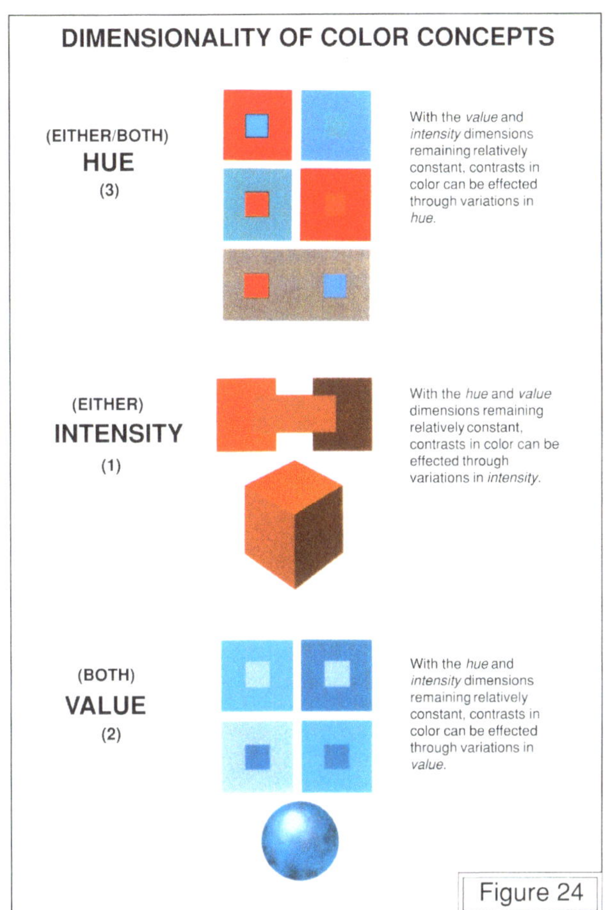

Figure 24

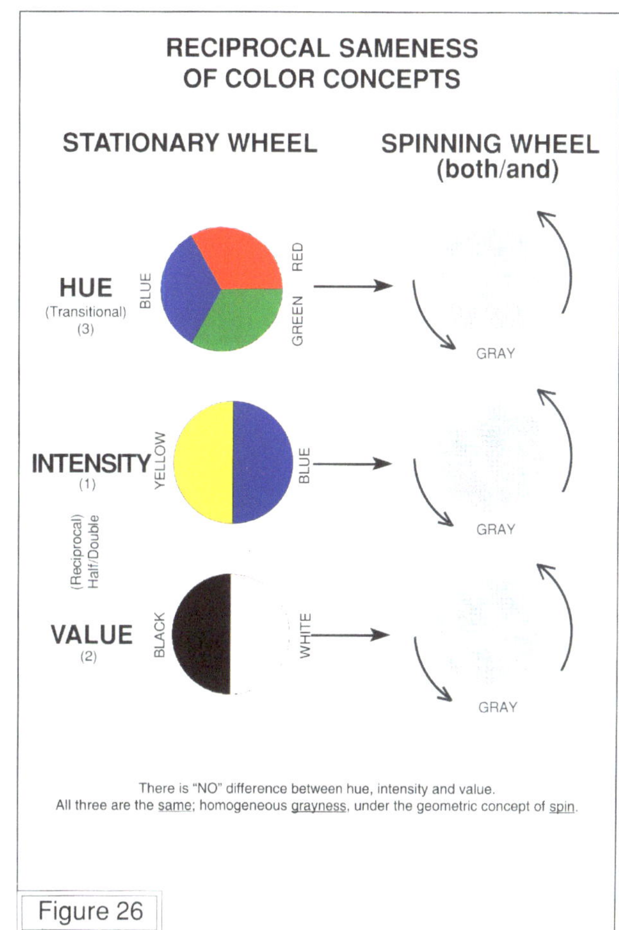

Figure 26

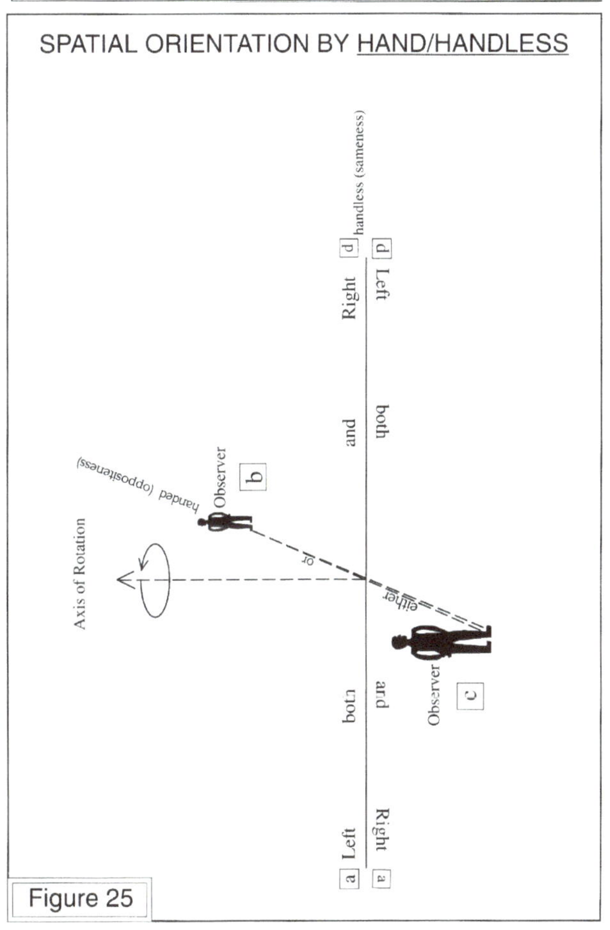

Figure 25

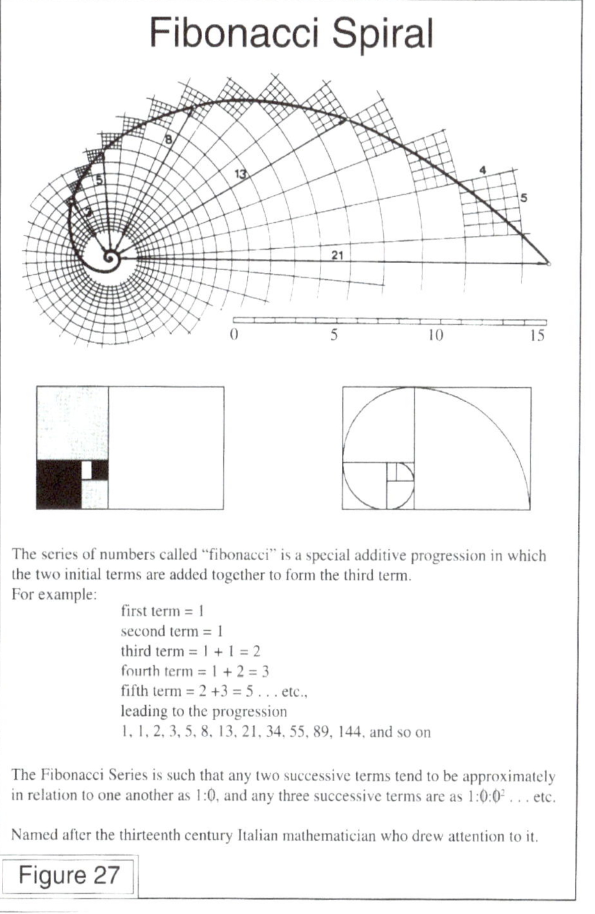

Figure 27

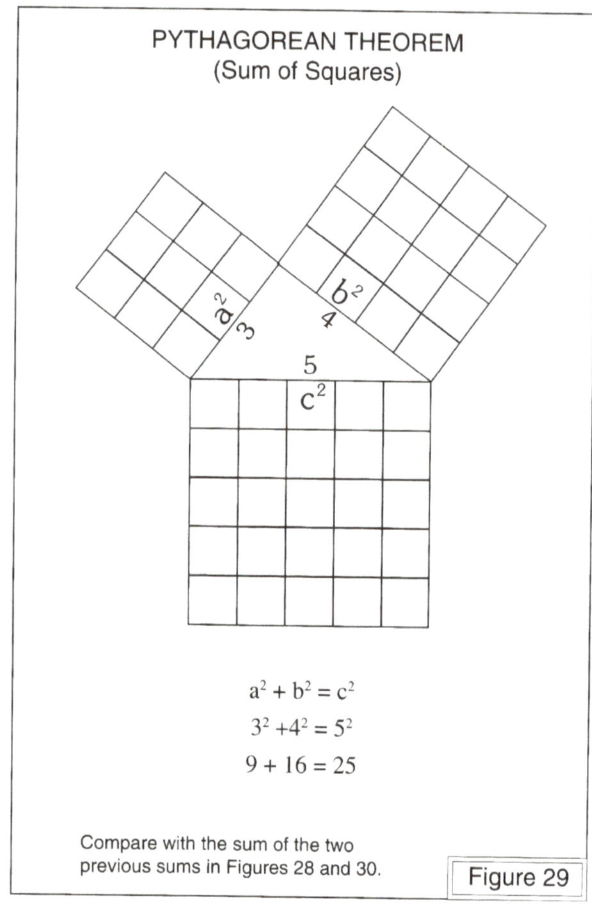

Compare with the sum of the two previous sums in Figures 28 and 30.

Figure 29

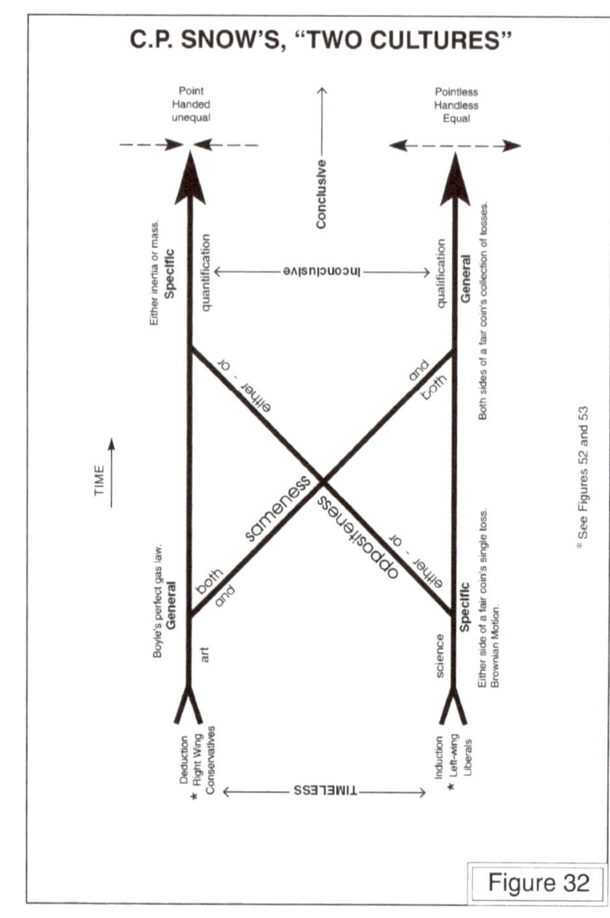

Figure 32

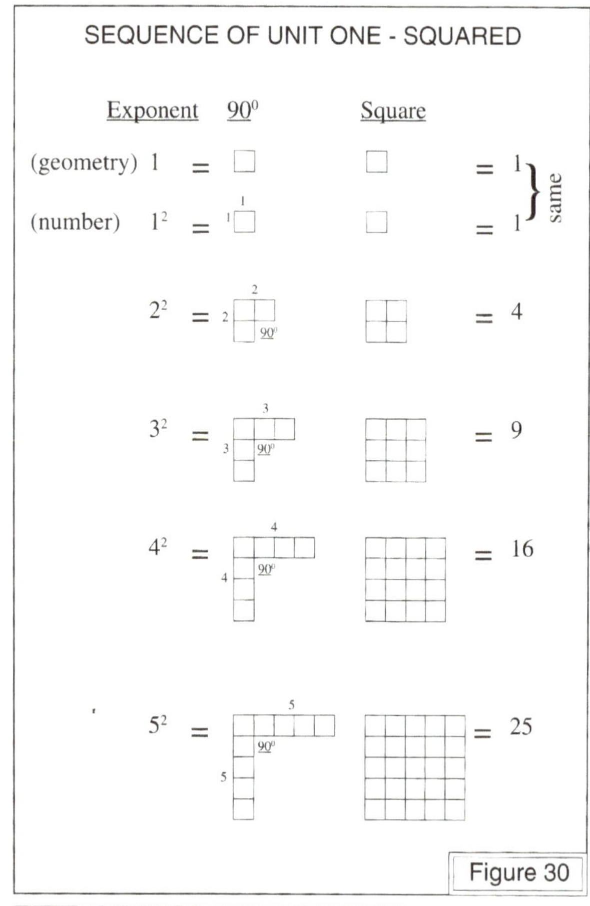

Figure 30

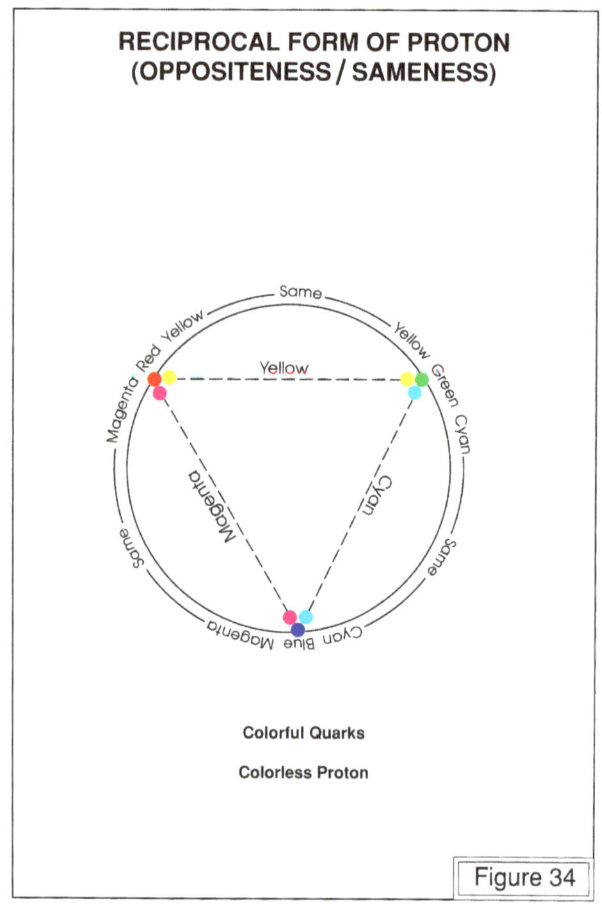

Figure 34

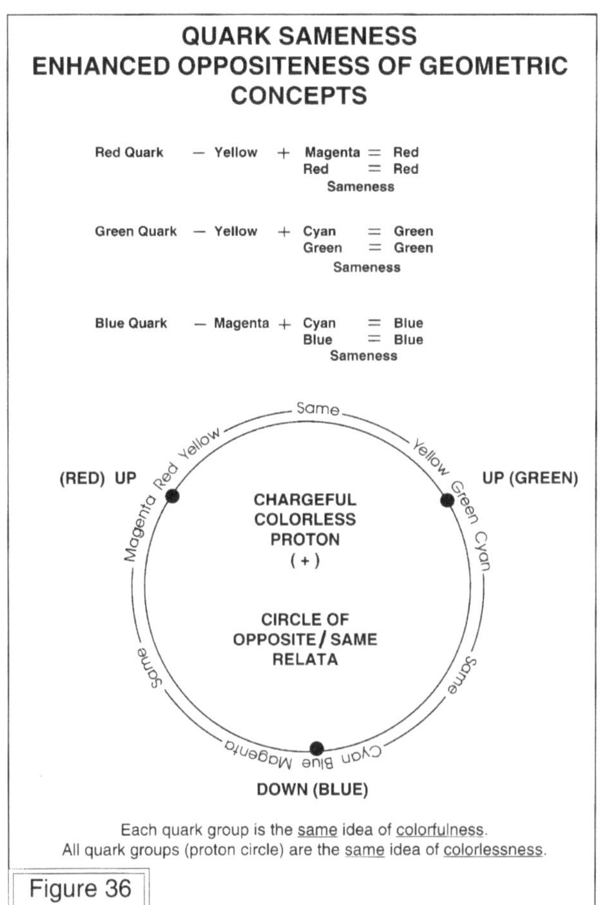

Figure 36

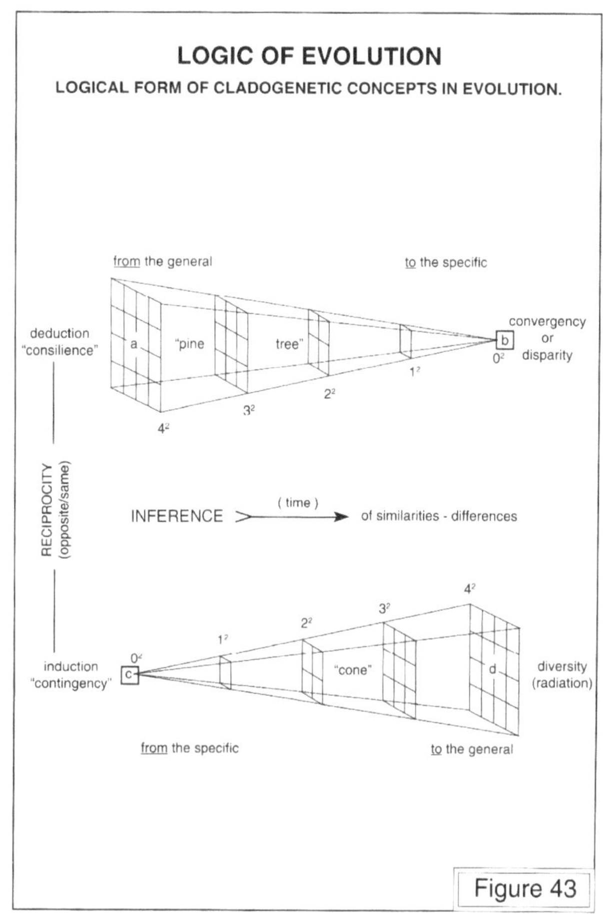

Figure 43

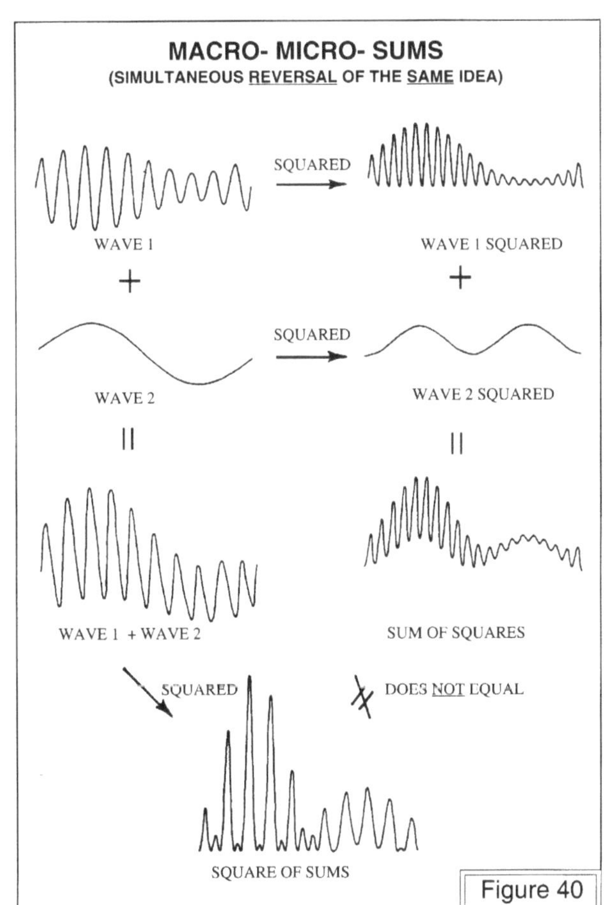

Figure 40

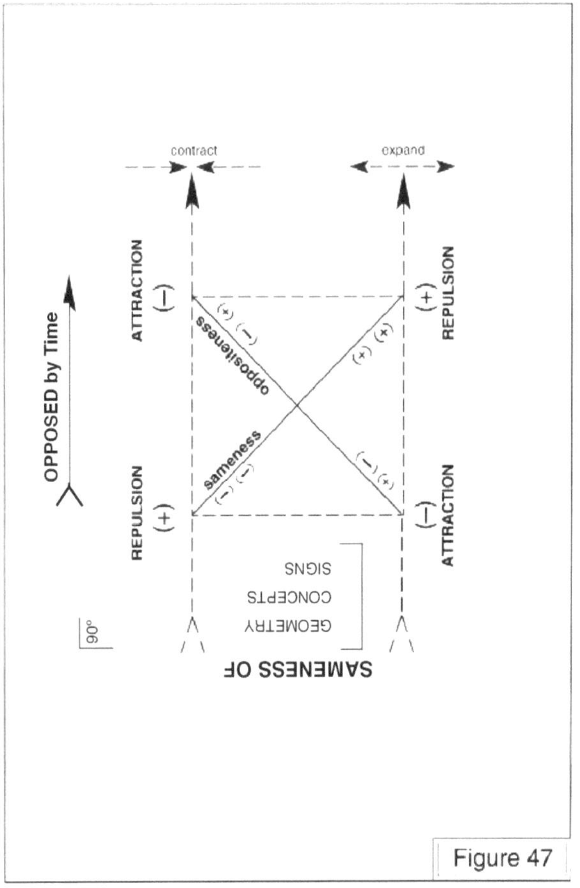

Figure 47

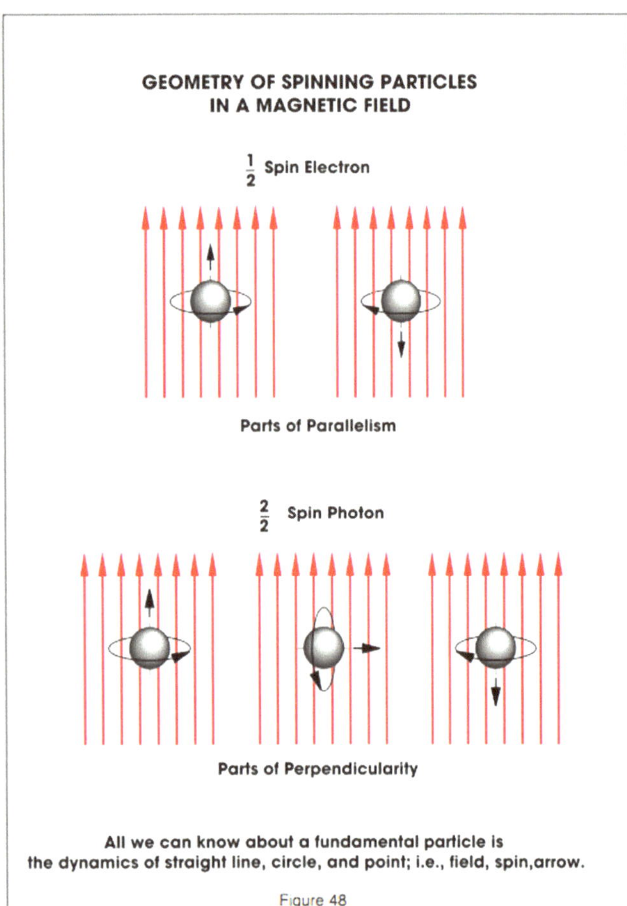

Figure 48

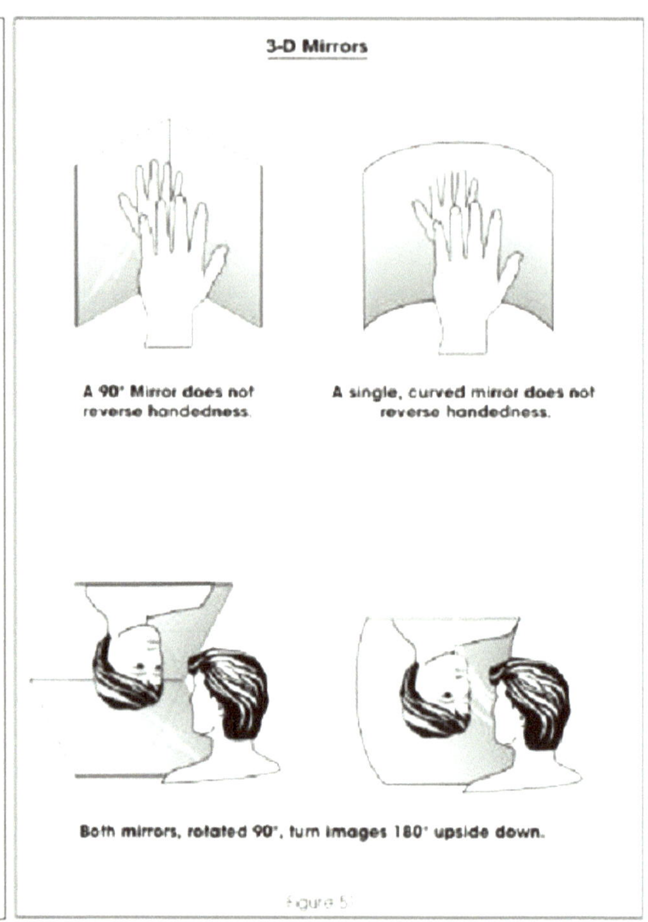

Figure 5?

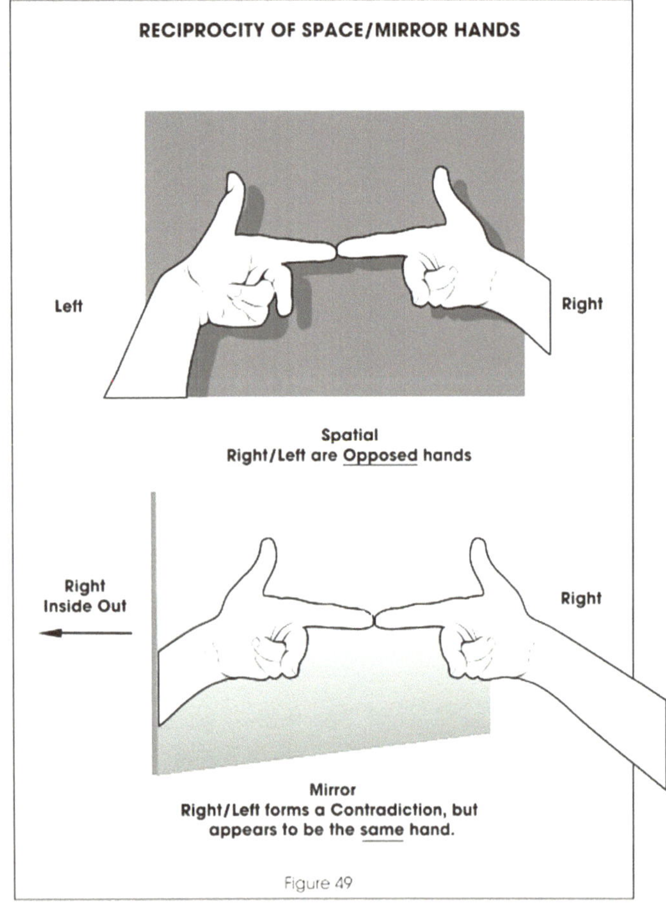

Figure 49

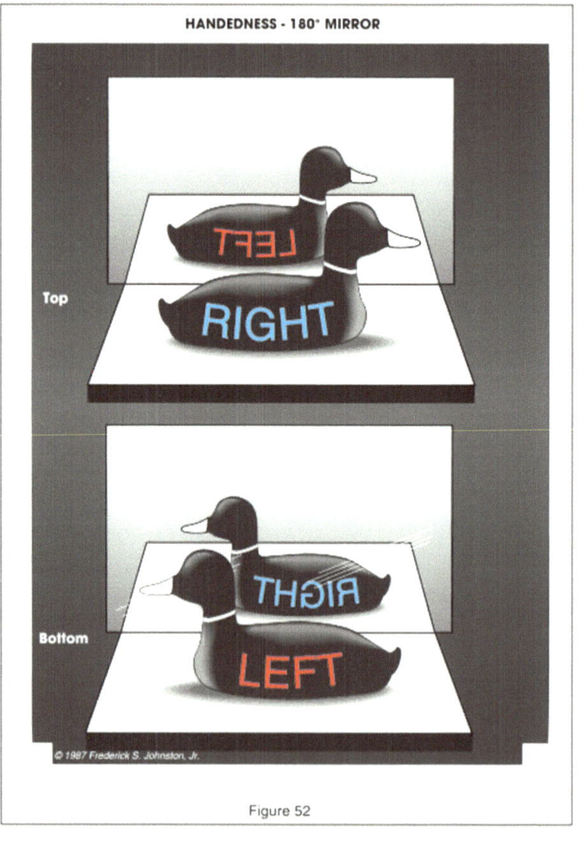

Figure 52

Current Logic of the mind's concepts

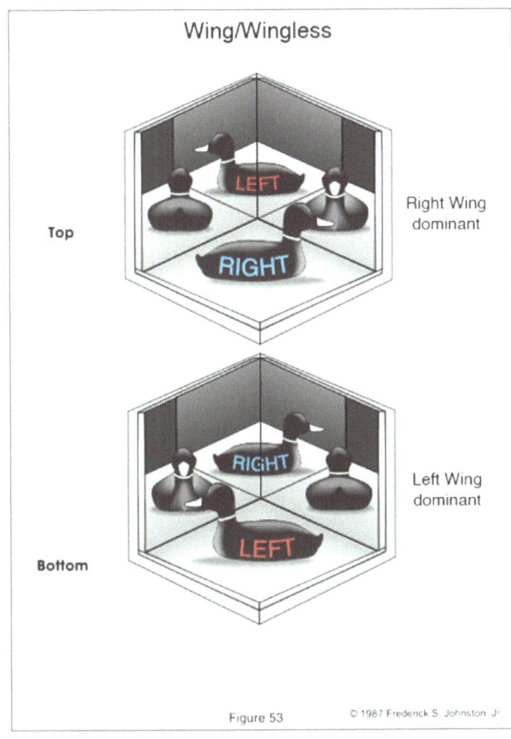

Wing/Wingless
Top
LEFT
RIGHT
Right Wing dominant
Bottom
RIGHT
LEFT
Left Wing dominant

Figure 53 © 1987 Frederick S. Johnston Jr.

The future of the minds more complex and enhance conceptual logic.

WAVELENGTH (ANGSTROM UNITS)
BALMER-ALPHA 6,563
BALMER-ALPHA 4,861
BALMER-ALPHA 4,340
BALMER-ALPHA 4,102

PATTERN OF DISTINCT LINES is formed when the light emitted by hydrogen atoms is dispersed according to wavelength The lines are those of the Blamer series and their approximate wavelengths are given in angstrom units (one angstrom is equal to 10 centimeters) The color photograph was made by Jon Brenneis and one of the authors (Schawlow) in the latter's laboratory at Stanford University The black-and-white spectrum was recorded in 1927 by Gerhard Herzberg who was then at the Darmstadt Technical University in Germany The intervals between the lines are different in the two photographs because Brenneis and Schawlow dispersed the light with a diffraction grating whereas Herzberg employed a prism the prism introduces a distortion because the reflective index of glass is a nonlinear function of wavelength In both cases the source of the spectrum was a gas-discharge tube in which hydrogen molecules (H) are dissociated into atoms by an electric potential In the color photograph the yellow and green bands and the continuous blue background at the right are not part of the atomic-hydrogen spectrum but result from molecular contaminants in the discharge tube Herzberg eliminated such contamination by cooling the tube in liquid nitrogen Lines to the right of Blamer delta which appear white or light blue in the color photograph are actually in the ultraviolet region of the electromagnetic spectrum

<u>The Spectrum of Atomic Hydrogen</u>
Theodore W. Hansch, Arthur L. Schwlow, and George W. Series

Scientific American
March 1979

W.H. Freeman and Co.

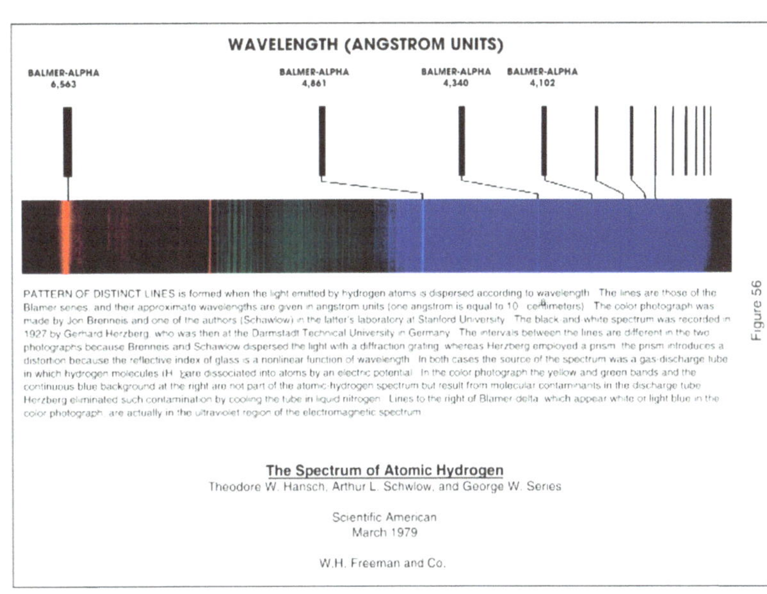

Figure 56

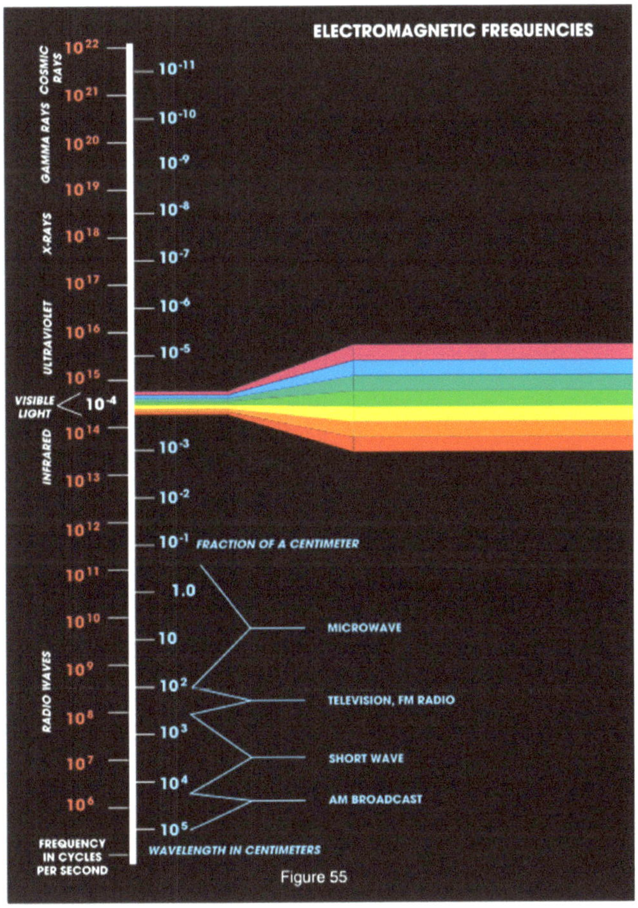

ELECTROMAGNETIC FREQUENCIES
COSMIC RAYS
GAMMA RAYS
X-RAYS
ULTRAVIOLET
VISIBLE LIGHT
INFRARED
RADIO WAVES
10²²
10²¹
10²⁰
10¹⁹
10¹⁸
10¹⁷
10¹⁶
10¹⁵
10¹⁴
10¹³
10¹²
10¹¹
10¹⁰
10⁹
10⁸
10⁷
10⁶
10⁻¹¹
10⁻¹⁰
10⁻⁹
10⁻⁸
10⁻⁷
10⁻⁶
10⁻⁵
10⁻⁴
10⁻³
10⁻²
10⁻¹ FRACTION OF A CENTIMETER
1.0
10
10²
10³
10⁴
10⁵
MICROWAVE
TELEVISION, FM RADIO
SHORT WAVE
AM BROADCAST
FREQUENCY IN CYCLES PER SECOND
WAVELENGTH IN CENTIMETERS

Figure 55

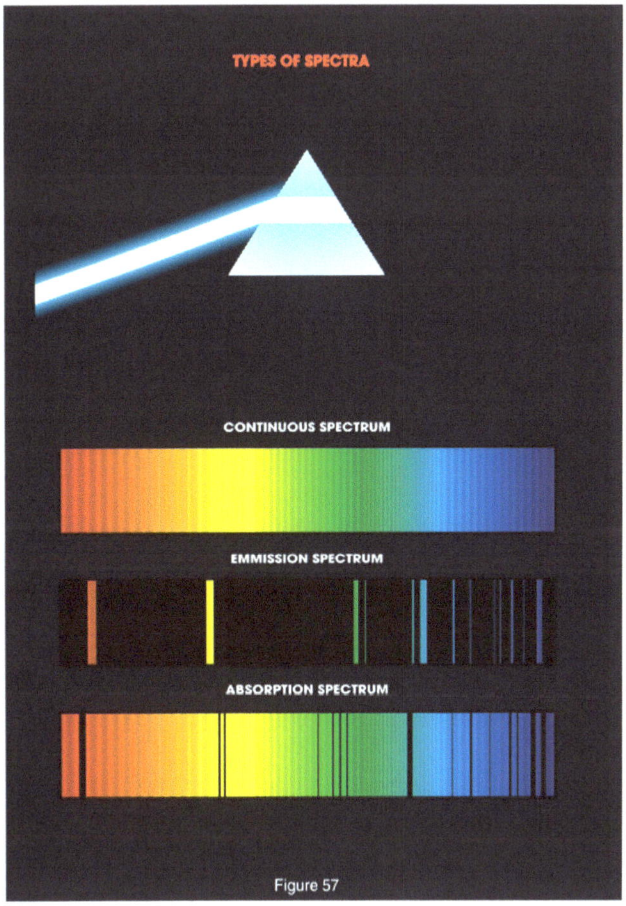

TYPES OF SPECTRA
CONTINUOUS SPECTRUM
EMMISSION SPECTRUM
ABSORPTION SPECTRUM

Figure 57

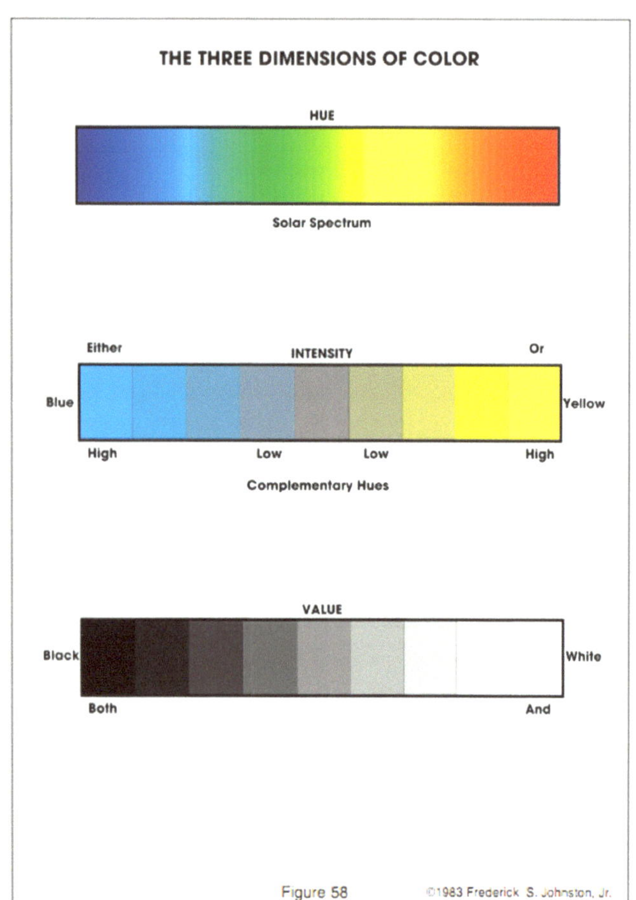

Figure 58

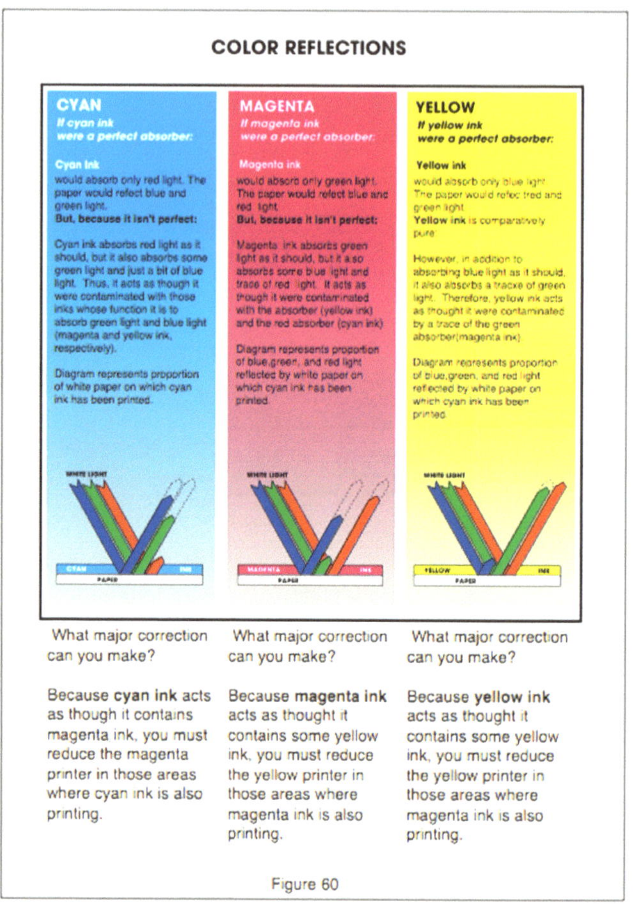

Figure 60

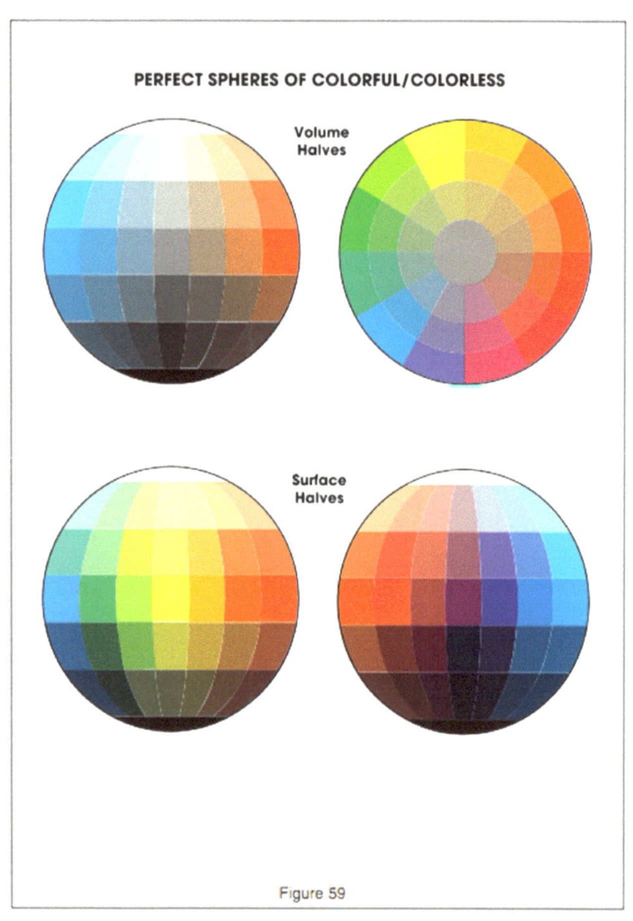

Figure 59

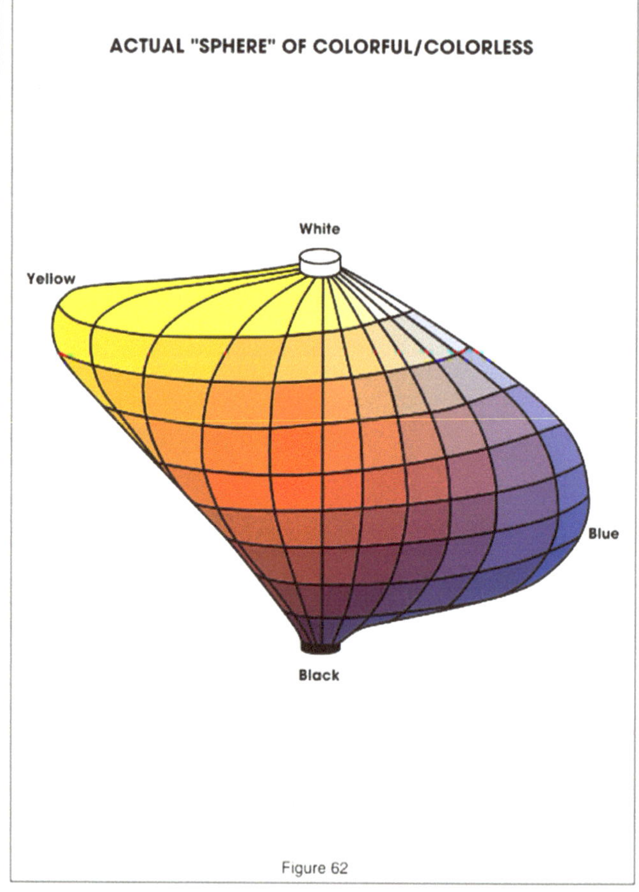

Figure 62

Encyclopaedia Britannica

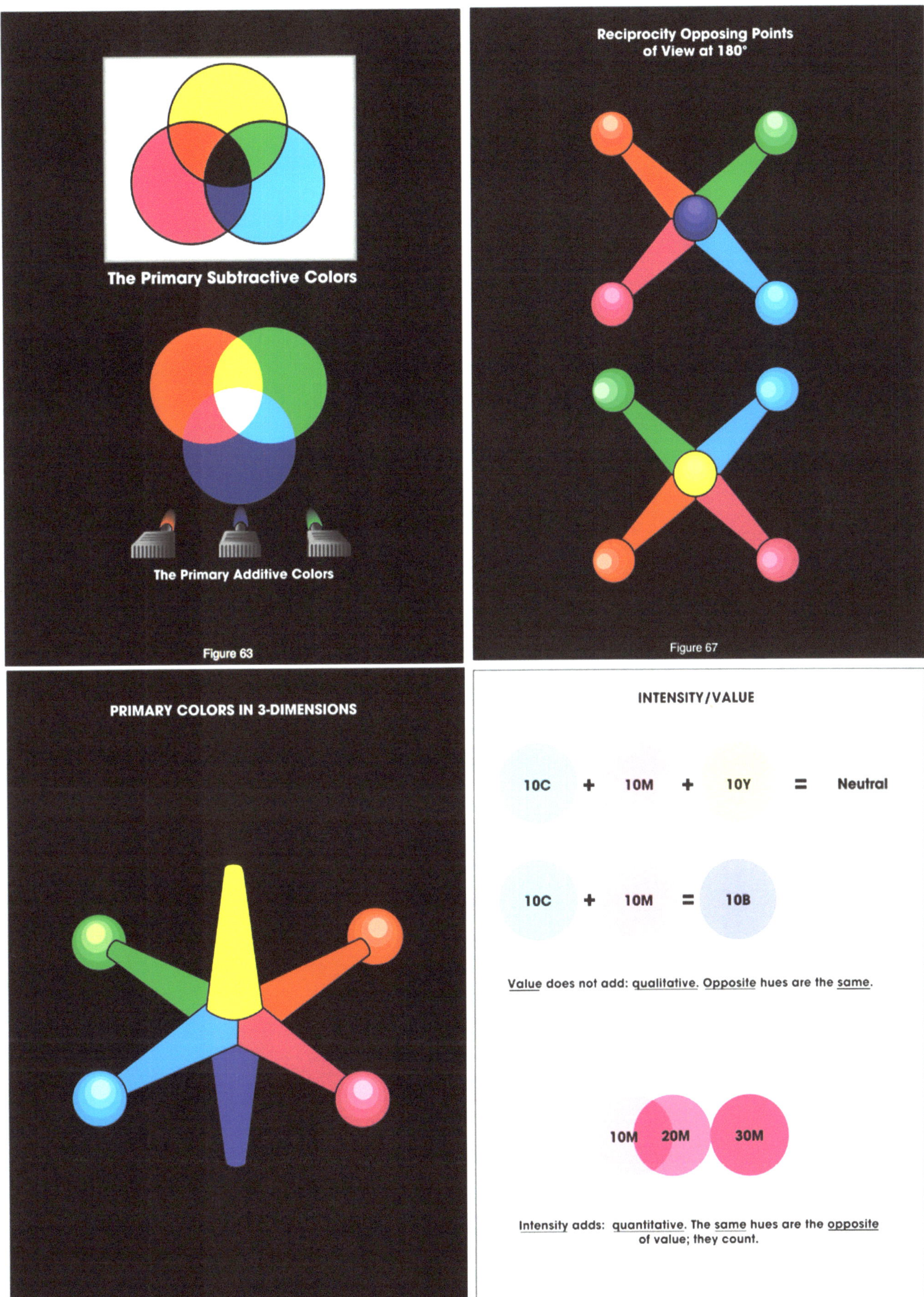

Figure 63

Figure 67

Figure 65

Figure 68

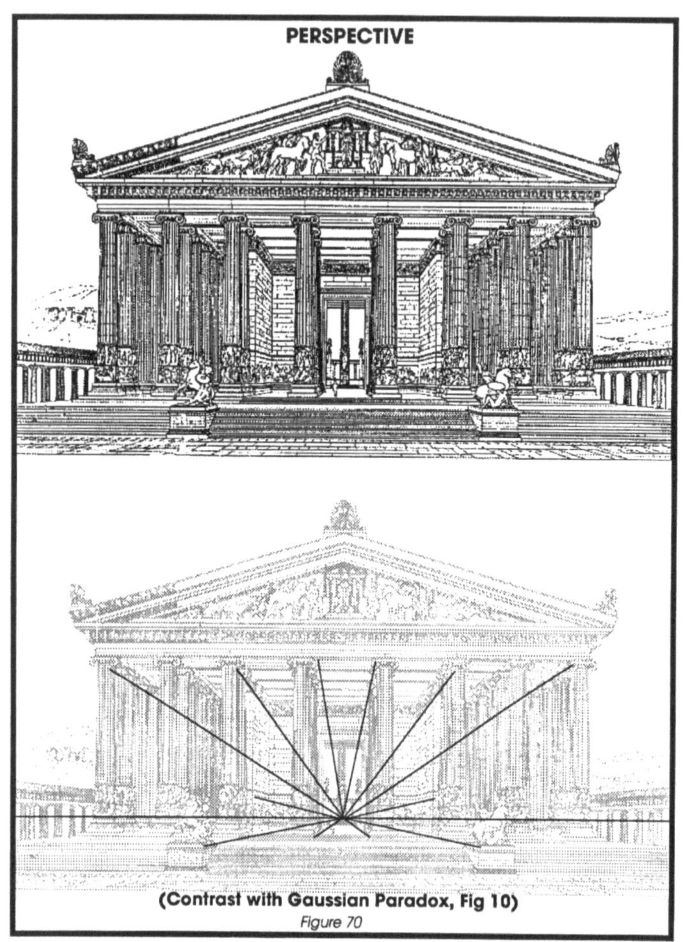

(Contrast with Gaussian Paradox, Fig 10)

Figure 70

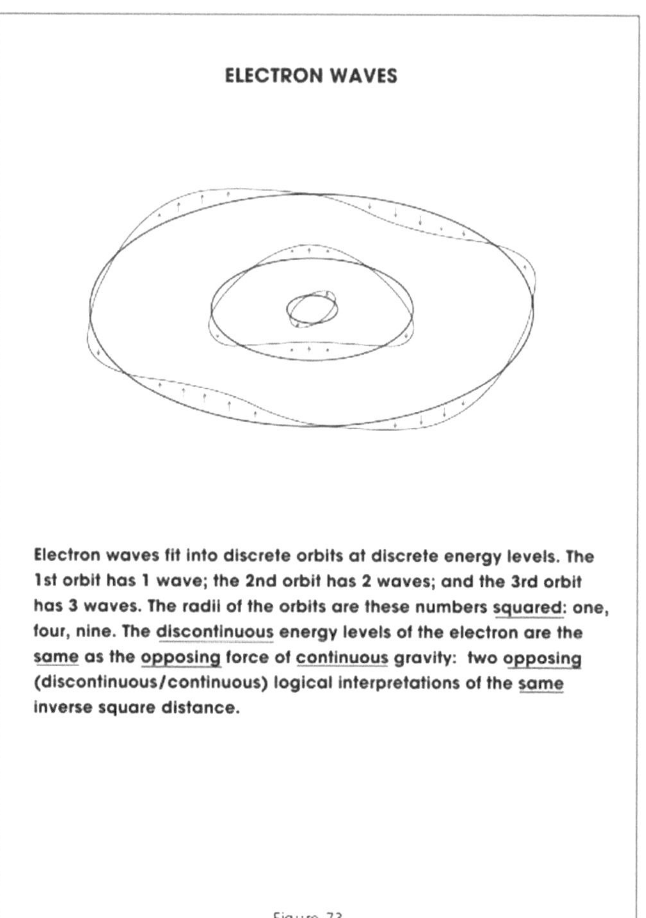

Figure 73

ELASTIC COLLISION BY RELECTION

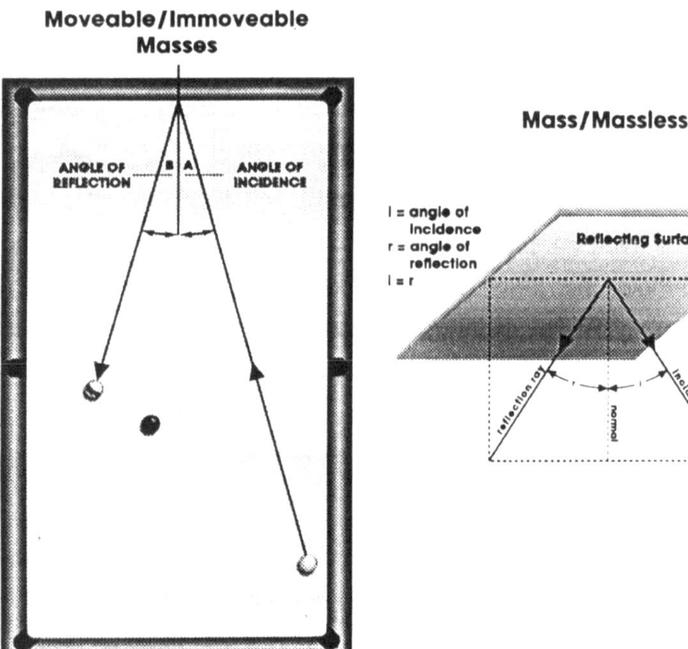

Laws of collision and laws of reflection are the same from the opposable idea of elasticity; or both are opposing parts of the same idea of elasicity.

Figure 71

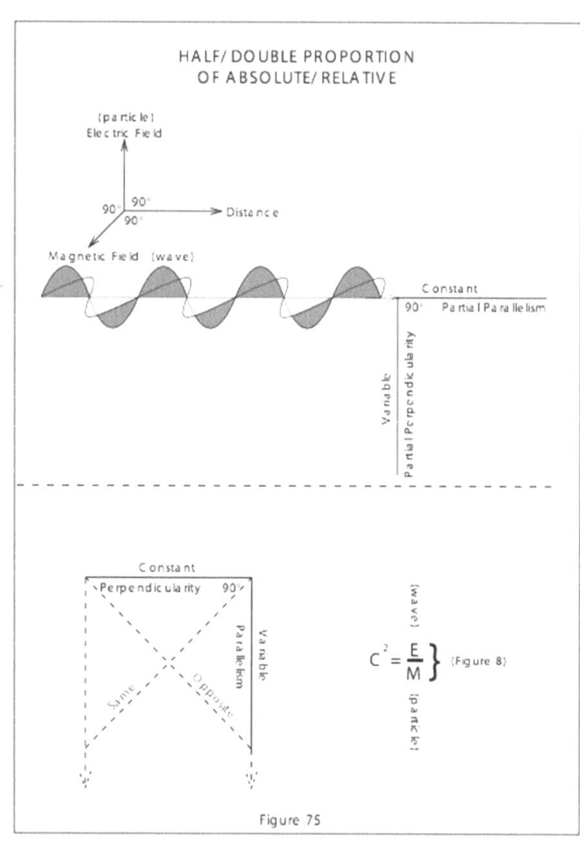

Figure 75

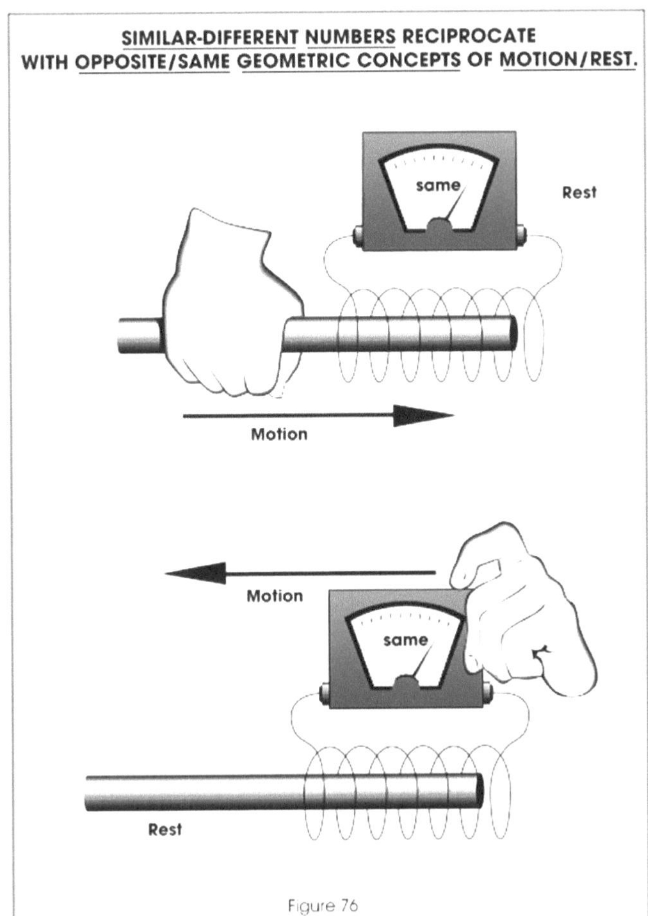

Figure 76

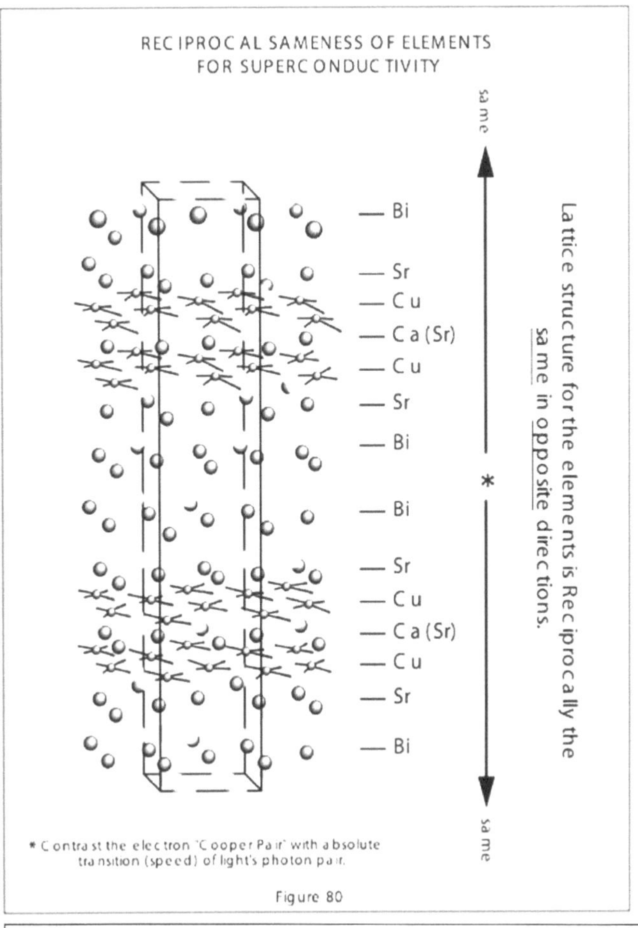

Figure 80

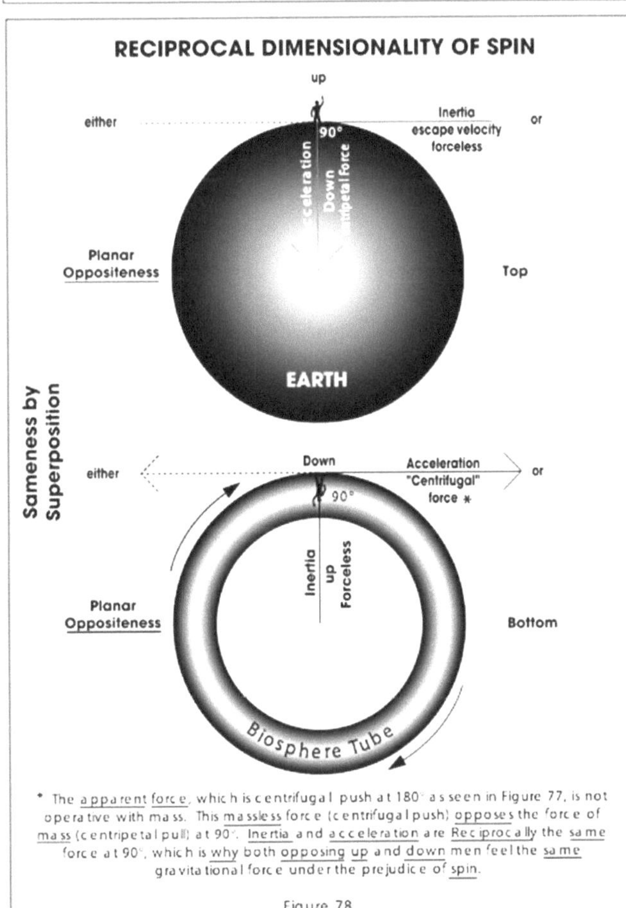

* The apparent force, which is centrifugal push at 180° as seen in Figure 77, is not operative with mass. This massless force (centrifugal push) opposes the force of mass (centripetal pull) at 90°. Inertia and acceleration are Reciprocally the same force at 90°, which is why both opposing up and down men feel the same gravitational force under the prejudice of spin.

Figure 78

ROMAN COIN OF JANUS

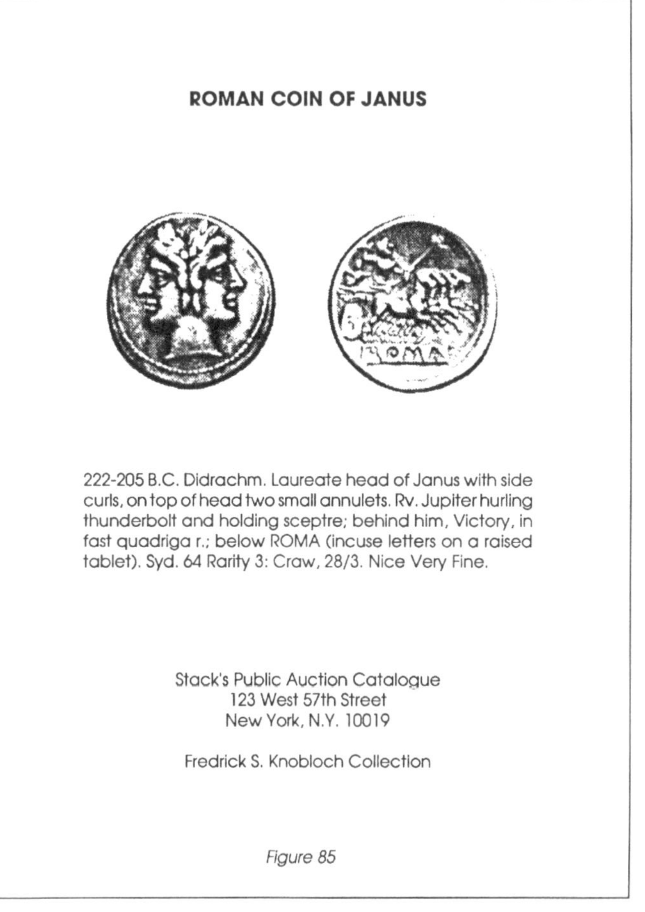

222-205 B.C. Didrachm. Laureate head of Janus with side curls, on top of head two small annulets. Rv. Jupiter hurling thunderbolt and holding sceptre; behind him, Victory, in fast quadriga r.; below ROMA (incuse letters on a raised tablet). Syd. 64 Rarity 3: Craw, 28/3. Nice Very Fine.

Stack's Public Auction Catalogue
123 West 57th Street
New York, N.Y. 10019

Fredrick S. Knobloch Collection

Figure 85

Figure 86

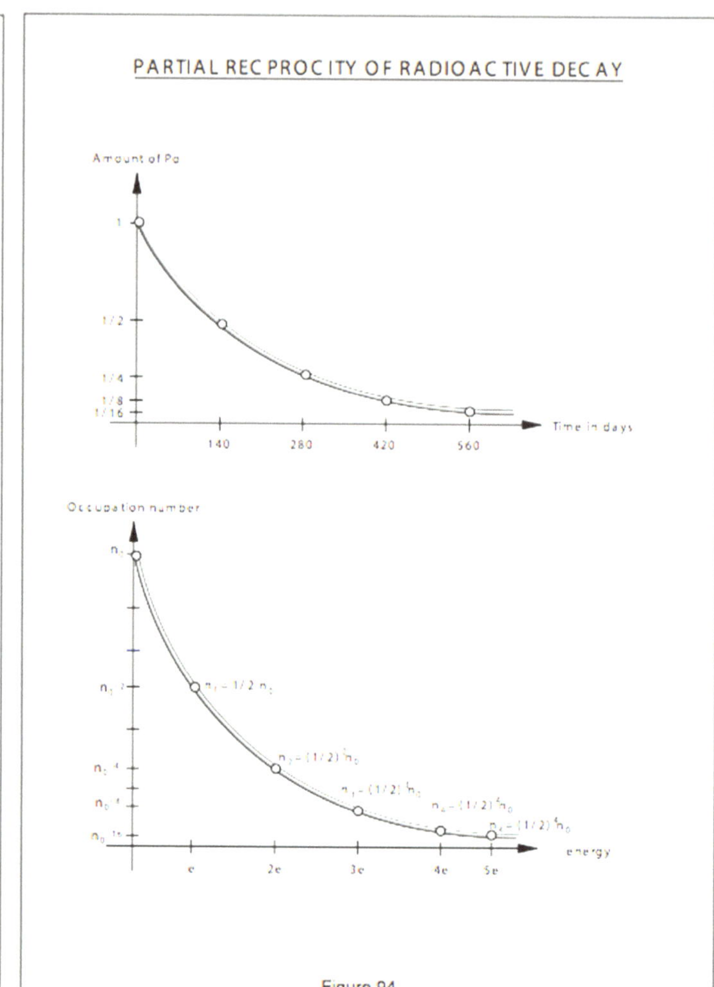

Figure 94

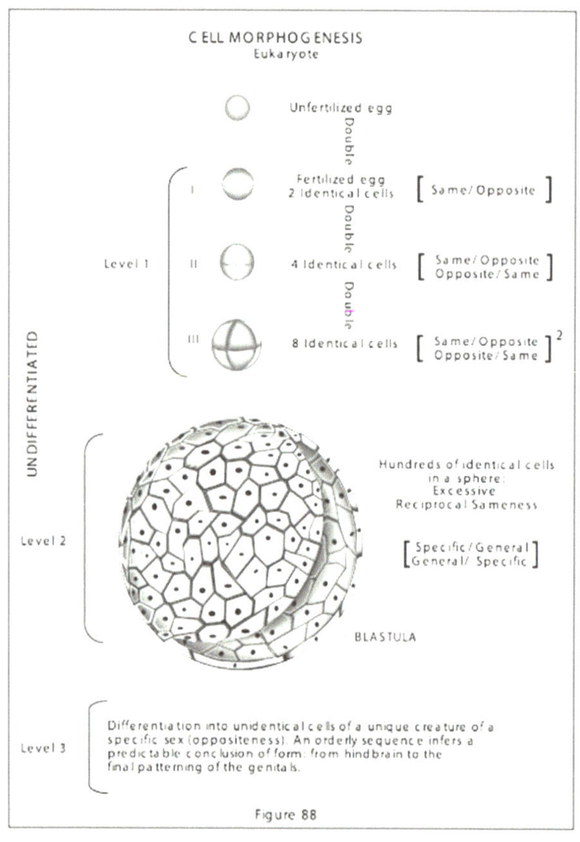

Figure 88

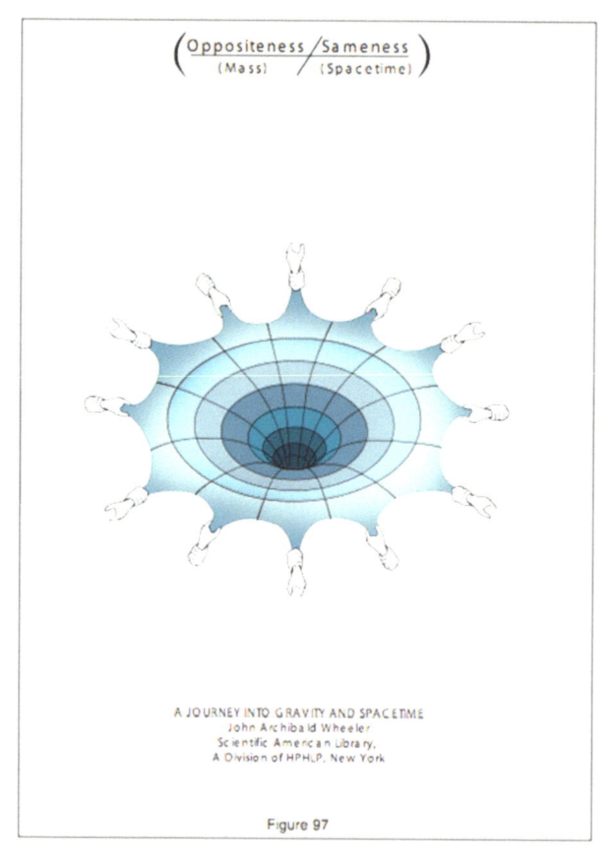

Figure 97

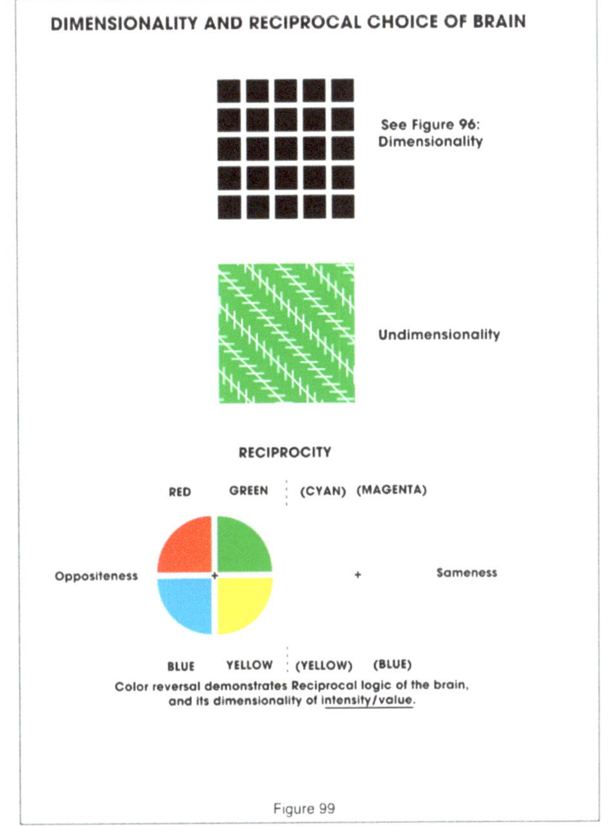

Figure 99

THE TRANSITIONAL CYCLOID

arc of A,B,C,D,E

* (Dynamic Interpretation of Figure 3's Mass Point of View)

* contrast with Figures 101 and 103

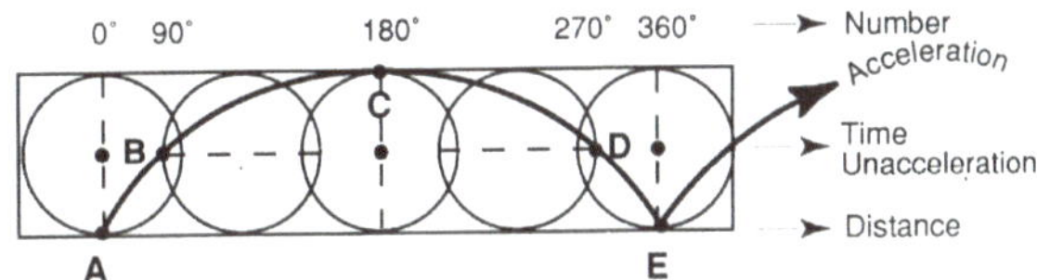

An infinite geometric progression of logarithmic proportion where the number e is the constant of change.

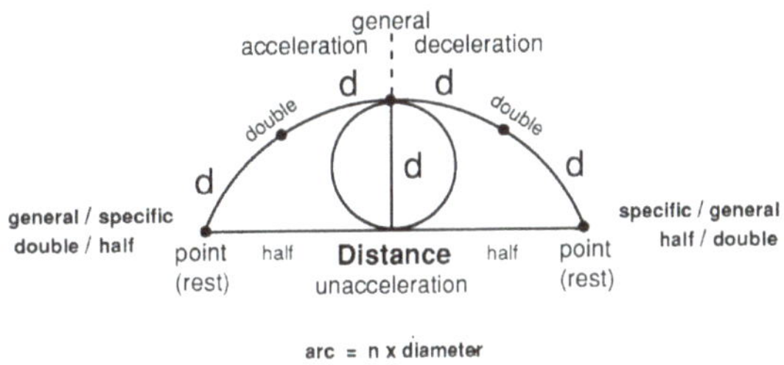

arc = n x diameter
Distance = 1/2 arc

Reciprocity of half / double area, unsquared.

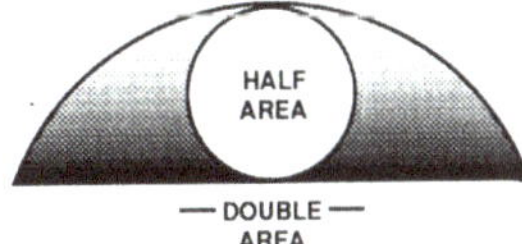

White circle is unaccelerated area (half).
Dark geometry is accelerated/decelerated area (double).

Figure 102

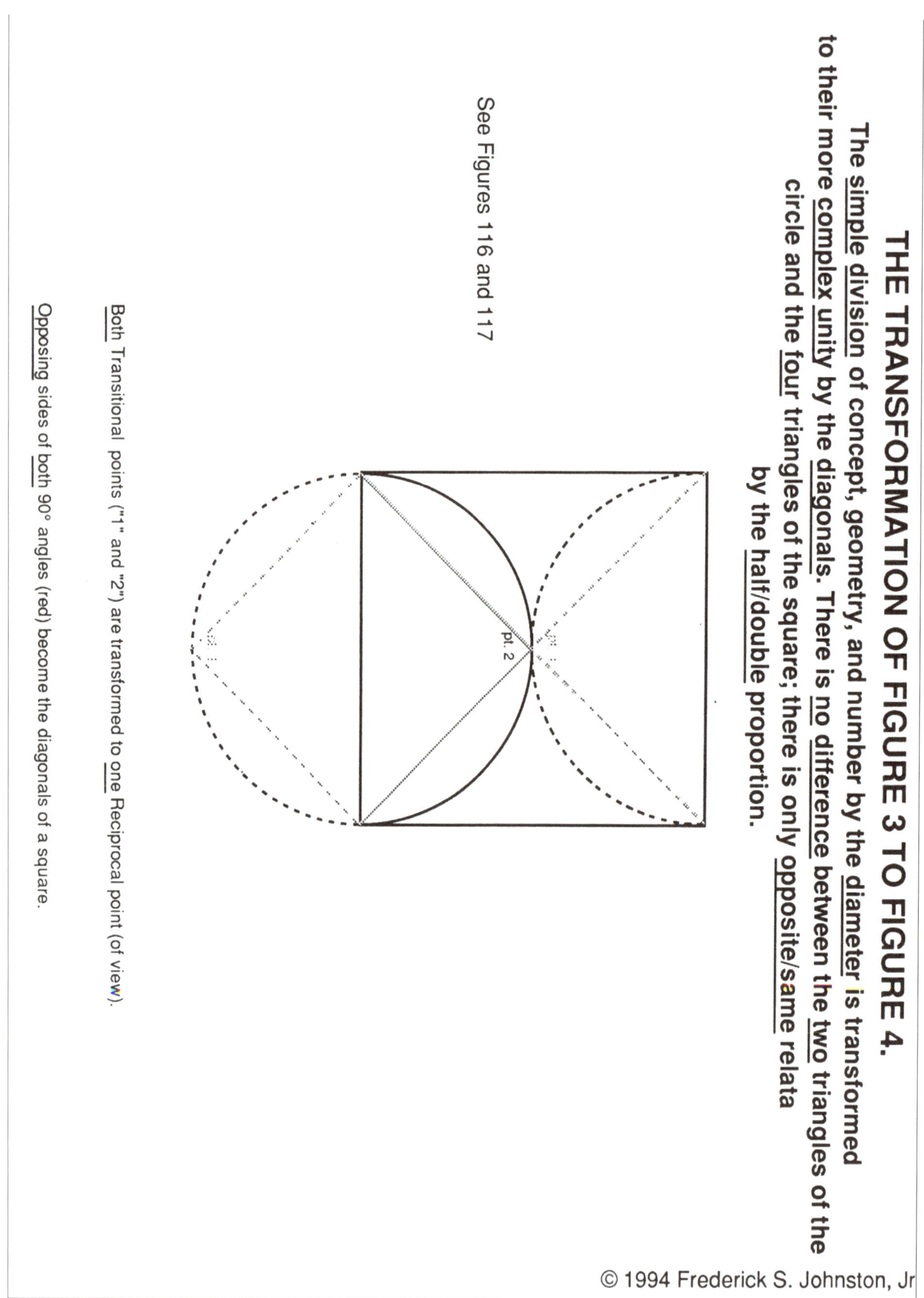
THE TRANSFORMATION OF FIGURE 3 TO FIGURE 4.
The simple division of concept, geometry, and number by the diameter is transformed to their more complex unity by the diagonals. There is no difference between the two triangles of the circle and the four triangles of the square; there is only opposite/same relata by the half/double proportion.
See Figures 116 and 117
pt. 2
Both Transitional points ("1" and "2") are transformed to one Reciprocal point (of view).
Opposing sides of both 90° angles (red) become the diagonals of a square.
© 1994 Frederick S. Johnston, Jr

Figure 104

THE FORCE OF GRAVITY
Action at a distance
d= 1/2 at²
Odd Even Odd Even Odd Even Odd Even
Time
in seconds squared on the real number line. See Figure 30.
1² 2² 3² 4² 5² 6² 7² 8²
Distance
in feet of successive squared numbers. See Figure 117.
(Half)
16 64 144 256 400 576 784 1024
Acceleration
in feet per second, per second.
(Double)
32 128 288 512 800 1152 1568 2048
Time squared makes distance and acceleration two reciprocally opposing (1/2 x 2/1) aspects of the gravitational force of sameness.
d= 1/2 at²
√1 → √4 → √9 → √16 → √25 → √36 → squares . . .
Mass Induction
from the specific to the general conclusion
expansion from a point
Gravity
Weaker
d= 2/1 at²
1² → 2² → 3² → 4² → 5² → 6² → squares. . .
Mass Deduction
from the general to the specific conclusion
contraction to a point
Gravity
Stronger
See Figures 8, 106 and 107.

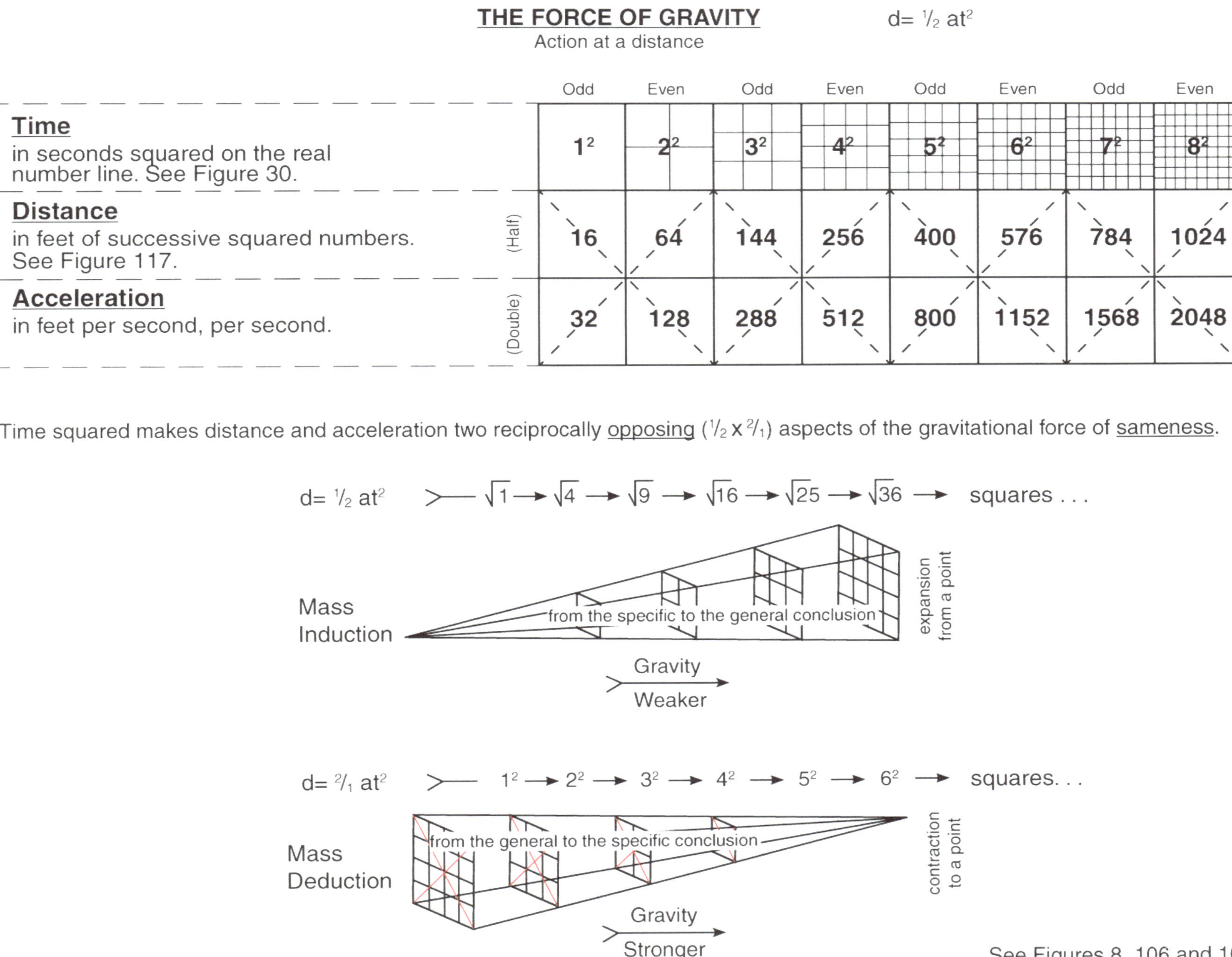

Figure 105

* SIMPLE MAGIC SQUARES

Reciprocal Sameness of Number, Geometry, & Concept.
(why the square is "magic")
Half the Distance from Figure 107

(specific/general)^2
-full / -less

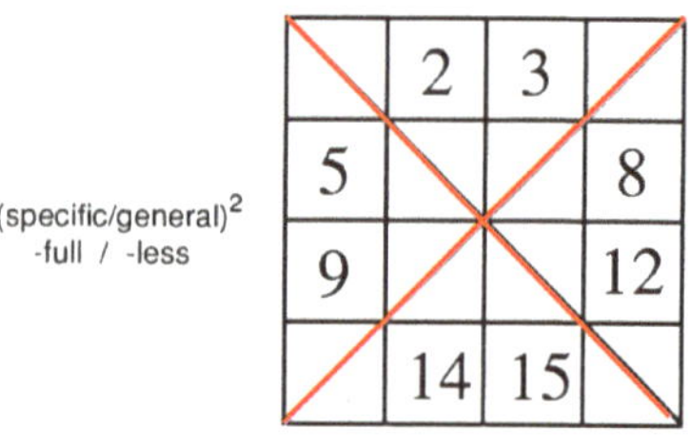

Numbers — parallel
Numberless — perpendicular

*** 90° to Figure 105 top**

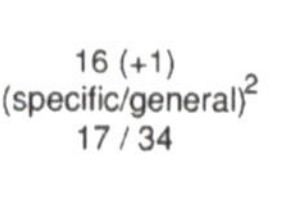

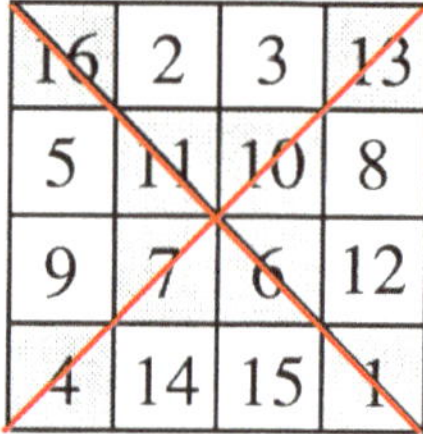

Numberless diagonals filled with successive numbers in reverse sequence

Figure 106

* COMPLEX MAGIC SQUARES

Enhancement of Complexity by Conserving the Simple Logical Order by Doubling the Distance from Figure 106

(specific/general)^2

	2	3			6	7	
9			12	13			16
17			20	21			24
	26	27			30	31	
	34	35			38	39	
41			44	45			48
49			52	53			56
	58	59			62	63	

***90° to Figure 105 top**

8^2 (+1)

64 (+1)
(specific/general)^2
65 /130

64	2	3	61	60	6	7	57
9	55	54	12	13	51	50	16
17	47	46	20	21	43	42	24
40	26	27	37	36	30	31	33
32	34	35	29	28	38	39	25
41	23	22	44	45	19	18	48
49	15	14	52	53	11	10	56
8	58	59	5	4	62	63	1

Figure 107

THE LOGICAL GEOMETRY OF PRIME NUMBERS ON THE PLANE OF COMPLEX NUMBERS

GAUSSIAN PRIMES

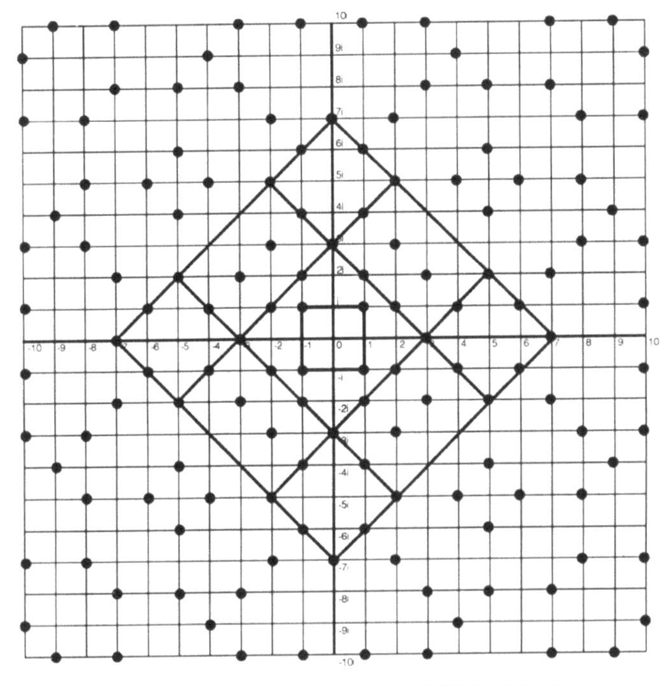

Unique prime numbers cannot be excluded from geometry.

What is the difference between the shaded/unshaded squares of these Gaussian Primes and the shaded/unshaded "magic squares" of Figure 107 bottom?

Figure 109

THE GEOMETRY OF PRIME NUMBERS SPIRALING ON A SIMPLE PLANE

100	99	98	97	96	95	94	93	92	91
65	64	63	62	61	60	59	58	57	90
66	37	36	35	34	33	32	31	56	89
67	38	17	16	15	14	13	30	55	88
68	39	18	5	4	3	12	29	54	87
69	40	19	6	1	2	11	28	53	86
70	41	20	7	8	9	10	27	52	85
71	42	21	22	23	24	25	26	51	81
72	43	44	45	46	47	48	49	50	83
73	74	75	76	77	78	79	80	81	82

Figure 111

UNSQUARED GEOMETRY OF REAL NUMBERS

(Momentum Infers a different Sum, Rather than Reciprocally the Same Squared and Accelerated Sum)

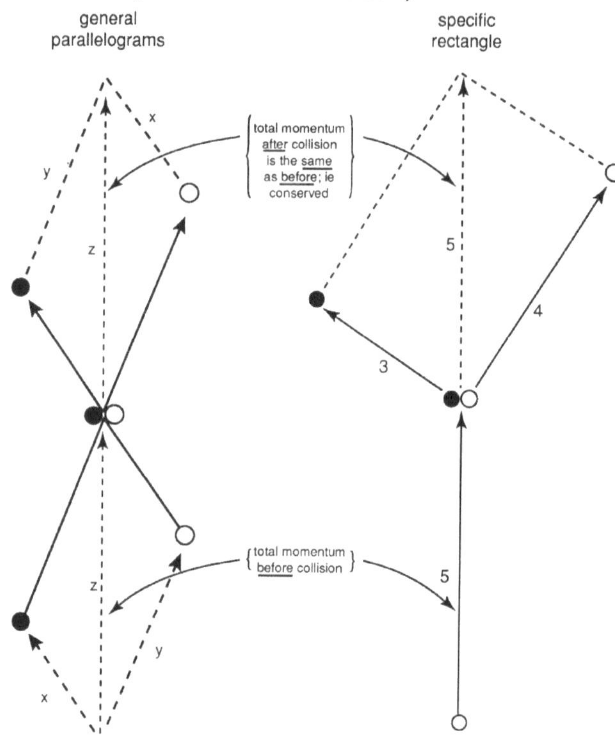

Contrast these unsquared sequences of real numbers that infer by parallelograms, with the squared simultaneity of natural numbers that Reciprocate by diagonals.

Figure 112

INVERSE SQUARE LAW OF "FREE FALL"

(Time in 1/4 seconds)

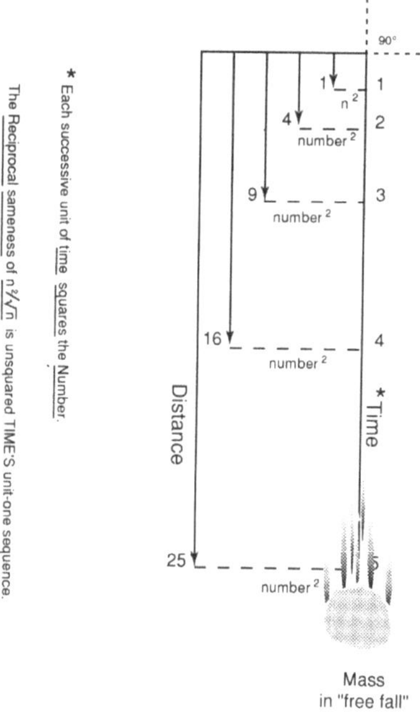

Figure 113

HARMONY OF FREQUENCIES

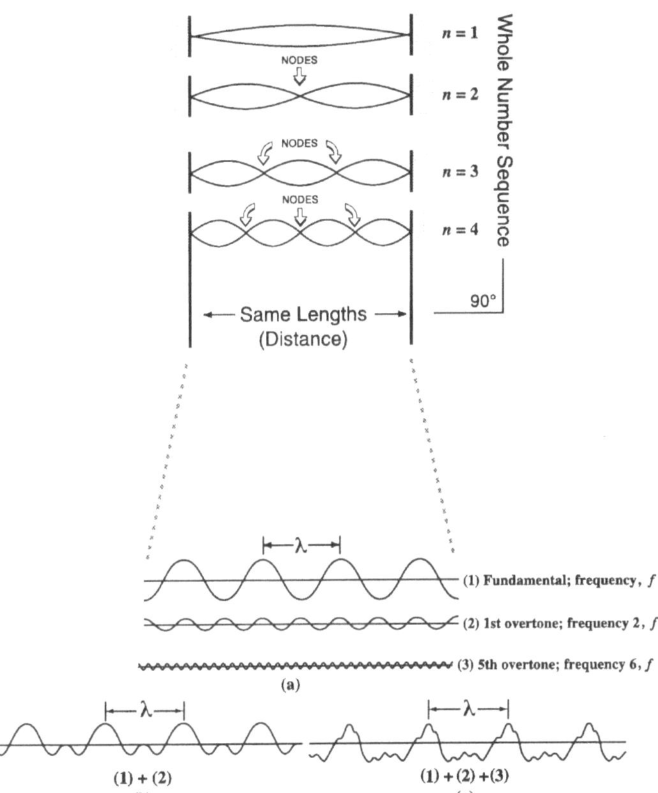

Figure 119

INTERVALS OF THE MUSICAL STAFF
(Unit-one Sequence)

(time/space Reciprocity)

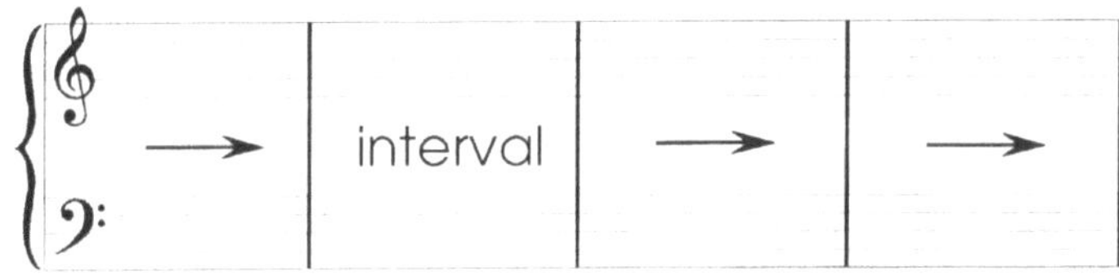

Each interval is an area of constraint for musical notation in time's unit-one periodicity.

Contrast with Figure 101

Figure 120

OCTAVES ON THE PIANO KEYBOARD
(Fundamental Harmony by doubling the distance)

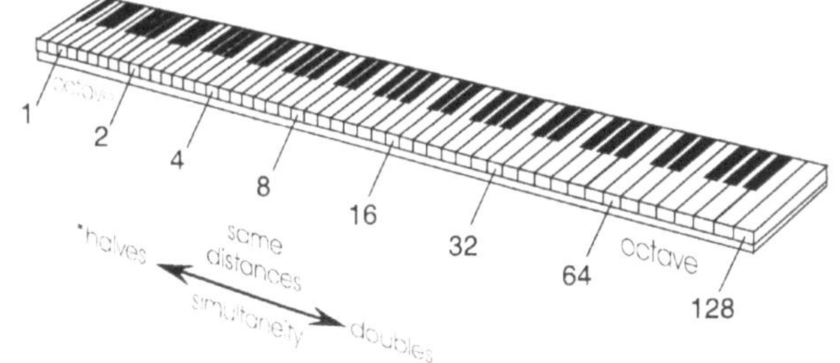

* These halvings are hidden by the expansion to higher frequences up the scale (right), which are harmonious by successive doubling of distance.

Figures 120 & 121

90° DISTANCE / ACCELERATION
(half/double)

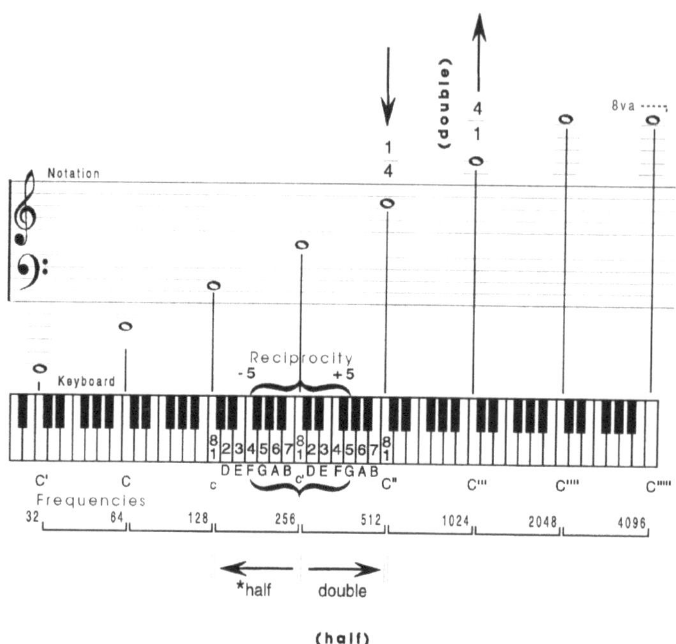

* This half is hidden from doubles's expansion to higher frequences from any note on the scale.

Figure 122

HARMONIOUS NOTES
(PIANO SCALE'S LOGICAL SEQUENCE)

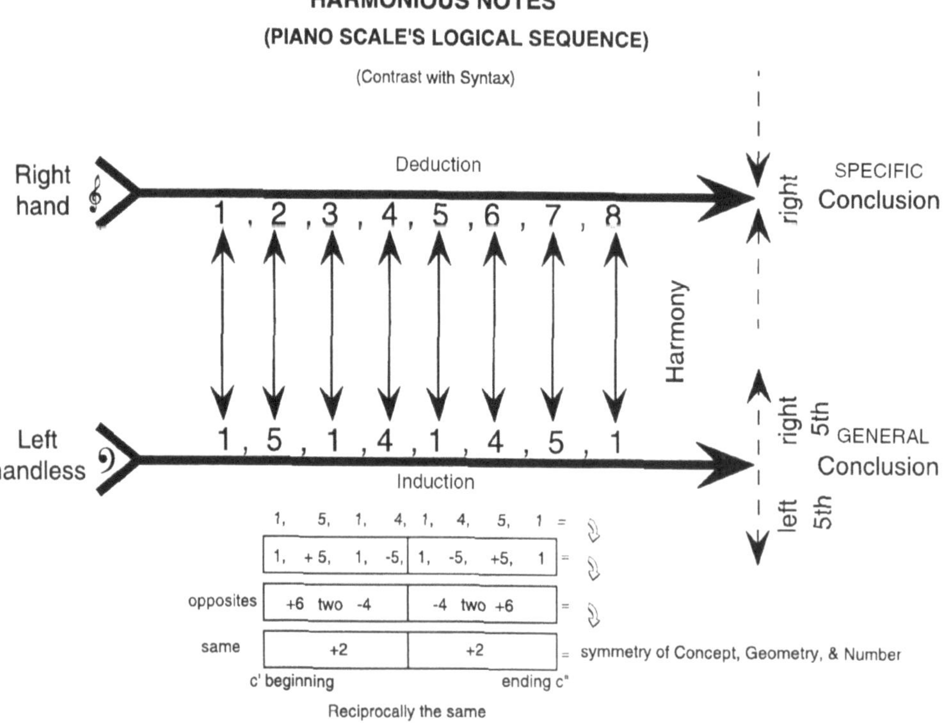

Figure 123

131

132

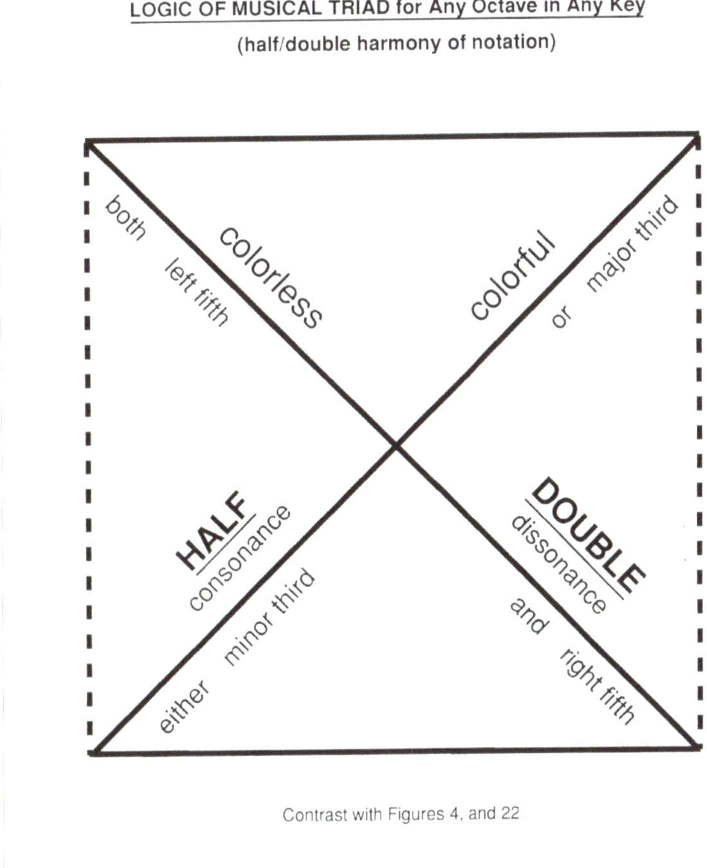

Figure 124

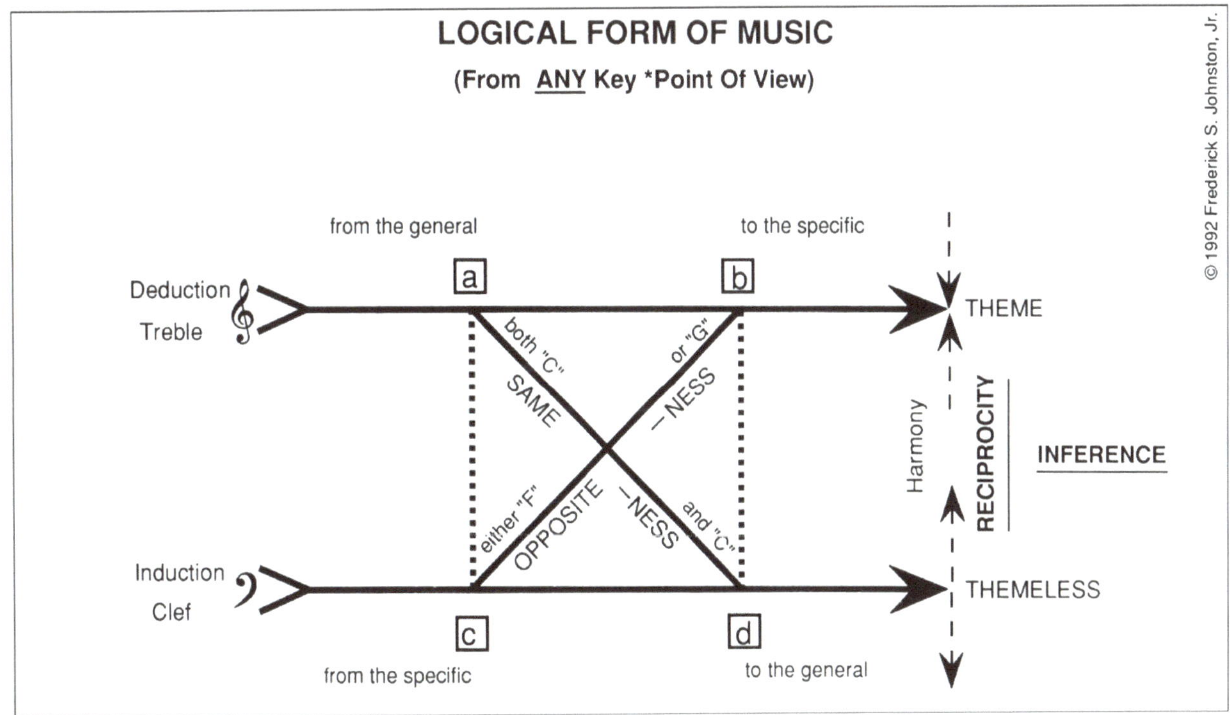

Figure 125

COLOR BALANCE

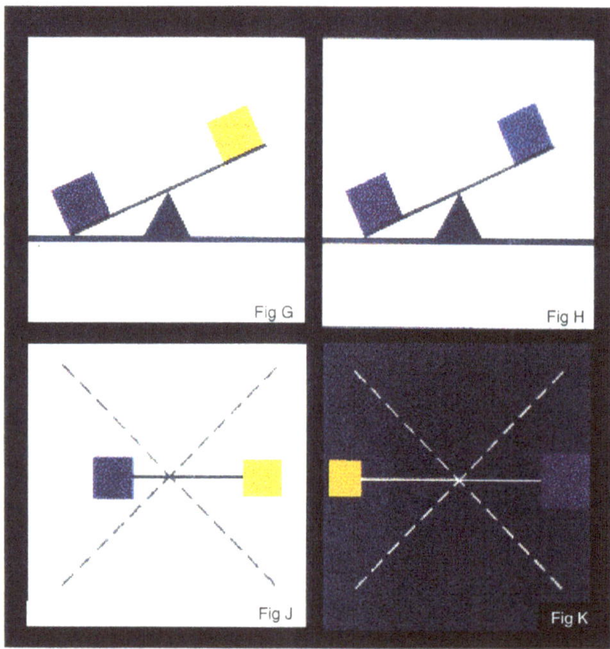

Fig.G—Two colors of equal size and of approximately the same intensity vary in "weight" due to difference in their value dimensions. Blue, being the darker in value, has a greater degree of contrast against a light background and is the dominant color.

Fig. H—Two colors of equal size and of approximately the same value vary in "weight" due to differences in their intensity. The more intense color has a greater ground and is the dominant color.

Fig J—Blue, the dominant color because of its greater contrast in value against a light background, tends to balance a color of equal size and less contrast through adjustments in position within the picture area.

Fig K—Yellow, the dominant color because of its greater contrast in value against a dark background, tends to balance a color of lesser contrast through adjustments in size, position, or both within the picture area.

Figure 126

COLOR INFRARED FILTER

Reprinted courtesy of Eastman Kodak Co., Rochester, N.Y.

Figure 127

Figure 128

PERCEPTUAL BALANCE/IMBALANCE
For Dynamic Interpretation

The Golden ratio: C divides the segment AB such that the whole segment is to the large part as the large part is to the small. If the whole segment is of unit length, we have $1/x = x/(1-x)$. This leads to the quadratic equation $x^2+x-1=0$, whose positive solution is $x=(-1+\sqrt{5})/2$, or about 0.61803. The golden ratio is the reciprocal of this number, or about 1.61803. Which segment is <u>specific</u> and which is <u>general</u>?

Figure 130

THE GOLDEN RECTANGLE

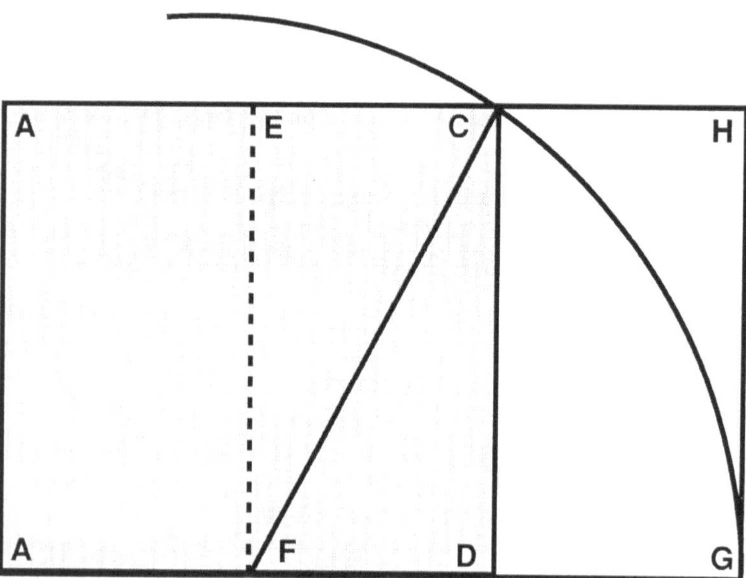

The geometric construction of a Golden Rectangle begins with a square, which is then divided in two equal parts by the dotted line EF. Point F now serves as the center of a circle whose radius is the diagonal FC. An arc of the circle is drawn (CG) and the base line AD is extended to intersect it. This becomes the base of the rectangle. The new side HG is now drawn at right angles to the new base, with the line BH brought out to meet it. The resultant Golden Rectangle has one unusual property: if the original square is taken away, what remains will still be a Golden Rectangle.

Figure 131

TIME AND AREA (SPATIAL PROPORTION) ARE RECIPROCALY THE SAME

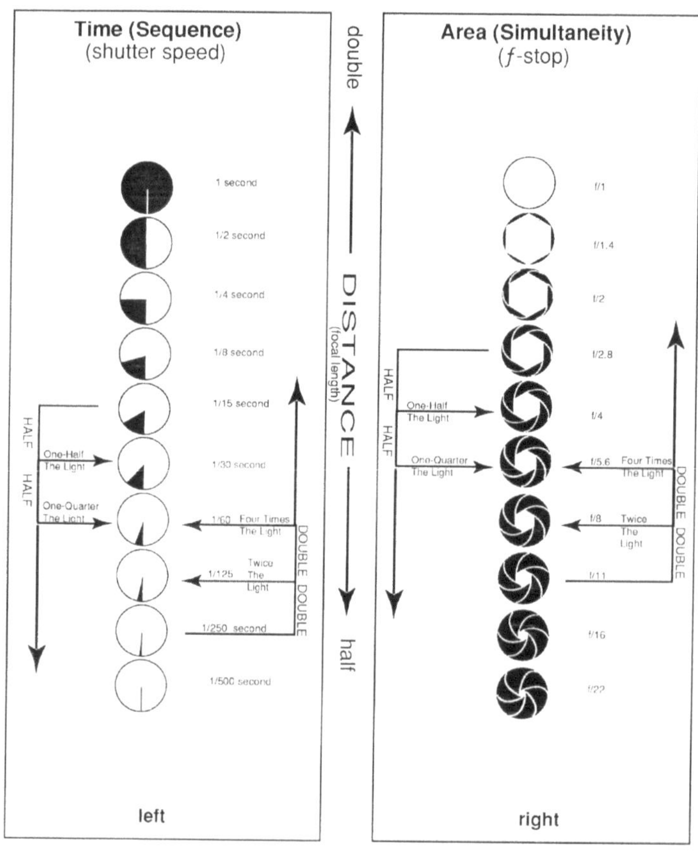

Figure 135

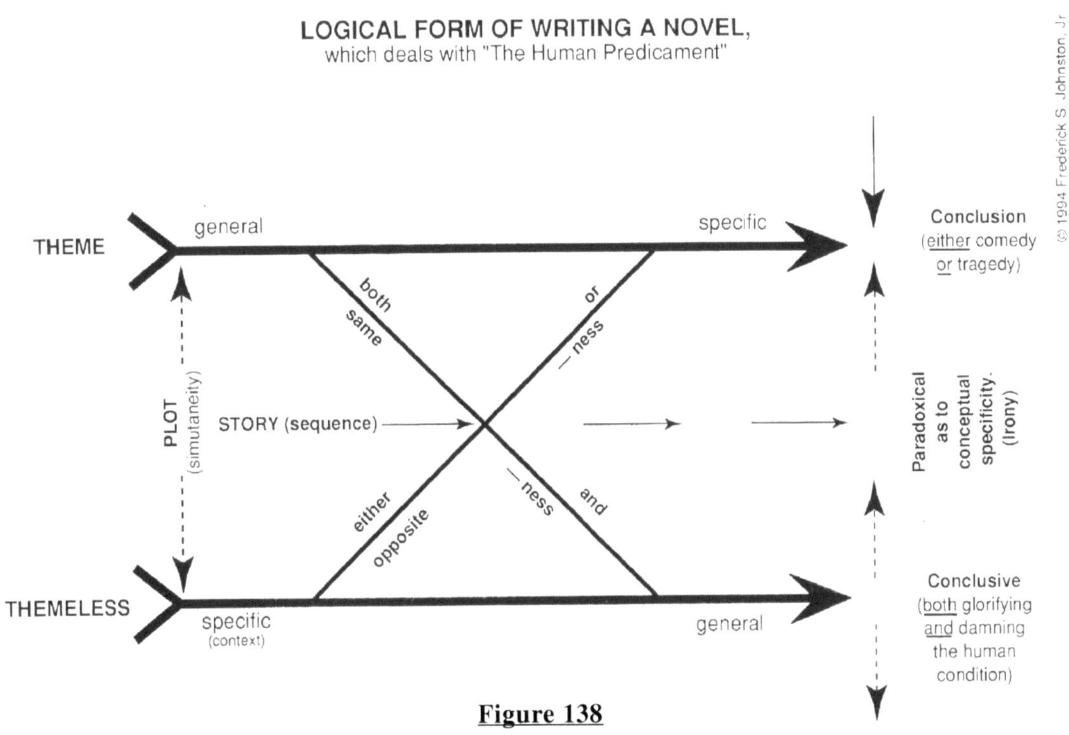

Figure 138

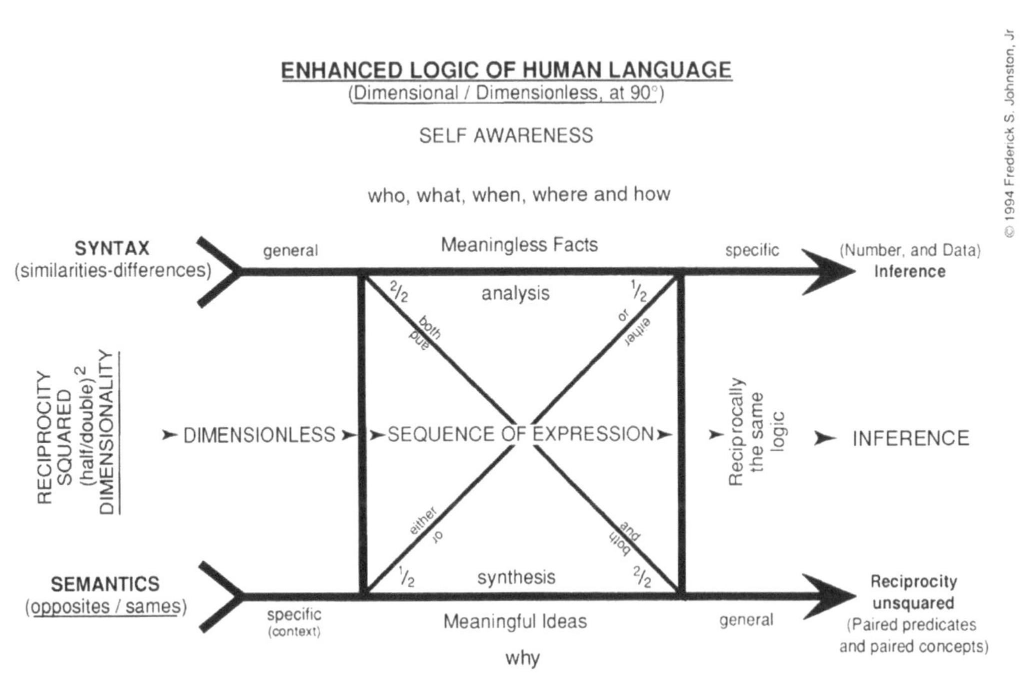

Figure 139

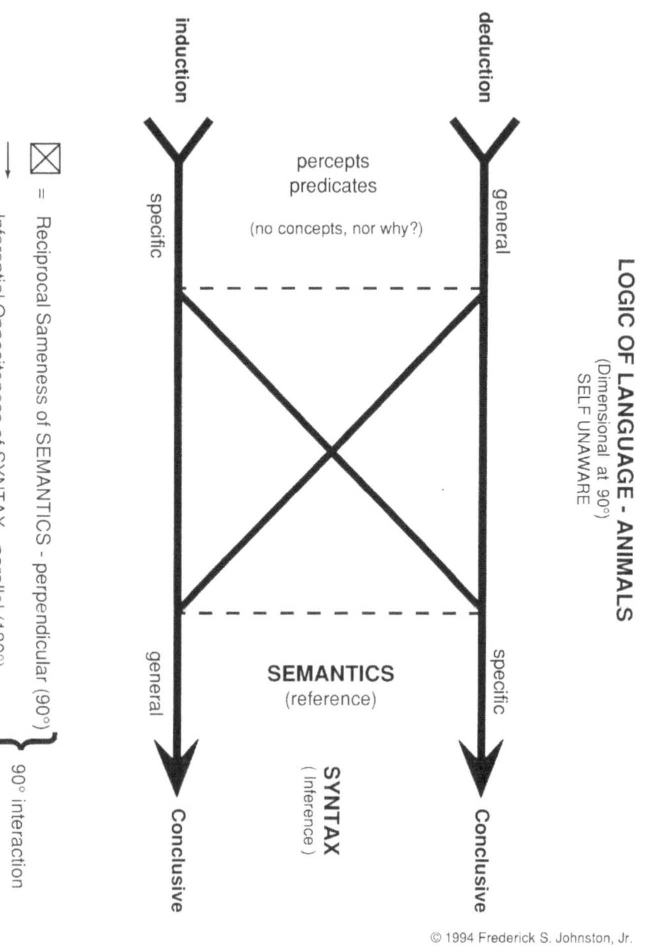

Figure 140

THE CALCULUS ARCH

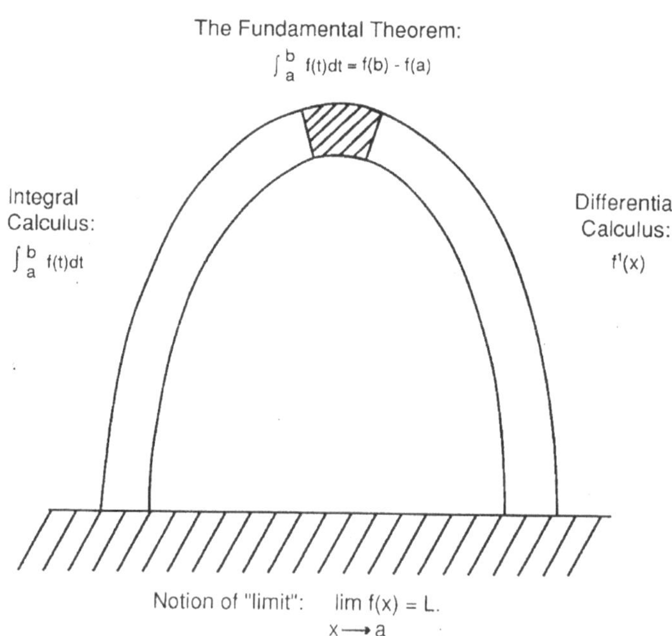

Figure 150

MENDELEEV'S PERIODIC TABLE OF THE ELEMENTS
(See Figure 149)

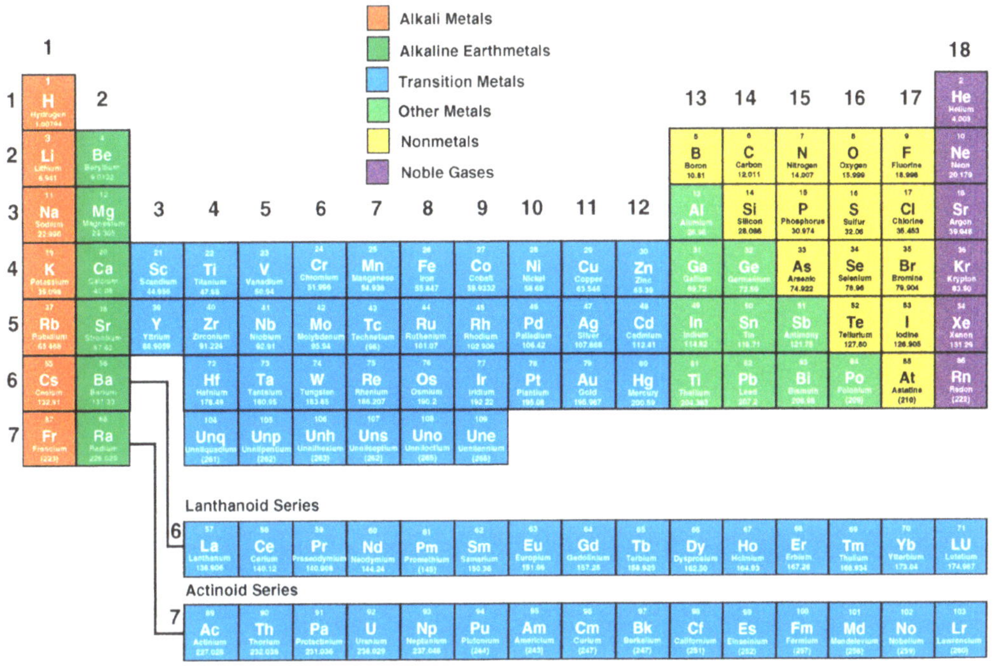

FIGURE 152

EARLY SEQUENCE OF POLYCHAETE WORM CELL DIVISION

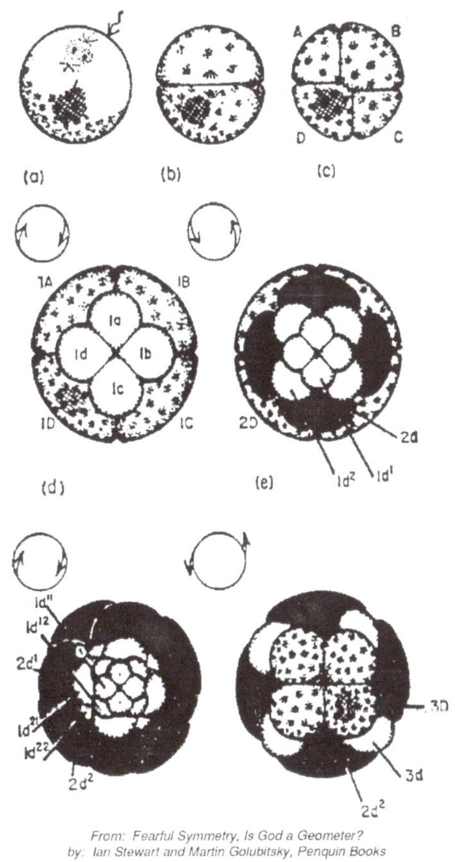

From: Fearful Symmetry, Is God a Geometer?
by: Ian Stewart and Martin Golubitsky, Penquin Books

Figure 153-a

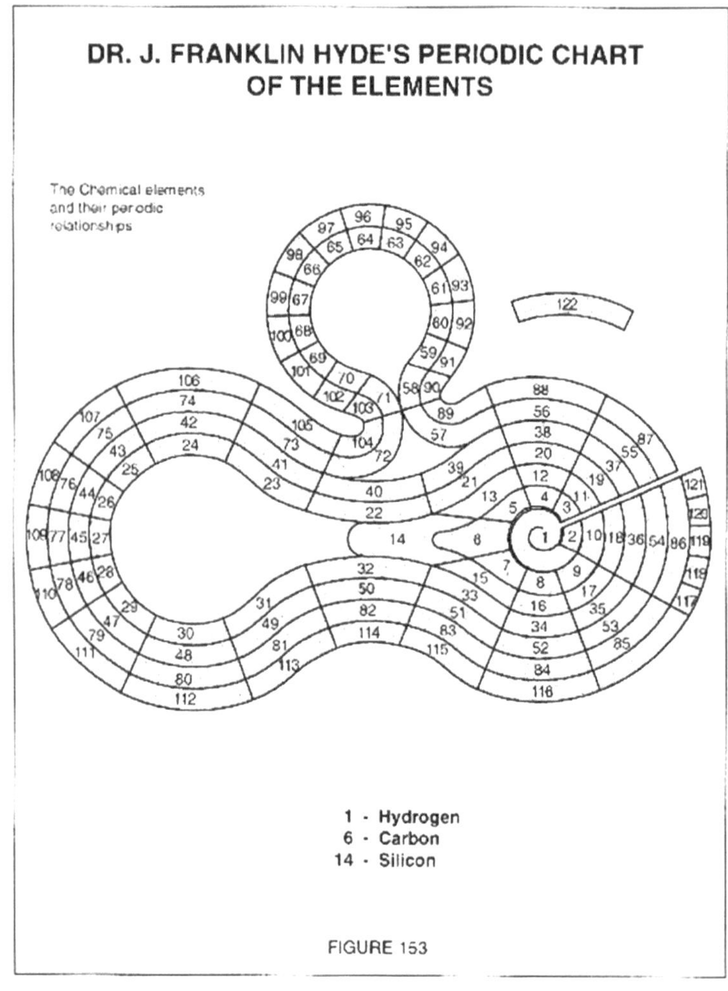

FIGURE 153

Reciprocal Functions of Sink-Faucet Handles

Right Handed
Simple Oppositeness

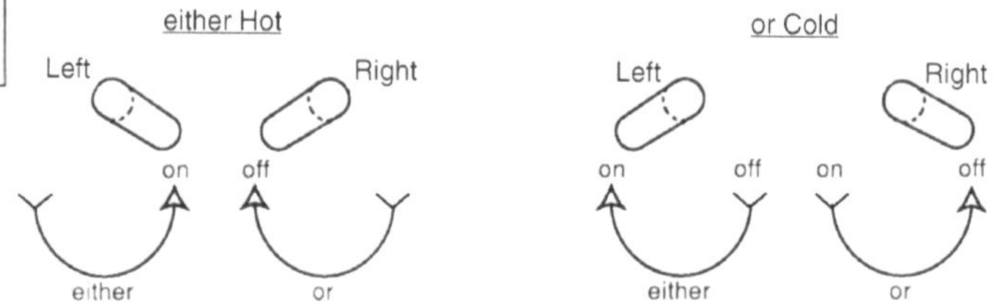

Right Handed
Complex Oppositeness

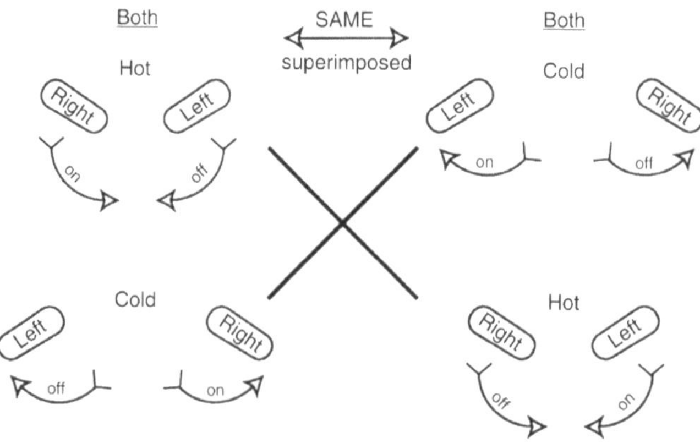

Figure 154-a

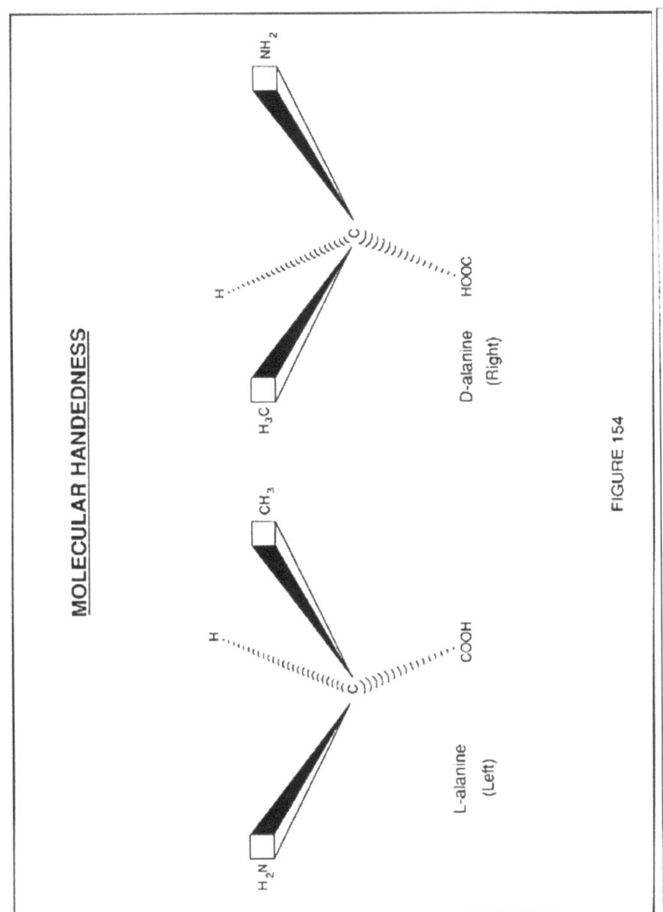

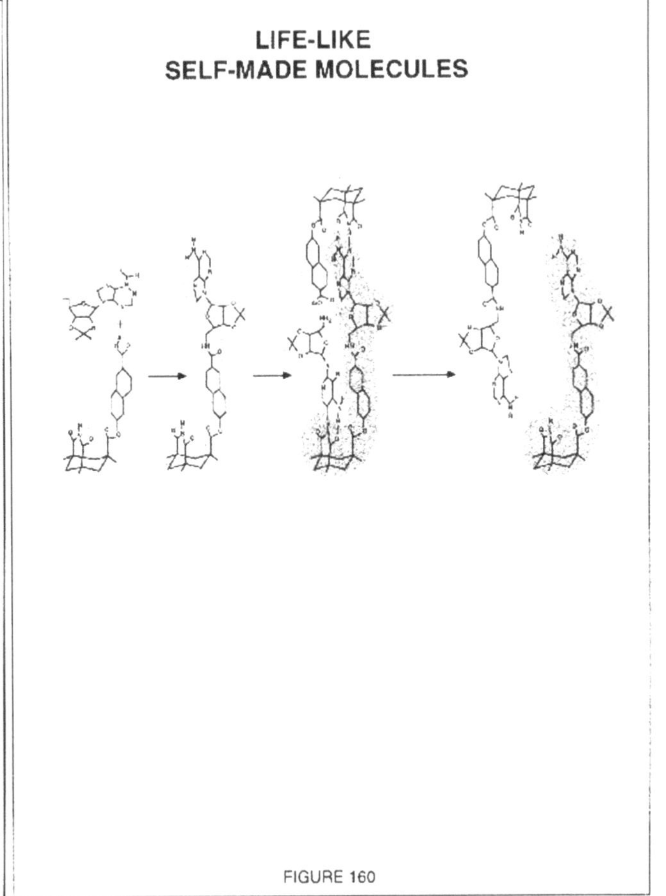

FIGURE 160

Figure 154-b

DNA

Reprinted with permission from George V. Kelvin.
Gary Felsenfeld, Scientific American, pp.60, October 1985

FIGURE 161

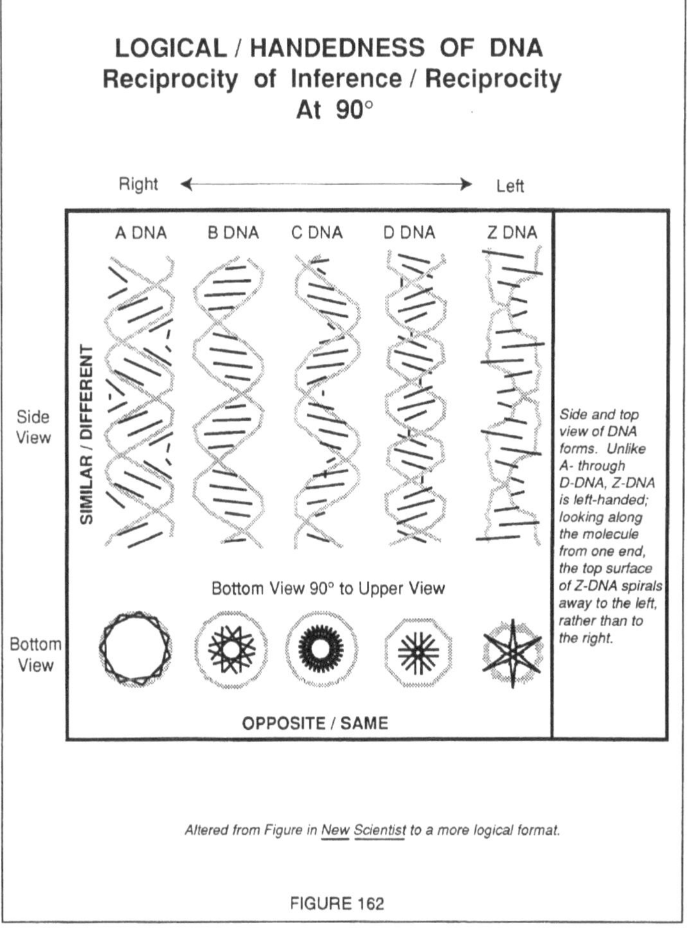

FIGURE 162

Reciprocal form of
DNA
Nucleotides
OPPOSITE / SAME
90°
SIMILAR - DIFFERENT
A
T
G
C
T
A
C
G
G
C
T
A
T
A
A
T
FIGURE 163

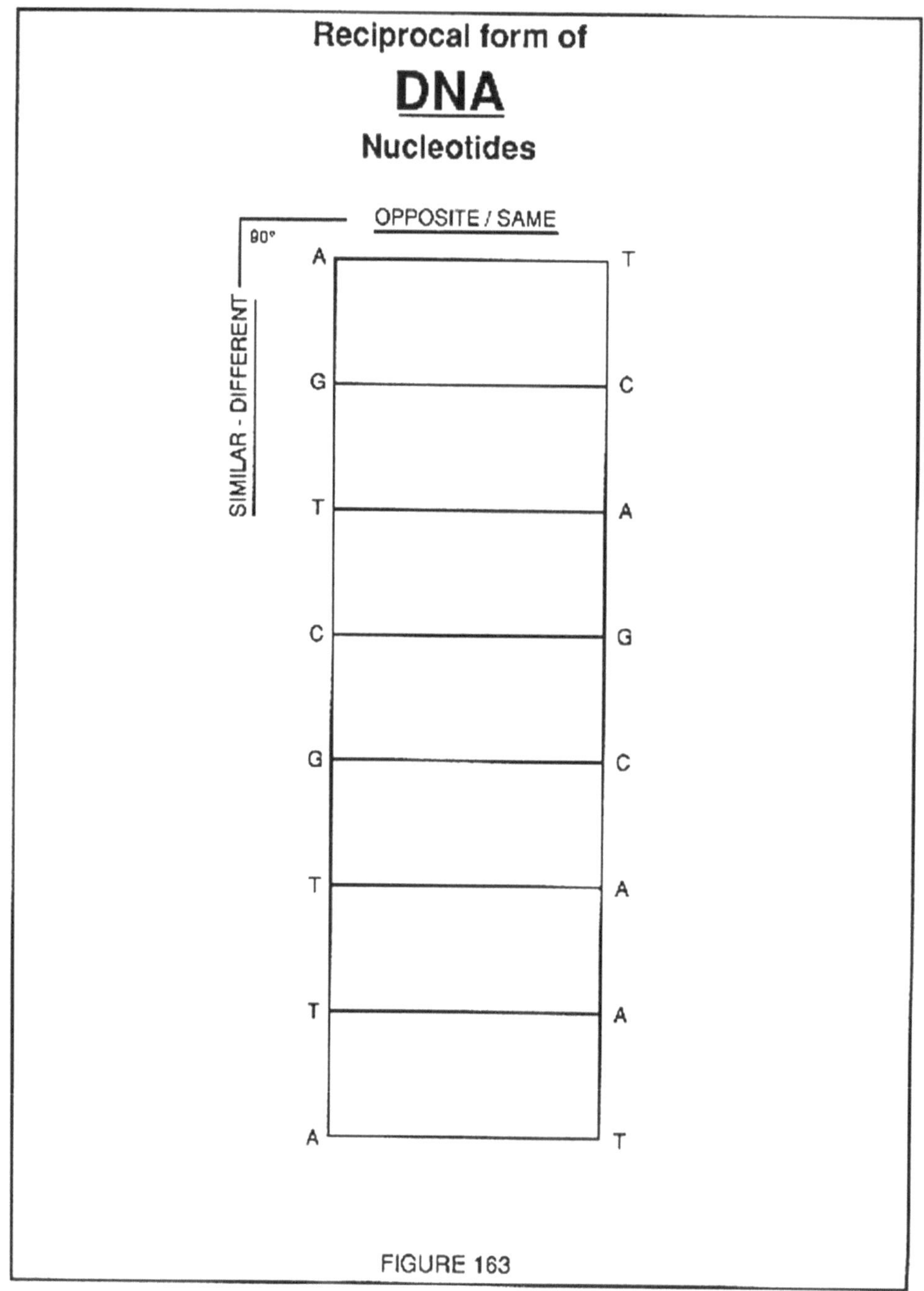

CONTRAST / COMPARISON OF CRYSTAL FORM AT 90°

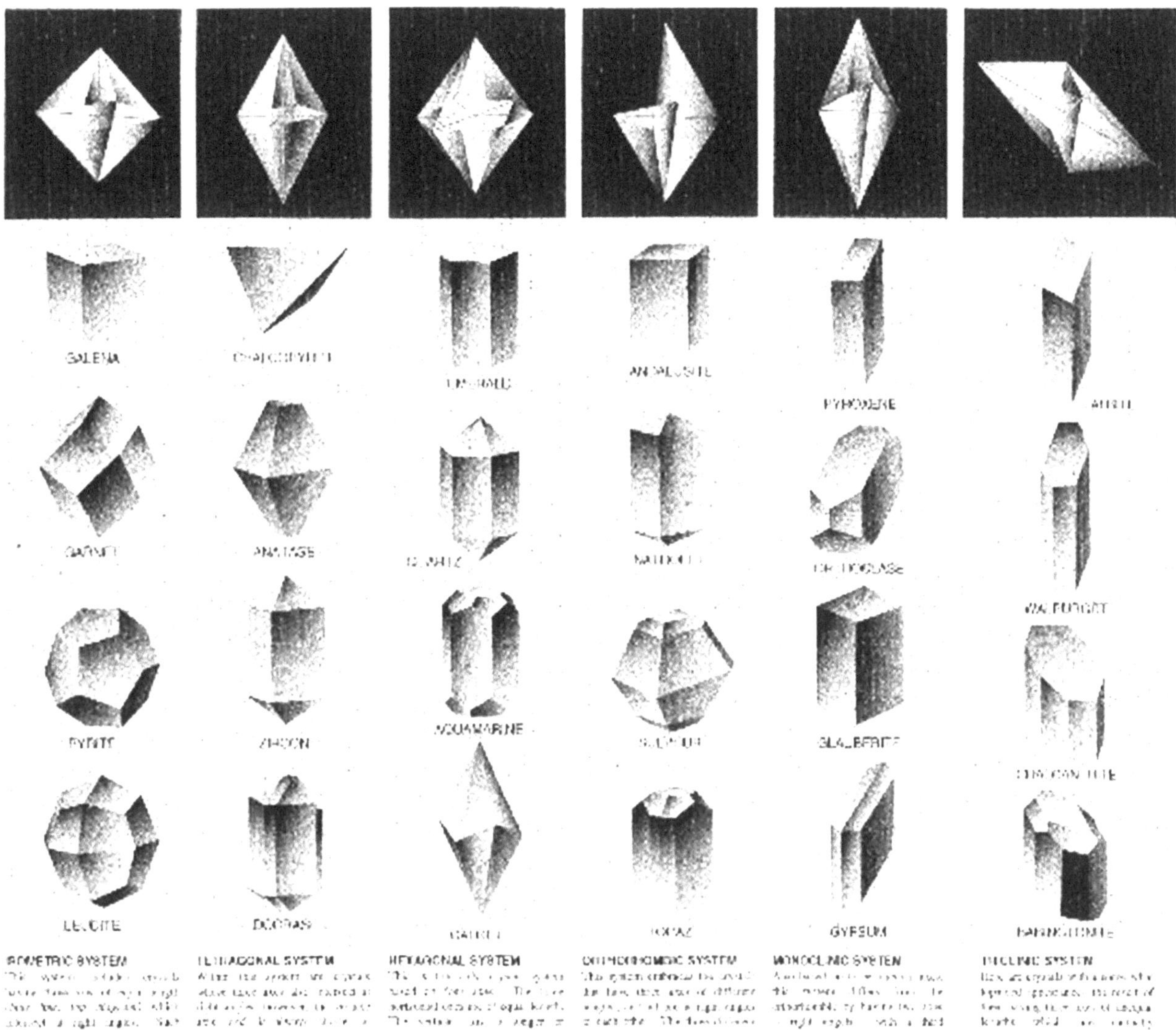

FIGURE 169

3 FORMS OF CARBON

THE THREE KNOWN FORMS OF CARBON

GRAPHITE

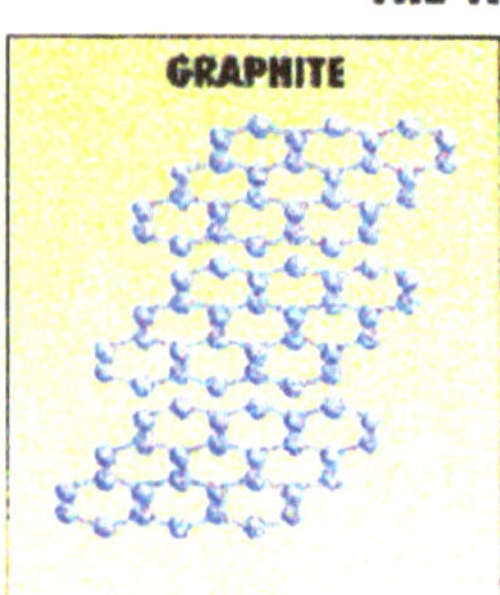

Graphite, the most common form of carbon, has molecules that form flat sheets of atoms arranged in a hexagonal pattern. Because the sheets can slide over one another, graphite is soft and slippery, almost greasy, making it an excellent lubricant. Its black color, combined with its spreadability, also makes graphite a natural for "lead" pencils. Graphite is resistant to heat and is a good conductor of electricity. Graphite fibers have very high tensile strength, so they are used in golf clubs to reduce breakage.

DIAMOND

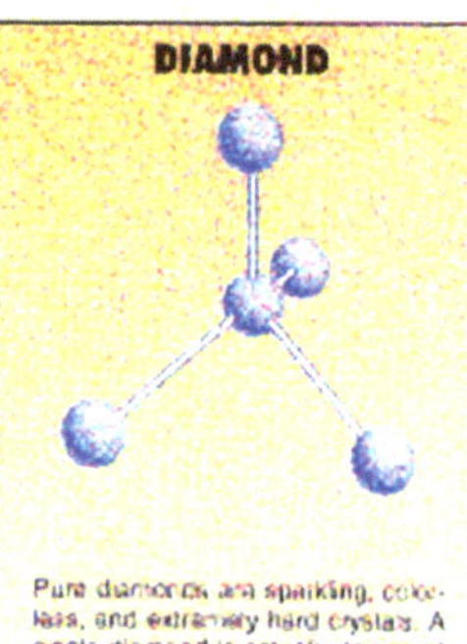

Pure diamonds are sparkling, colorless, and extremely hard crystals. A single diamond is actually one giant molecule consisting of nothing but carbon atoms. Each atom is bonded to four others that form the corners of the pyramidal structure called a regular tetrahedron. This very rigid structure gives diamond its remarkable hardness. The nature and strength of its chemical bonds make it the hardest substance known. Diamond anvils are used in high-pressure research.

BUCKYBALL

Buckminsterfullerene, unknown until recently, has a molecule with 60 carbon atoms that are arranged in regular pentagons and hexagons to form a hollow cage. This form of carbon is named in recognition of the late Buckminster Fuller's geodesic dome, which has a similar structure. This unique structure gives buckyball a whole range of properties that are different from either diamond or graphite, such as superconductivity and the ability to trap other chemicals.

FIGURE 170

PROTEIN FOLDING

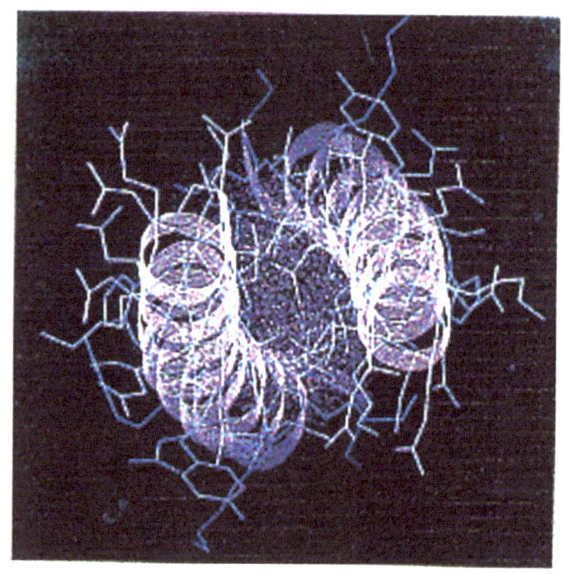

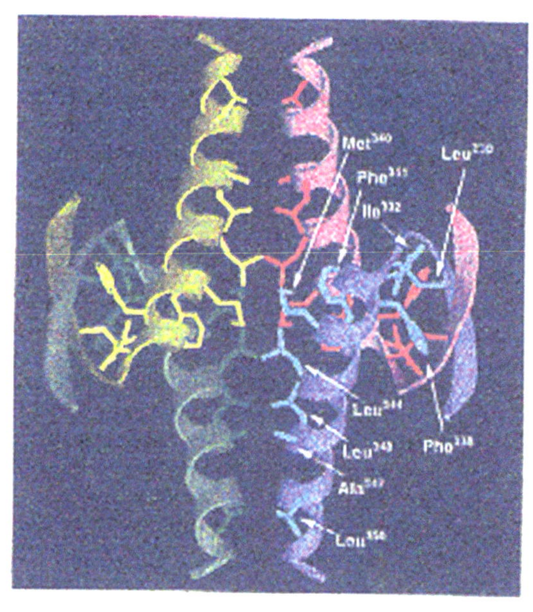

FIGURE 171

RECIPROCAL SAMENESS OF DIAGONAL, DIAMETER, & HYPOTENUSE

(See Figure Inside Front Cover Of Addendum II)

The logical basis for the transformations seen in Figures 3,4,& 5, and the constant changes of Figure 105.

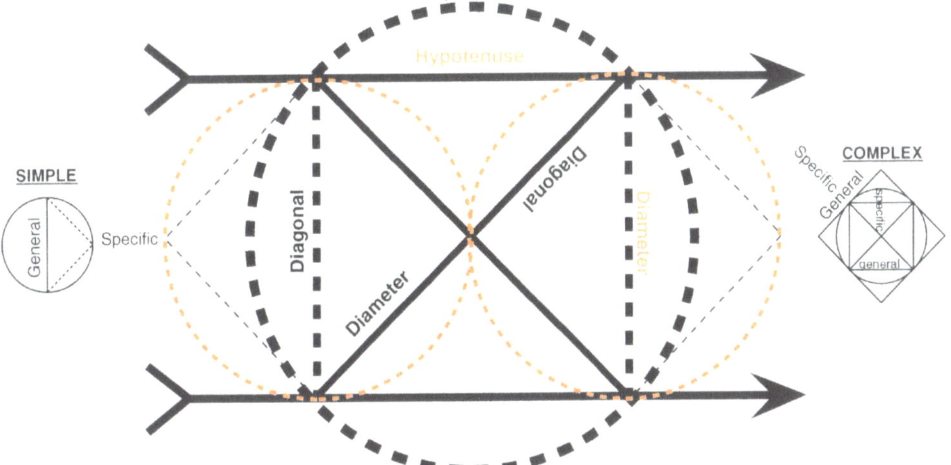

The solid black diagonal of the black larger square inside the dotted black circle is the same length as the diameter of the dotted black circle outside the square.

The hypotenuse (orange) of the larger black triangle is the same length as the diagonal (black) of the smaller black square.
The hypotenuse (orange) of the larger black triangle is the same length as the diameter (orange) of the smaller orange circle.

FIGURE 175

DENDRITIC, OR FRACTAL, INCREASE IN COMPLEXITY

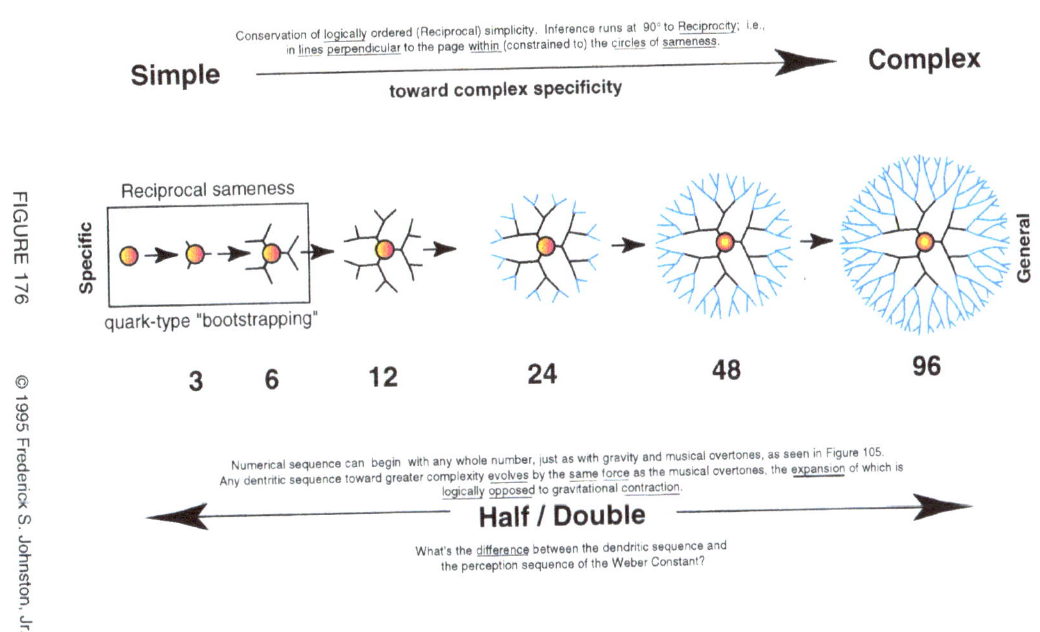

RECIPROCITY OF BOTH PARALLELISM AND PERPENDICULARITY OF THE RHOMBUS

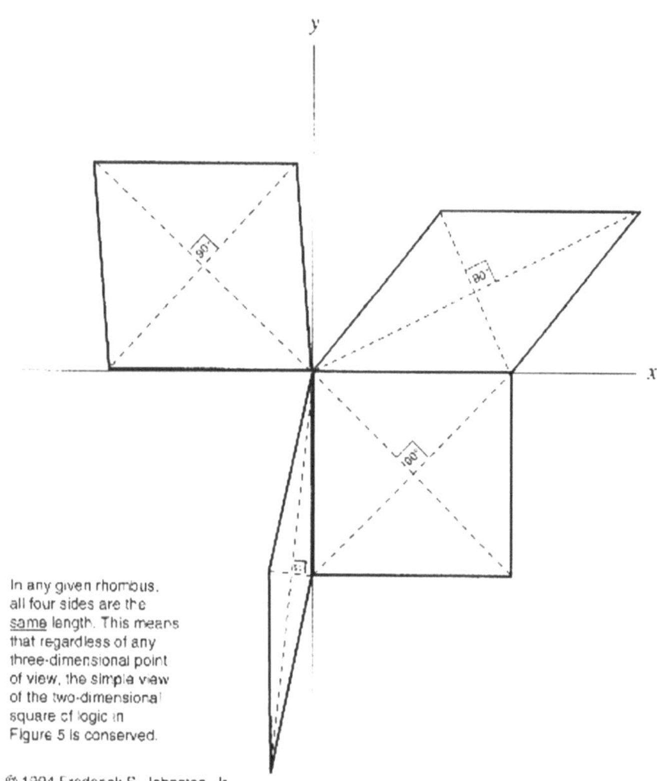

FIGURE 177

RECIPROCAL SAMENESS OF COLLIDING ANGLES

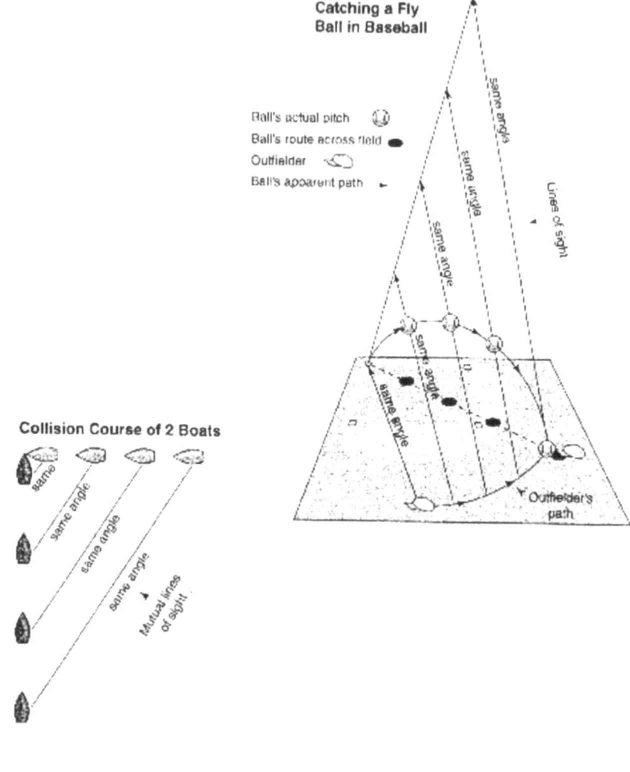

FIGURE 179

RECIPROCALLY THE SAME DIAGONALS CREATE PERPENDICULARITY

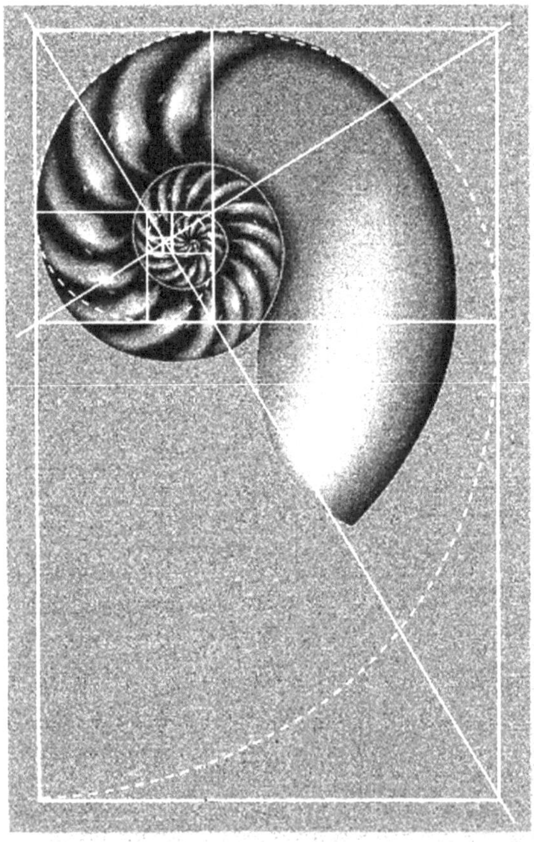

FIGURE 181

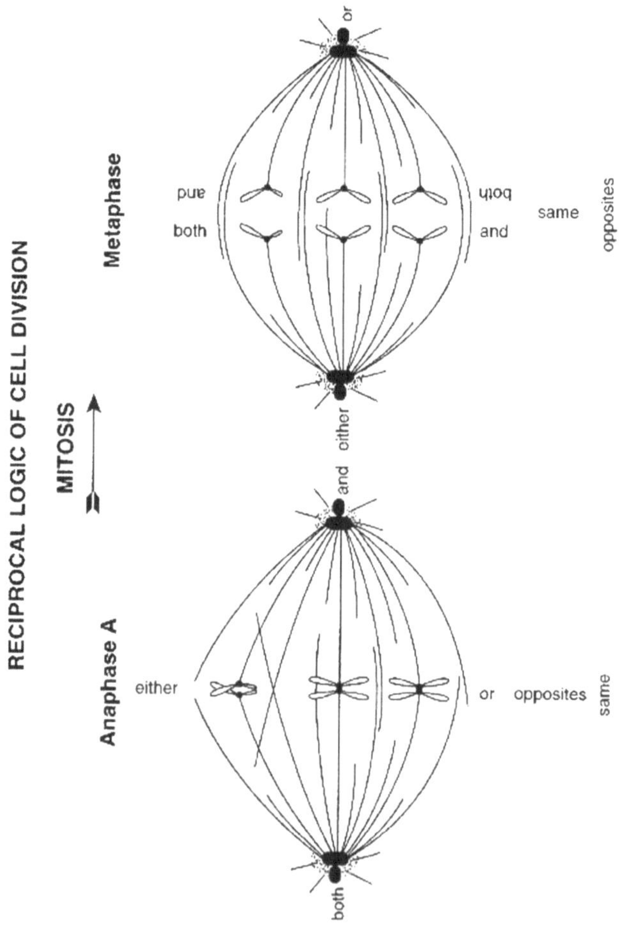

RECIPROCAL LOGIC OF CELL DIVISION

MITOSIS

Metaphase

or

both and

and both

same

opposites

either

and

Anaphase A

either

or opposites

same

both

FIGURE 182

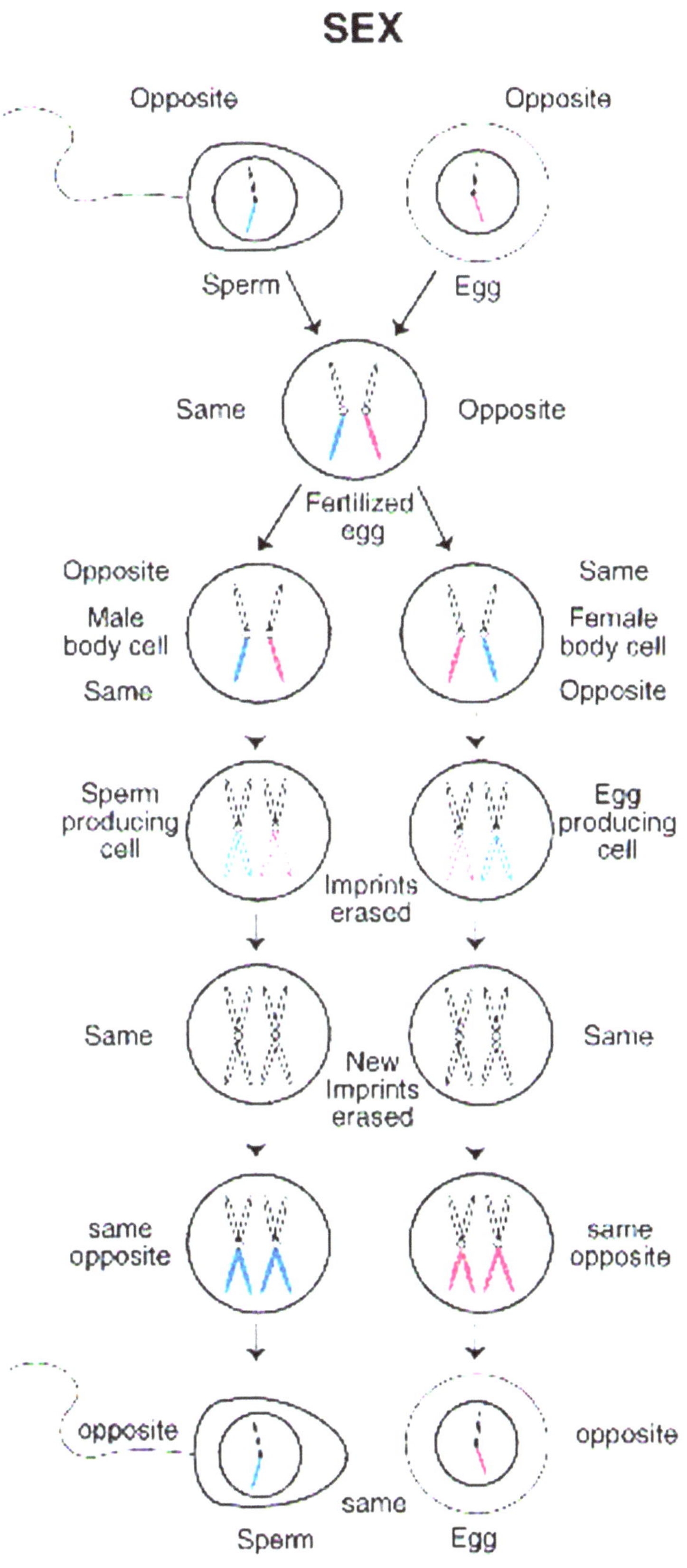

SEX

Opposite
Opposite
Sperm
Egg
Same
Opposite
Fertilized egg
Opposite
Male body cell
Same
Same
Female body cell
Opposite
Sperm producing cell
Egg producing cell
Imprints erased
Same
New Imprints erased
Same
same opposite
same opposite
opposite
opposite
same
Sperm
Egg

FIGURE 184

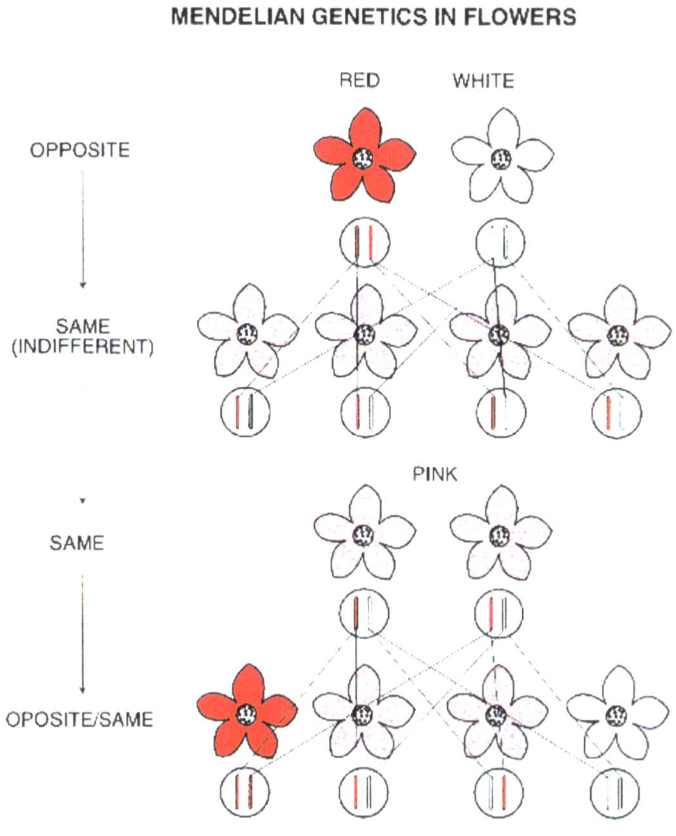

MENDELIAN GENETICS IN FLOWERS

RED
WHITE
OPPOSITE
SAME (INDIFFERENT)
PINK
SAME
OPOSITE/SAME

If sexual reproduction is Reciprocal, then the male/female chromosome should be reversible; i.e., the same chromosome with opposite male/female sex. In all birds the male is xx and the female is xy!

FIGURE 185

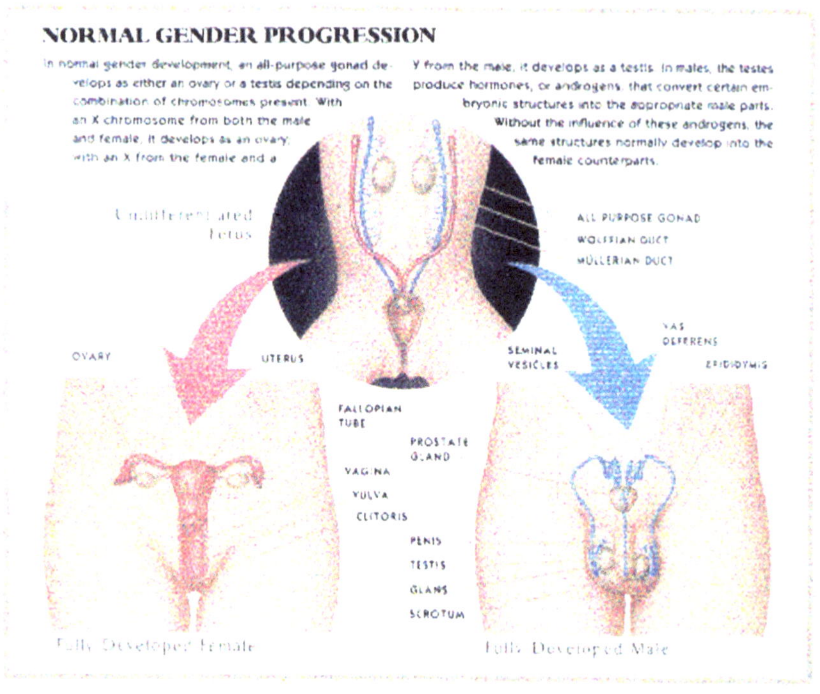

FIGURE 186

THE EVOLUTION OF COMPLEXITY

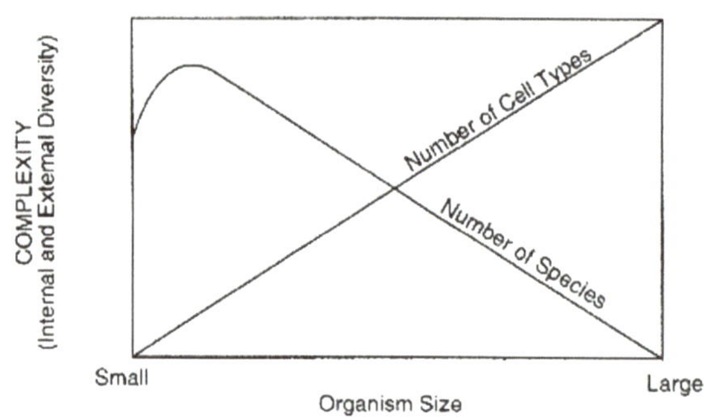

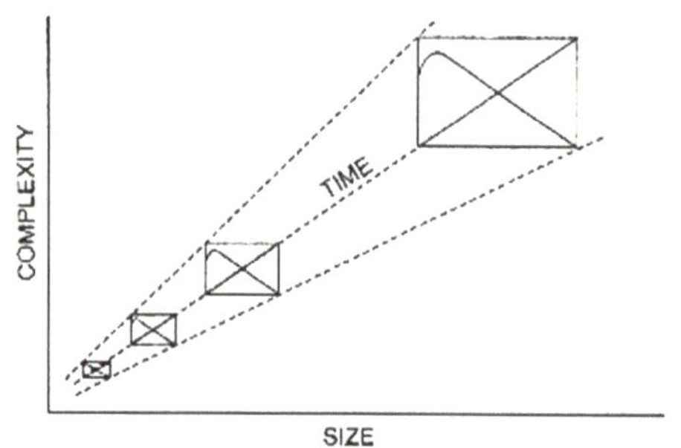

A three-dimensional graph showing how internal and external complexity and size increase with time during the course of evolution.

The Evolution of Complexity, John Tyler Bonner, Princeton University Press © 1988, Princeton University Press.

FIGURE 191

2-DIMENSIONAL ELECTRO/MAGNETISM STATIC GEOMETRIC CONCEPTS

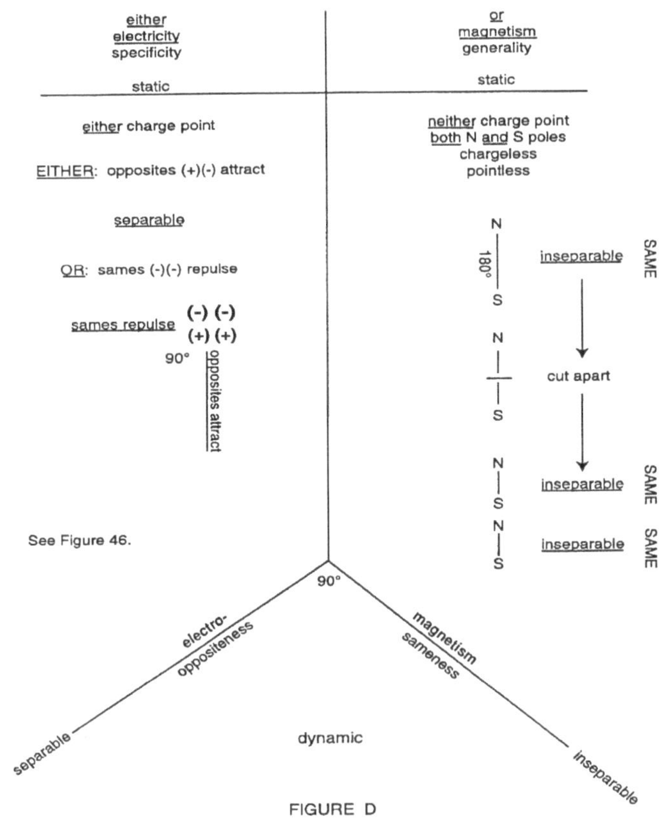

FIGURE D

2-D DYNAMIC GEOMETRIC CONCEPTS ELECTRO/MAGNETISM

circle of sameness

both — and

vectors of oppositeness

and — both

circle of sameness

either

or

opposite - ness

same- - ness

charge specific

magne - tism

charge - less

general

perpendicularity of geometric concepts

Neither electro-oppositeness nor magnetic-sameness moves unless both move simultaneously. If either moves, both move. Neither can move unless either/both functions Reciprocate at 90°.

FIGURE E

HANDEDNESS OF ELECTROMAGNETISM

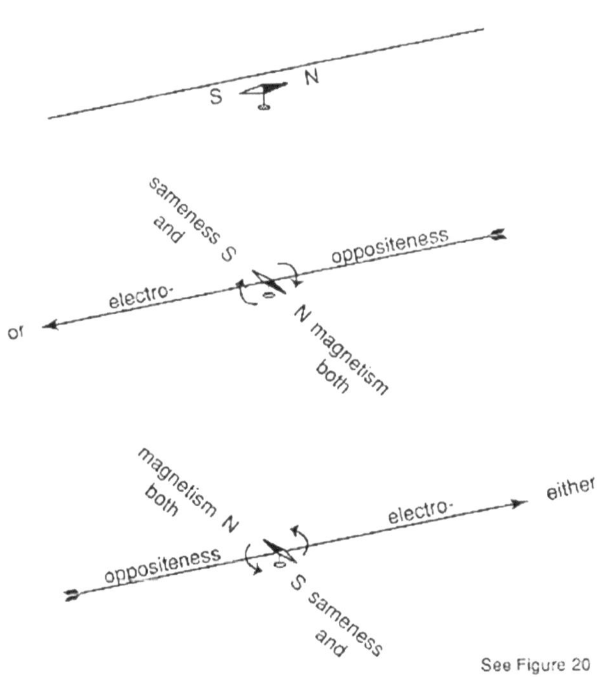

Hans Oersted discovered that instead of parallelism, there must be a perpendicularity of either charge and both poles. Notice how the geometric concepts of Reciprocal logic are naturally formed at 90°.

FIGURE F

RECIPROCAL OF FIGURE E

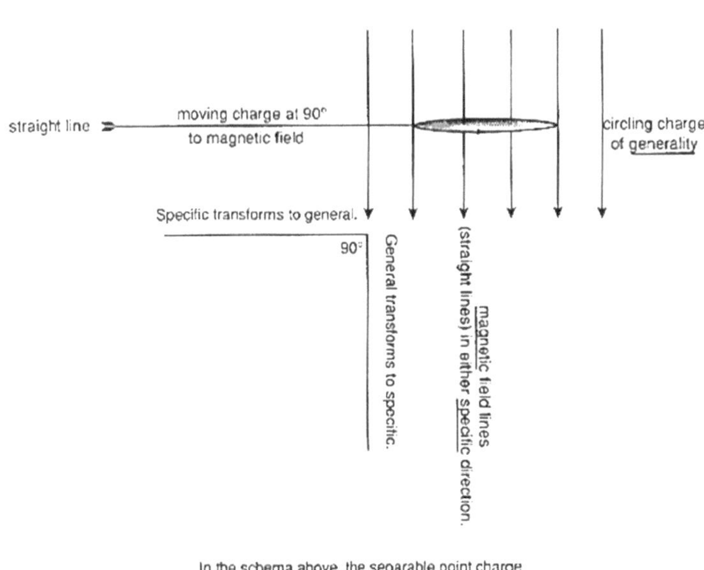

In the schema above, the separable point charge transforms into an inseparable circle of charge when entering a strong magnetic field of straight lines at 90°.

FIGURE G

SAMENESS OF MAGNETIC FIELD LINES

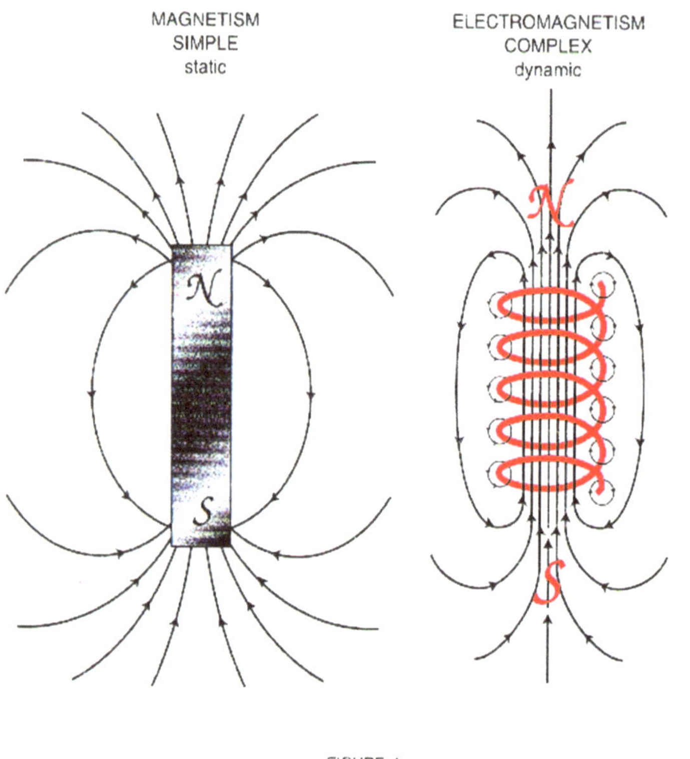

FIGURE I

INFERENTIAL ORDER
From the General to the Specific.

INFERENTIAL CHAOS (Reciprocity)

(see Figure 8)

0

Stable solutions

bifurcation point

bifurcation point

Distance from equilibrium

Continuum

Chaos

The bifurcation cascade to chaos in a simple nonlinear system, also known as period doubling.

Frontiers of Complexity, The Search for Order in a Chaotic World, Peter Coveney and Roger Highfield, Fawcett Columbine, New York

INFERENTIAL ORDER
From the Specific to the General.

FIGURE K-1

THE LOGIC OF LIFE'S EVOLUTION

from the specific to the general, with each successive Generality creating a more complex Specificity

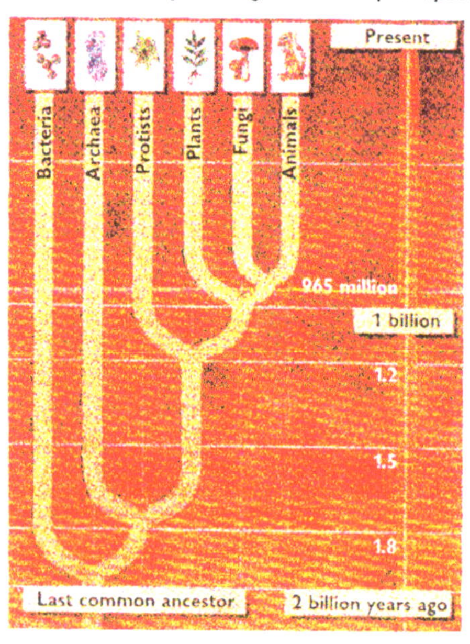

FIGURE K-2

LIFE/DEATH
Change/Constant

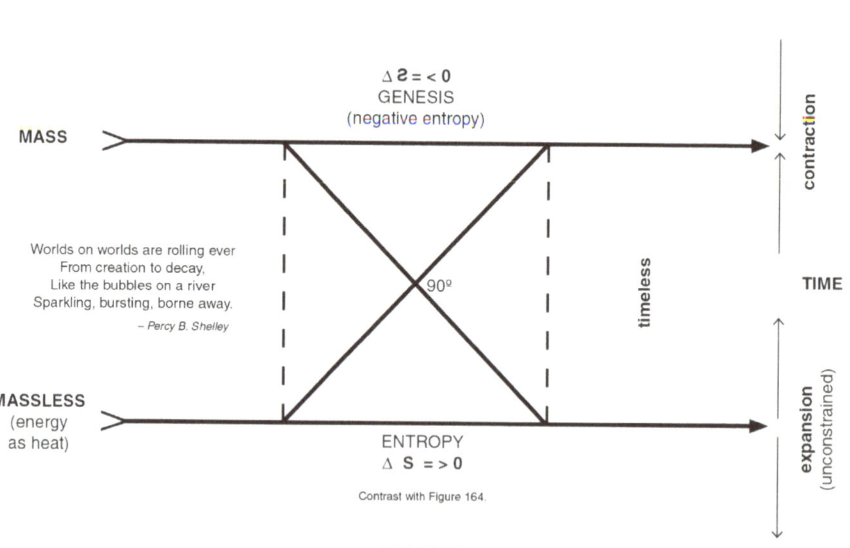

Contrast with Figure 164.

FIGURE M

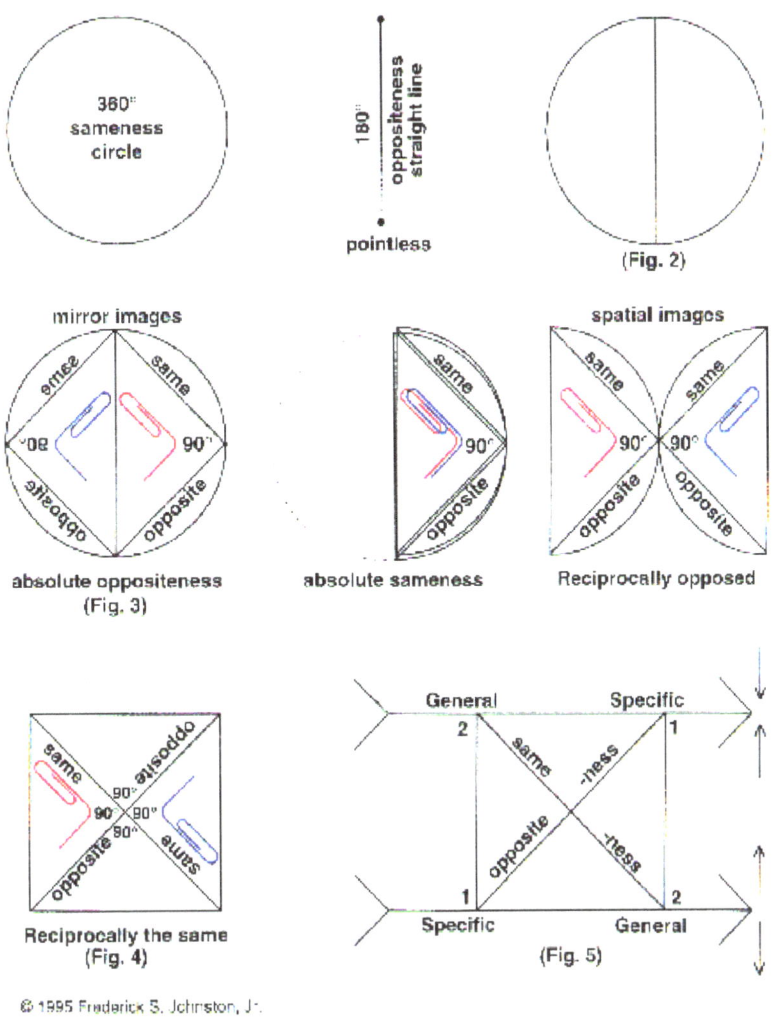

FIGURE N

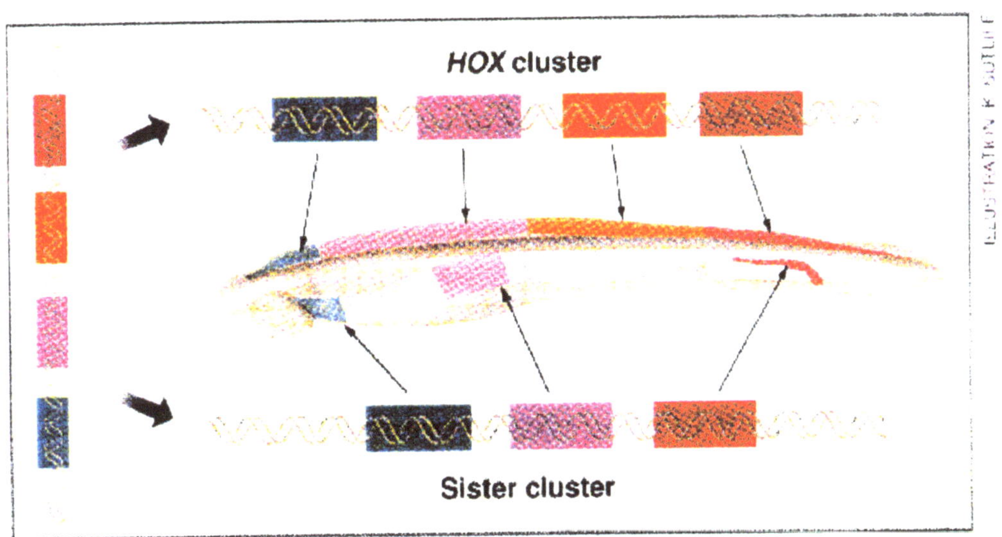

Division of labor. The descendants of a proto-*HOX* gene cluster (left) may help shape the *Amphioxus* ectoderm (above) and the gut (below).

Figure 194

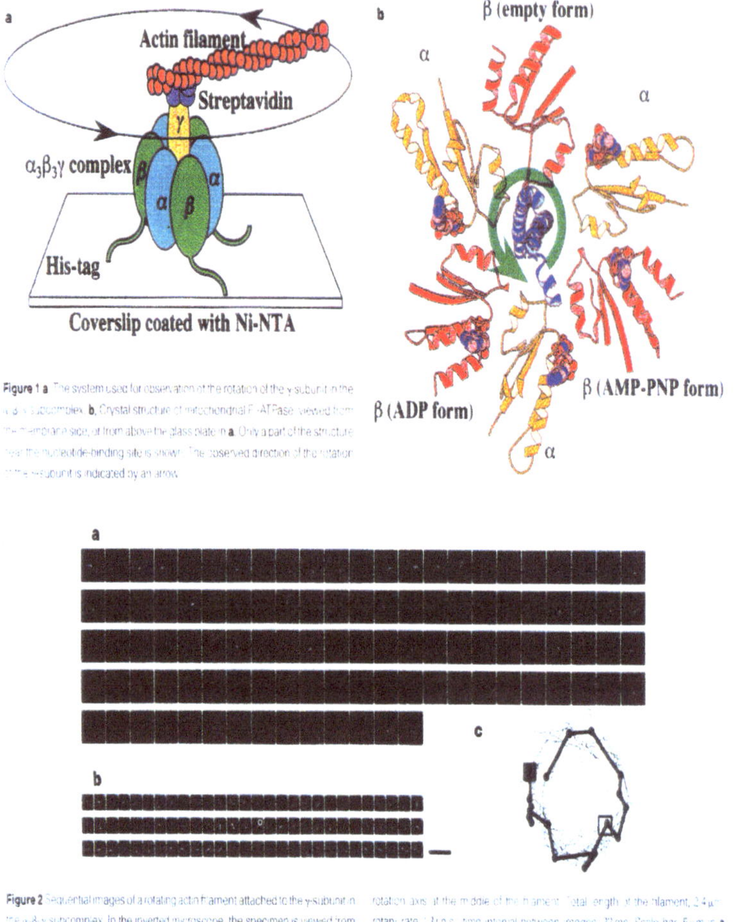

Figure 195

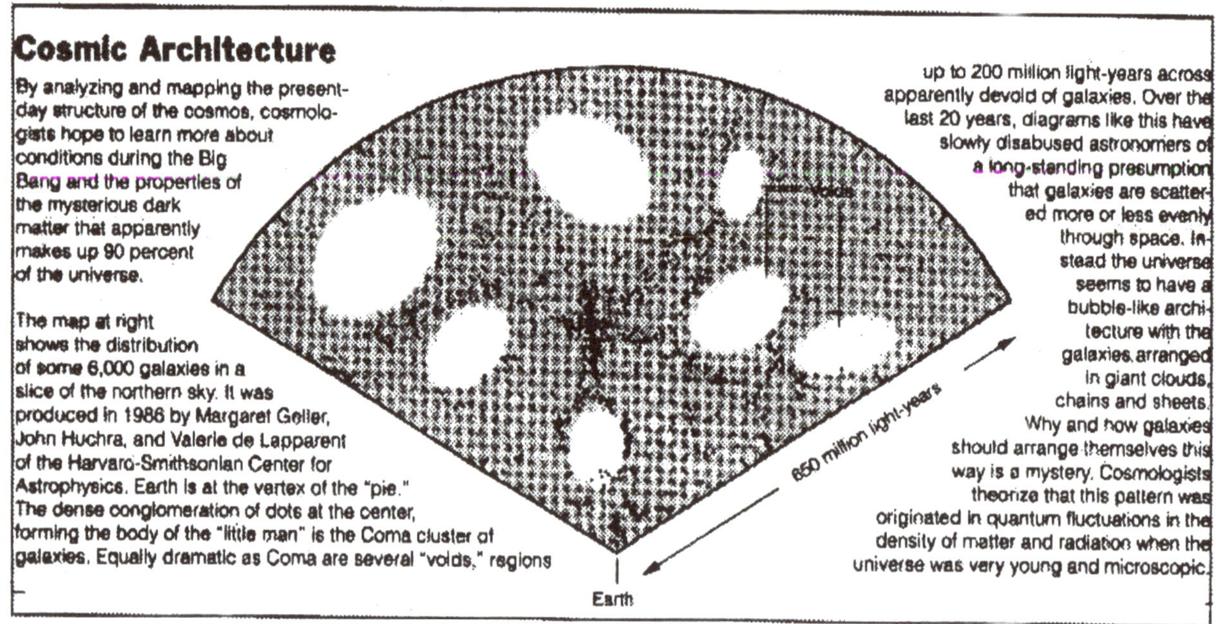

Cosmic Architecture

By analyzing and mapping the present-day structure of the cosmos, cosmologists hope to learn more about conditions during the Big Bang and the properties of the mysterious dark matter that apparently makes up 90 percent of the universe.

The map at right shows the distribution of some 6,000 galaxies in a slice of the northern sky. It was produced in 1986 by Margaret Geller, John Huchra, and Valerie de Lapparent of the Harvard-Smithsonian Center for Astrophysics. Earth is at the vertex of the "pie." The dense conglomeration of dots at the center, forming the body of the "little man" is the Coma cluster of galaxies. Equally dramatic as Coma are several "voids," regions up to 200 million light-years across apparently devoid of galaxies. Over the last 20 years, diagrams like this have slowly disabused astronomers of a long-standing presumption that galaxies are scattered more or less evenly through space. Instead the universe seems to have a bubble-like architecture with the galaxies arranged in giant clouds, chains and sheets. Why and how galaxies should arrange themselves this way is a mystery. Cosmologists theorize that this pattern was originated in quantum fluctuations in the density of matter and radiation when the universe was very young and microscopic.

Figure 202

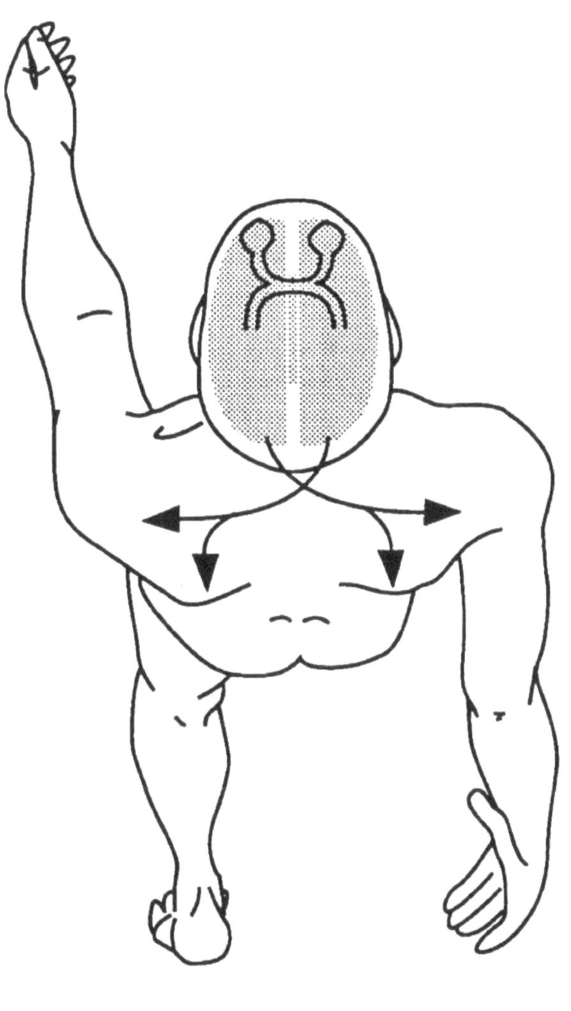

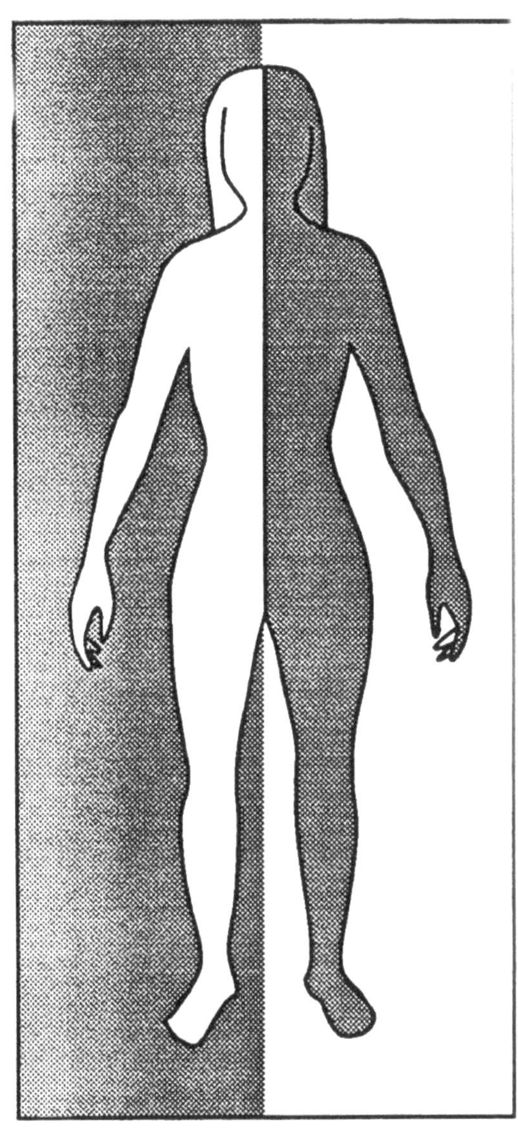

Copied from Robert E. Ornstein's
Psychology of Consciousness
W.H. Freeman and Co., Publishers

Copied from John M. Bailey's
Liberal Art's Physics
W.H. Freeman and Co., Publishers

Figure 204

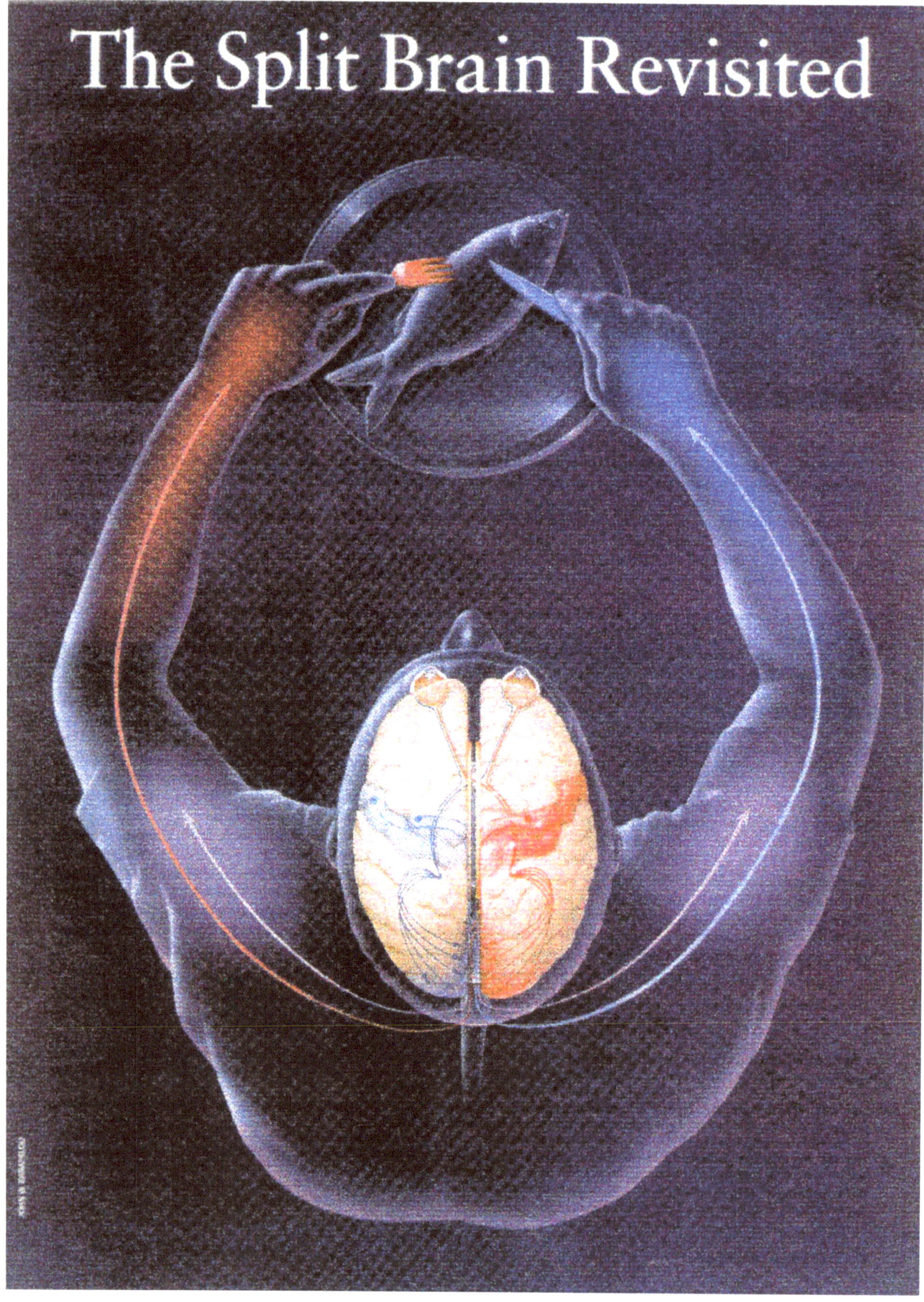

"The Split Brain Revisited," by Michael S. Gazzaniga, *Scientific American*, July 1998.

Figure 205

THE RECIPROCAL HEMISPHERES

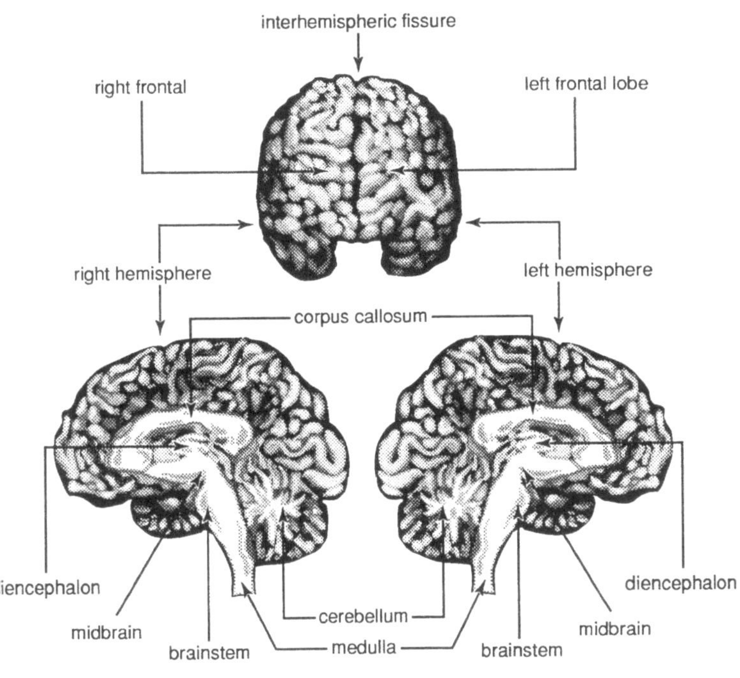

Two Reciprocal (either/both) functions of the same brain.

Figure 206

A Comparison of Left-Mode and Right-Mode Characteristics

L -MODE	R -MODE
Verbal: Using words to name, describe, define.	Nonverbal: Awareness of things, but minimal connection with words.
Analytic: Figuring things out step-by-step and part-by-part.	Synthetic: Putting things together to form wholes.
Symbolic: Using a symbol to stand for something. For example, the drawn form stands for eye, the sign + stands for the process of addition.	Concrete: Relating to things as they are, at the present moment.
Abstract: Taking out a small bit of information and using it to represent the whole thing.	Analogic: Seeing likenesses between things; understanding metaphoric relationships.
Temporal: Keeping track of time, sequencing one thing after another: Doing first things first, second things second, etc.	Nontemporal: Without a sense of time.
Rational: Drawing conclusions based on reason and facts.	Nonrational: Not requiring a basis of reason or facts; willingness to suspend judgement.
Digital: Using numbers as in counting.	Spatial: Seeing where things are in relation to other things, and how parts go together to form a whole.
Logical: Drawing conclusions based on logic: one thing following another in logical order – for example, a mathematical theorem or a well-stated argument.	Intuitive: Making leaps of insight, often based on incomplete patterns, hunches, feelings, or visual images.
Linear: Thinking in terms of linked ideas, one thought directly following another, often leading to a convergent conclusion.	Holistic: Seeing whole things all at once; perceiving the overall patterns and structures, often leading to divergent conclusions.

Dichotomies

Intellect	Intuition
Convergent	Divergent
Intellectual	Sensuous
Deductive	Imaginative
Rational	Metaphoric
Vertical	Horizontal
Discrete	Continuous
Abstract	Concrete
Realistic	Impulsive
Directed	Free
Differential	Existential
Sequential	Multiple
Historical	Timeless
Analytic	Holistic
Explicit	Tacit
Objective	Subjective
Successive	Simultaneous

Figure 207

THE HANDEDNESS TEST

Indicate your preferences in the use of the hands in the following activities by putting + in the appropriate column. Where the preference is so strong that you would never try to use the other hand unless forced to, put ++. In any case in which you are indifferent put + in both columns.

	Left	Right
1. Writing	☐	☐
2. Drawing	☐	☐
3. Throwing	☐	☐
4. Scissors	☐	☐
5. Toothbrush	☐	☐
6. Knife (without fork)	☐	☐
7. Spoon	☐	☐
8. Broom (upper hand)	☐	☐
9. Striking match (match)	☐	☐
10. Opening box (lid)	☐	☐

To find your laterality quotient, add up the number of +s in each column. Subtract the number under LEFT from the number under RIGHT, divide by the total number of +s, and multiply by 100.

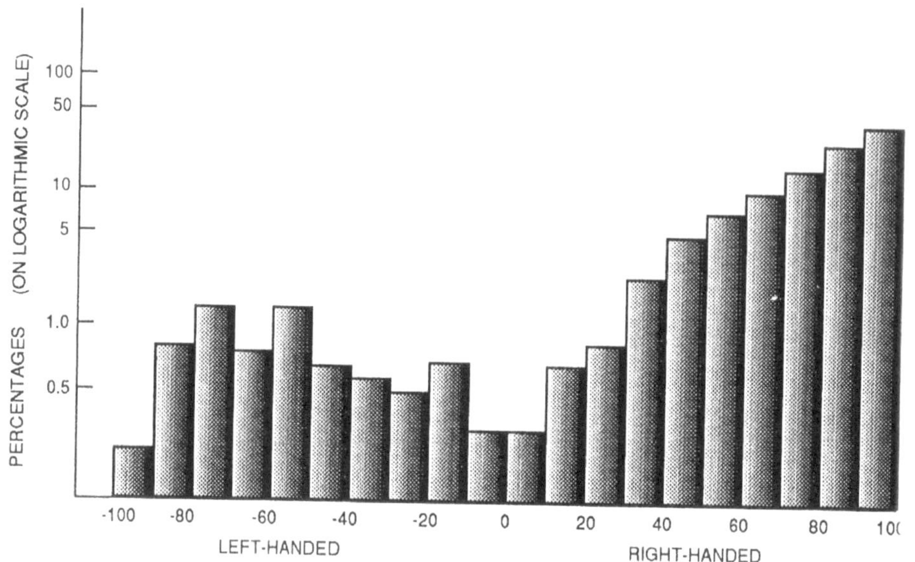

Are you 100 percent left- or right-handed? Probably not–most people perform some tasks with one hand and some with the other. By taking the Edinburgh Handedness Inventory (left), a standard test, you can find your own laterality quotient; the chart (above) illustrates an average distribution of scores. Note how right-handers seem to be more strongly right-handed than left-handers are left-handed; they tend to cluster toward the higher end of the scale, while those who are left-handed are more evenly distributed between ambidexterity (0) and total left-handedness. Findings like these have led some scientists to conclude that left-handedness can best be described as the lack of a predisposition toward being right-handed.

Figure 208

HUMAN BRAIN
PREFRONTAL CORTEX

Oppositeness	Sameness
either	both (neither)
imbalance	balance
half	double
choice	reconciliation
charge	chargeless
asymmetry	symmetry
hand	handless
point	pointless
sequential relations	simultaneous relations (interactions)
happy	sad

Figure 209

THE TRUNE BRAIN (Vertical)

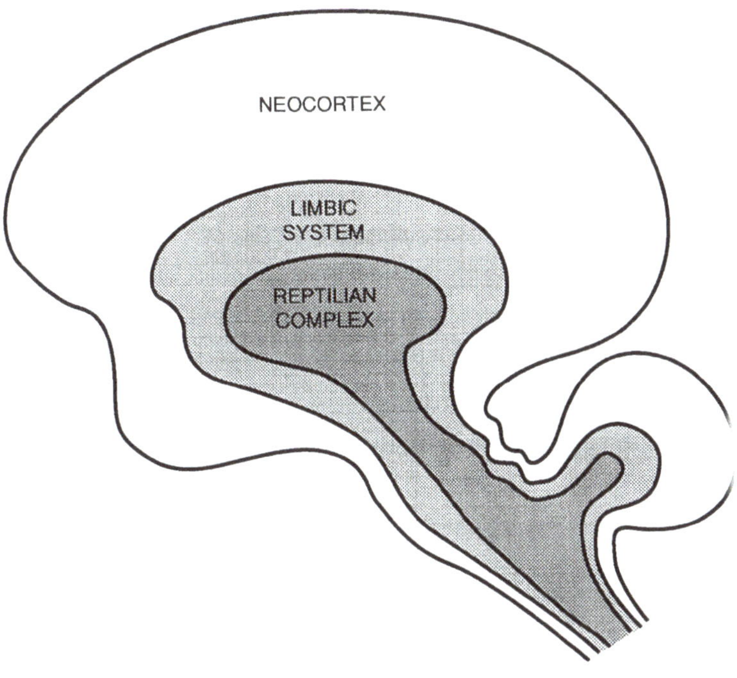

The three brains are anatomically separable and chemically distinct.

Figure 215 See also Figure 306

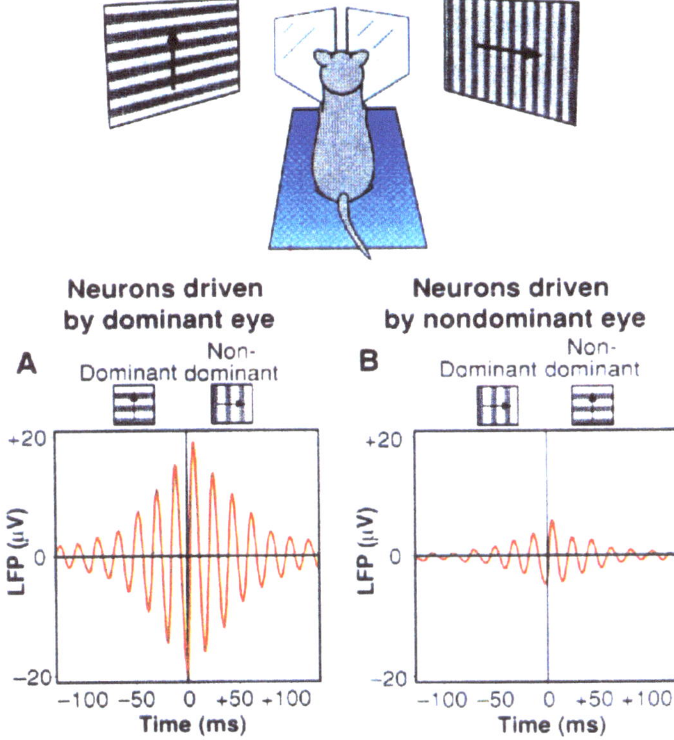

To see or not to see. Brain neurons that respond to the image that a cat perceives (A) show a high degree of synchronous firing (represented by the amplitude of the waves); less synchrony is shown by neurons responding to the eye whose image is not perceived (B).

"Listening In on the Brain," by Marcia Barinaga, *Science*, April 17, 1998.

Figure 217

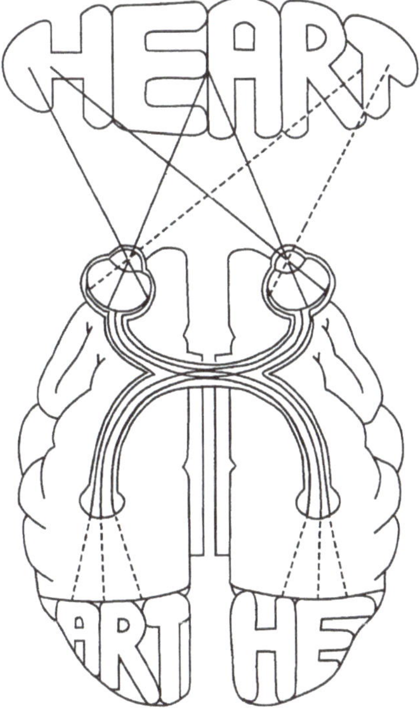

Copied from Robert E. Ornstein's The Psychology of Consciousness W.A. Freeman and Co., Publisher

Figure 221

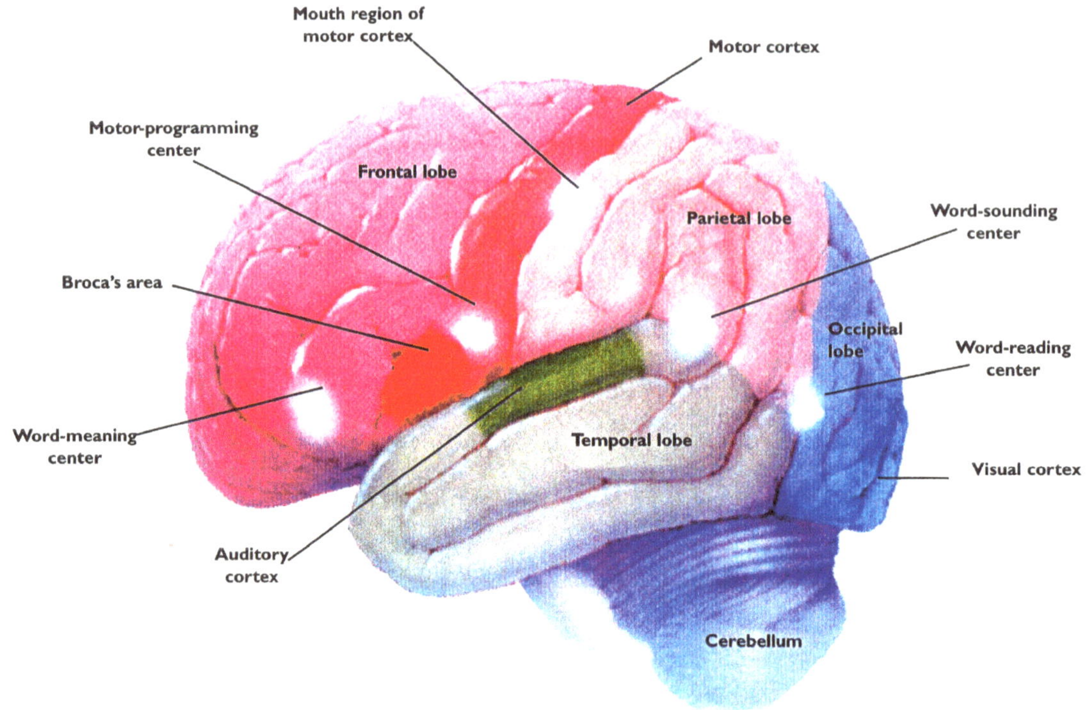

Figure 222

MYELIN SHEATH

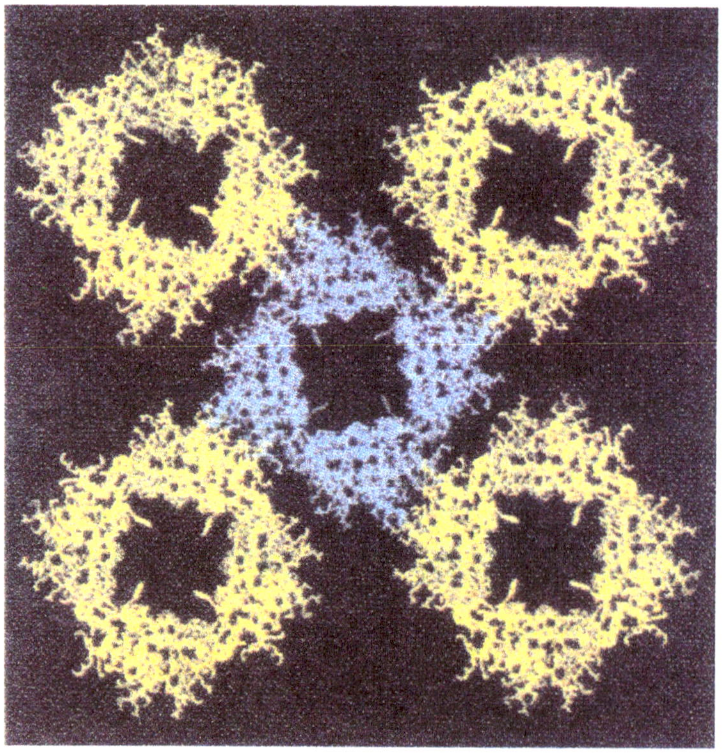

Figure 225

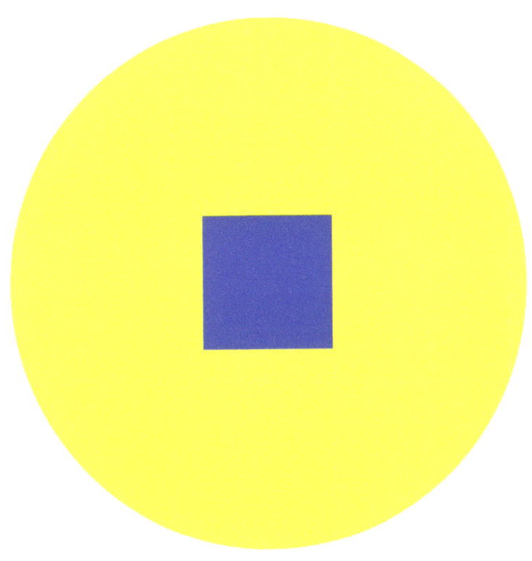

Colorful blue center, colorful yellow surrounding

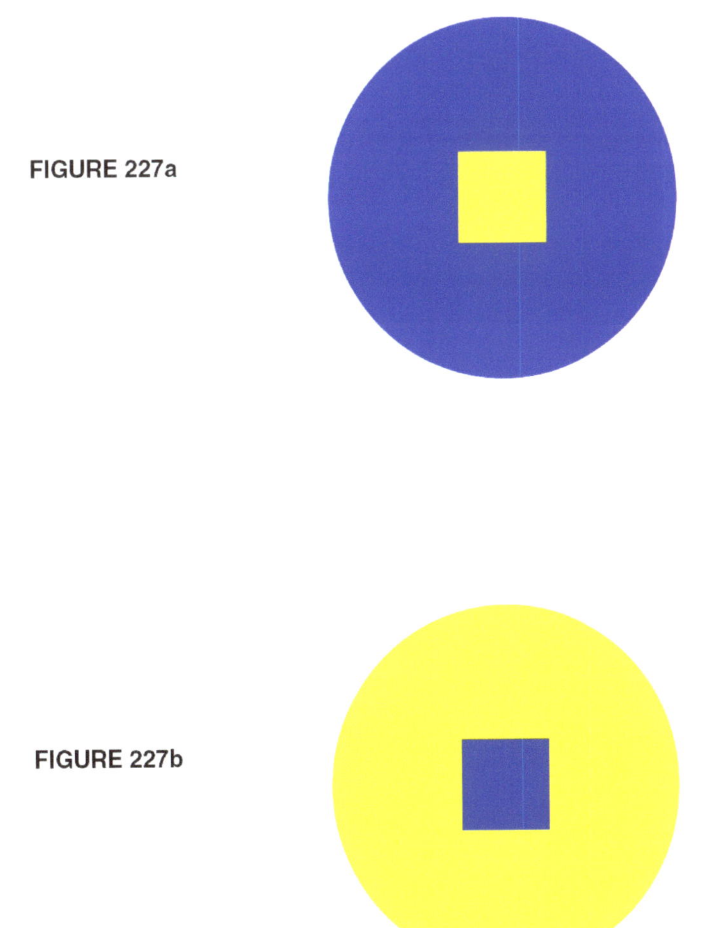

FIGURE 227a

FIGURE 227b

THE LAND EFFECT
(two 90° Colors)

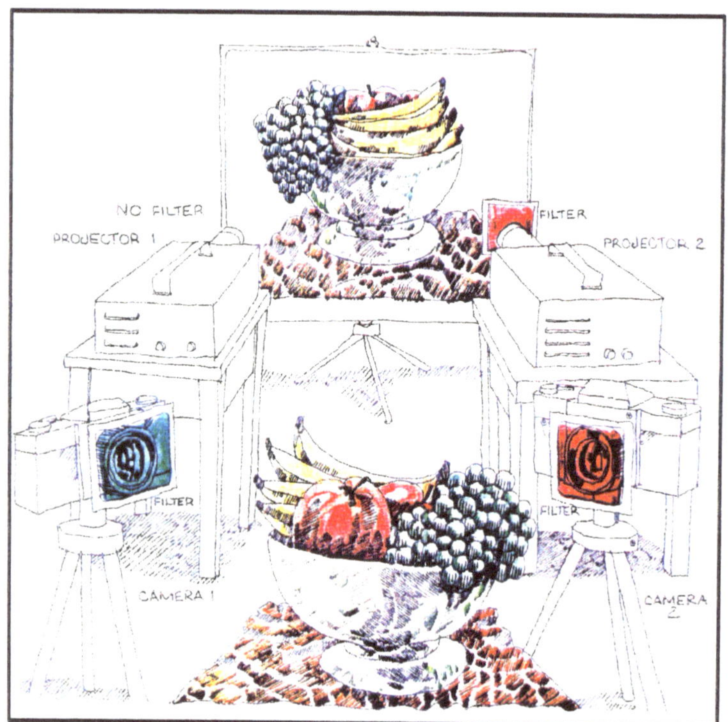

Figure 230

THE LAND EFFECT

(colorful/colorless)

Figure 231

REVERSE LAND EFFECT

HIGH INTENSITY GREEN AND BRIGHT LIGHT
IN A LOW WHITE LIGHT CONDITION

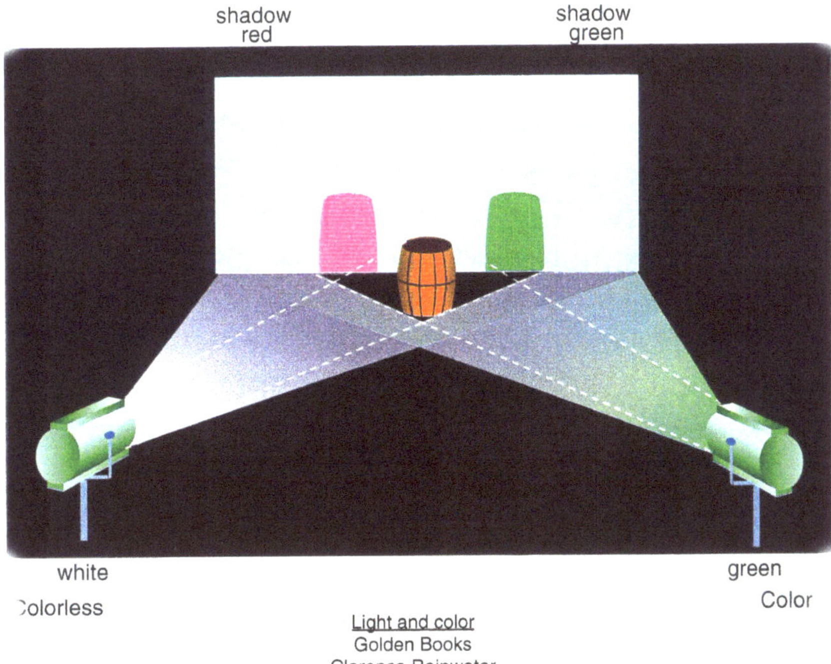

Light and color
Golden Books
Clarence Rainwater

Color/Colorless create two shadows
of 90° Reciprocal colors (red/green).

Figure 232

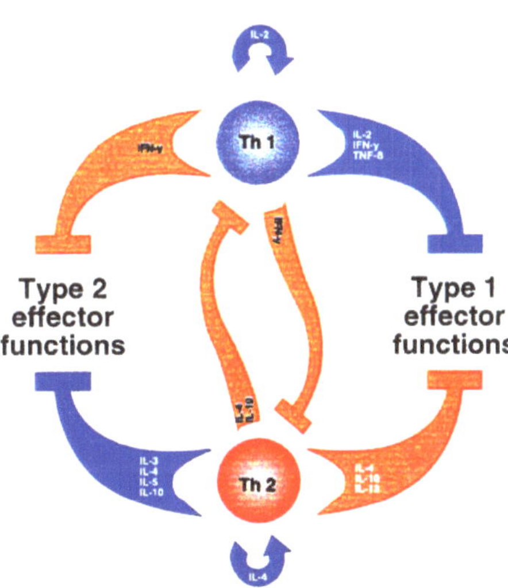

The cytokines produced by Th1 and Th2 lymphocyte populations determine their opposing effector (blue) and inhibitory (pink) functions. The Th1 and Th2 pathways are symmetrical, each controlling a unique set of immune responses, and each augmenting the **development** of cells of the same subset while suppressing the expansion and/or effector functions of the other subset. IL-2 and IL-4 are shown as autocrine growth factors for Th1 and Th2 cells, respectively.

Taken from Abul K. Abbas, Kenneth M. Murphy, and Alan Sher: Functional diversity of helper T lymphocytes. Nature, Volume 383. October 31, 1996.

Figure 234

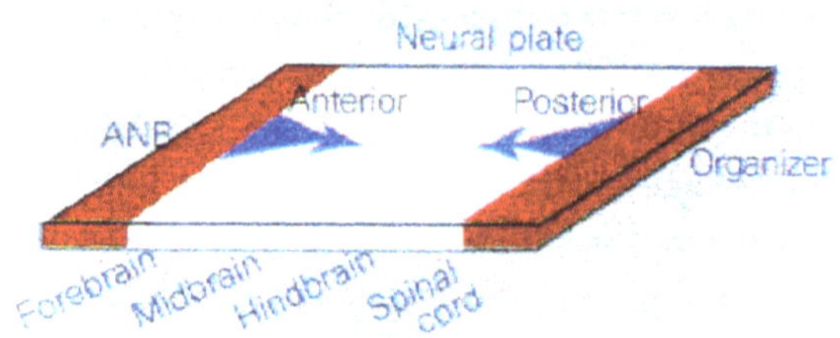

Figure 2 **Planar model for initial anteroposterior patterning in the vertebrate neural plate.** Early anteroposterior patterning seems to occur by the action of planar signals from the anterior[3] (the ANB) and the posterior (the organizer) boundary regions, with the neural plate behaving as a developmental field[4].

Figure 235 A

GEOMETRIC CONCEPTS OF BRAIN CEREBRAL LATERALIZATION

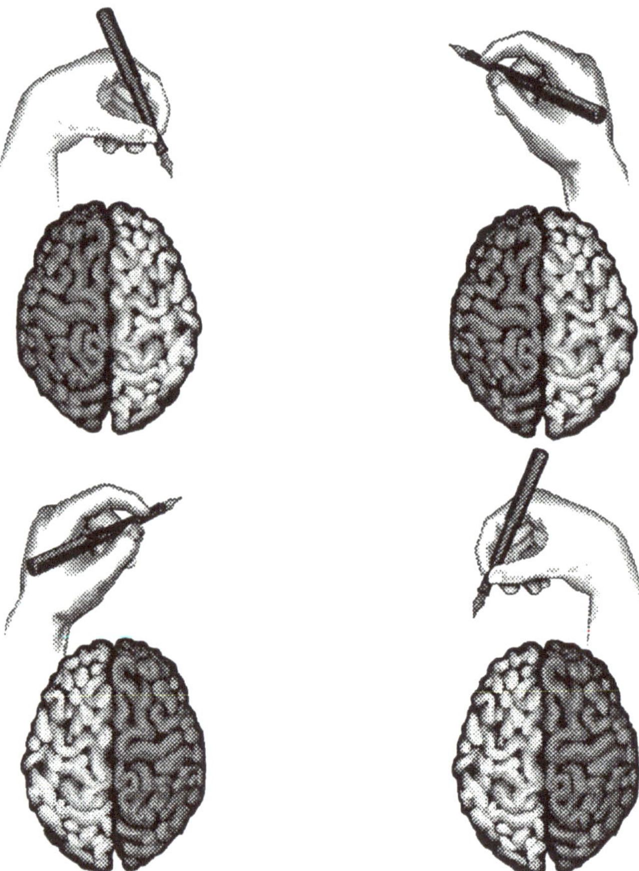

December, Science 80
Figure 236

Scientists are trying to determine if hand and writing posture indicate which brain hemisphere is used for language. Initial results show that left-handers who use the hooked position for writing process language on the left side of their brains, as do the majority of right-handers who use the straight hand position. But left-handers who use a straight hand posture seem to process language in the right hemisphere, as do the small minority of right-handers who use a hooked position.

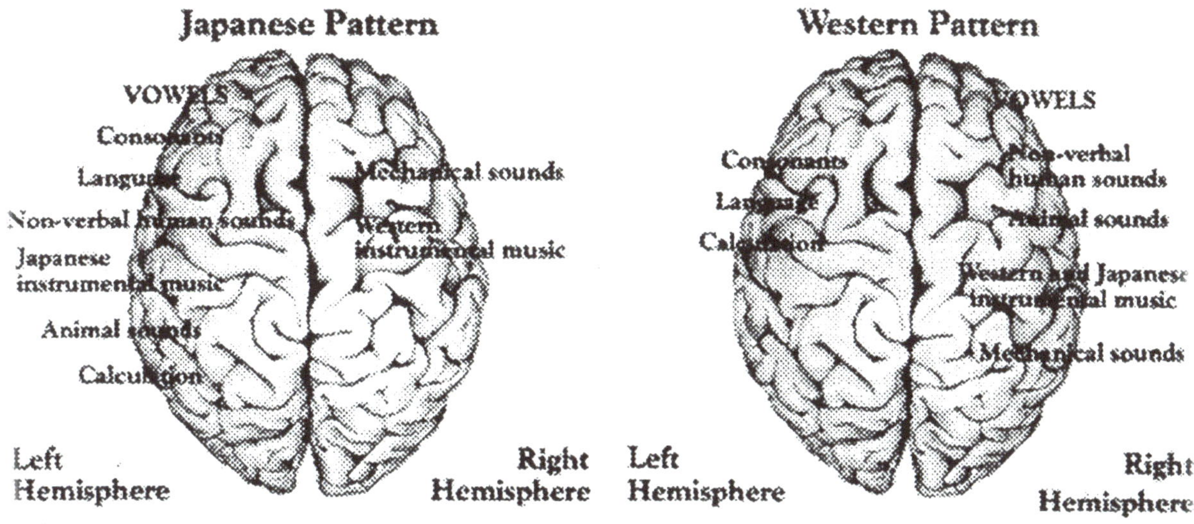

Tadanobu Tsunoda, a specialist in hearing difficulties, believes that the two halves of the Japanese brain divide up the labor of processing sounds in a way that differs from Western brains. The key to those differences, he postulates, is that the Japanese deal with all vowels in the left hemisphere, while Westerners handle isolated vowels in the right hemisphere.

Figure 237

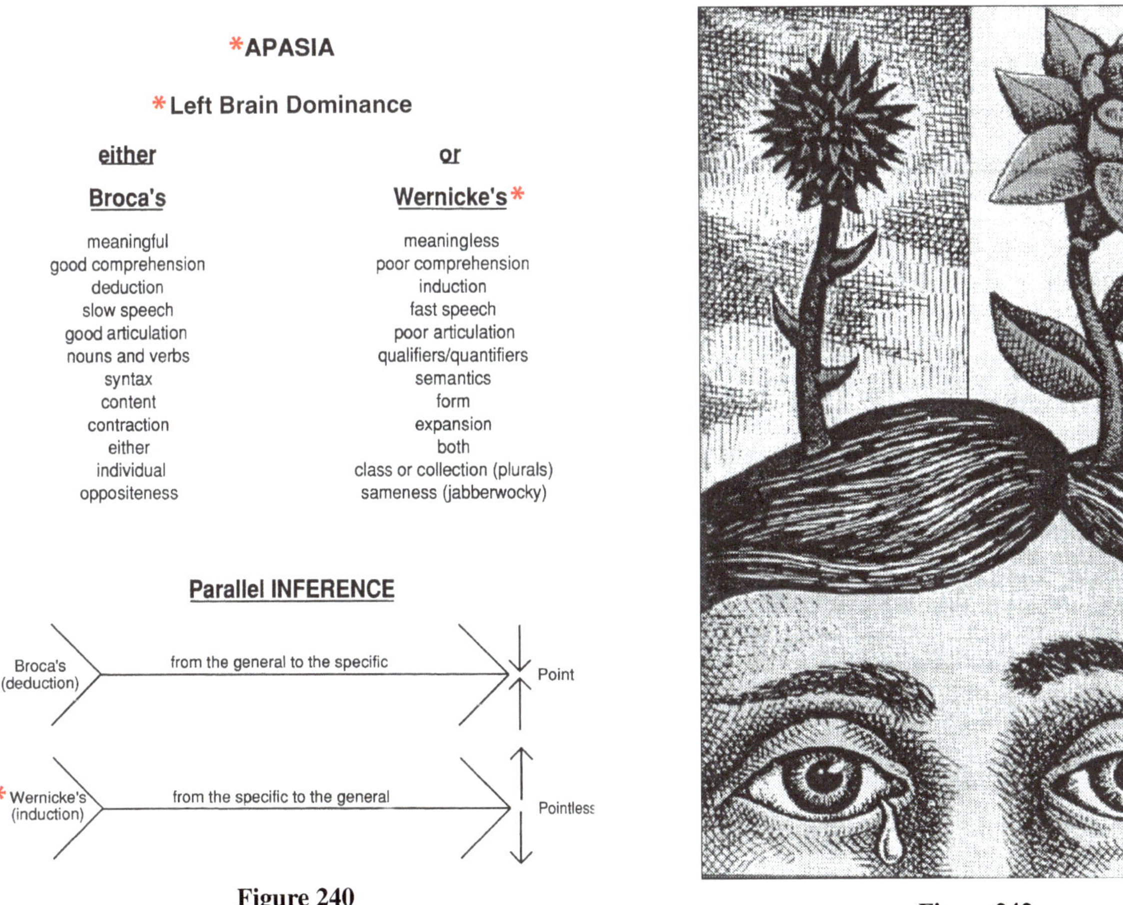

Figure 240

Figure 242

THE RECIPROCAL LOGIC of HAPPY/SAD

How the Brain Computes Tears and Laughter

The brain handles happiness and sadness in different areas, not in a single emotional center as was thought. New fast scanning methods show that happiness is marked by a decrease in activity in the cortex, in areas responsible for forethought and planning. Sadness is associated with enhanced activity in regions of the limbic system. PET scans below show changes in activity of subjects experiencing the two emotional states.

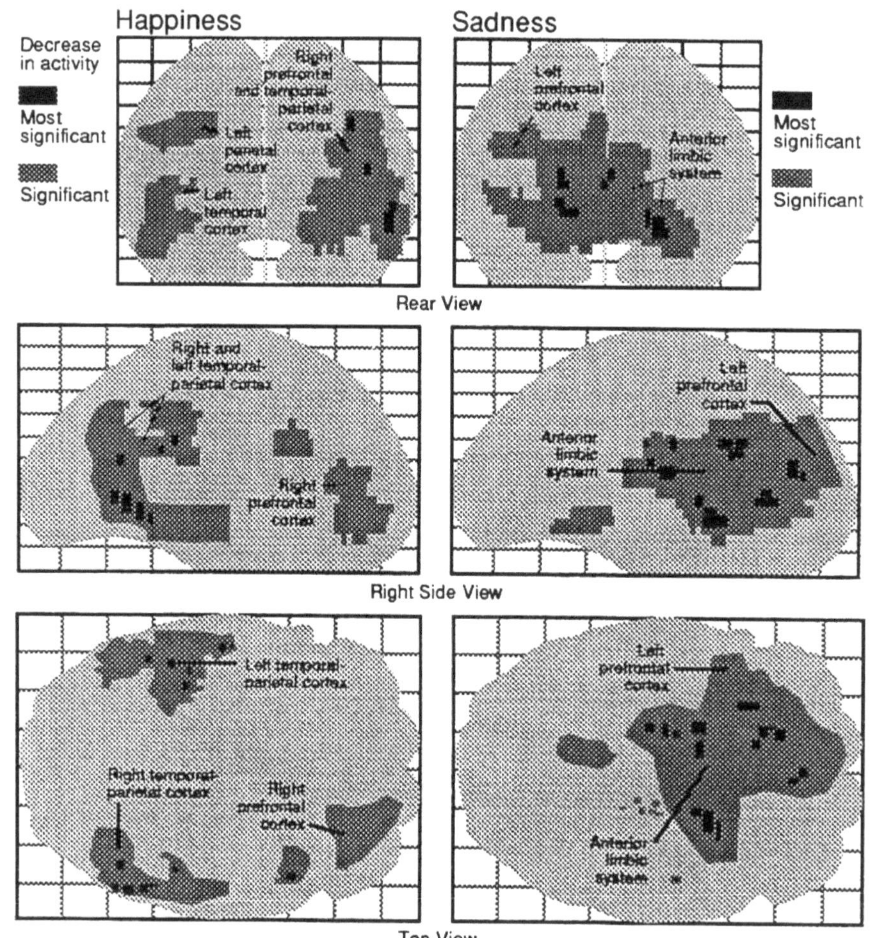

Taken from Daniel Goleman: The Brain Manages Happiness and Sadness in Different Centers. The New York Times, Tuesday, March 28, 1995. Illustration adapted from Frank O'Connell.

Figure 243

BRAIN'S EMOTIONAL SEPARATION
through Reciprocal oppositeness

ZIGGY By Tom Wilson

Figure 246

nature's code

AAA	AAC	ACA	ACC	CAA	CAC	CCA	CCC
AAG	AAU	ACG	ACU	CAG	CAU	CCG	CCU
AGA	AGC	AUA	AUC	CGA	CGC	CUA	CUC
AGG	AGU	AUG	AUU	CGG	CGU	CUG	CUU
GAA	GAC	GCA	GCC	UAA	UAC	UCA	UCC
GAG	GAU	GCG	GCU	UAG	UAU	UCG	UCU
GGA	GGC	GUA	GUC	UGA	UGC	UUA	UUC
GGG	GGU	GUG	GUU	UGG	UGU	UUG	UUU

"The Invention of the Genetic Code," by Brian Hayes,
American Scientist, January–February 1998
Figure 248

Horizon Magazine, May, 1977.

Figure 253

Figure 254

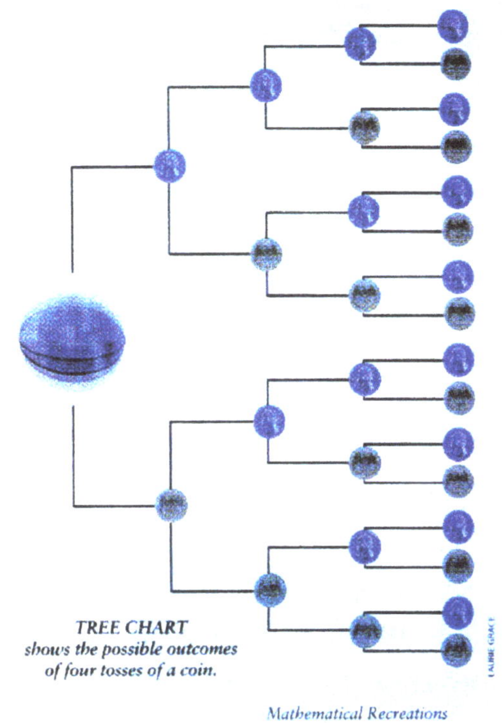

Figure 255 A

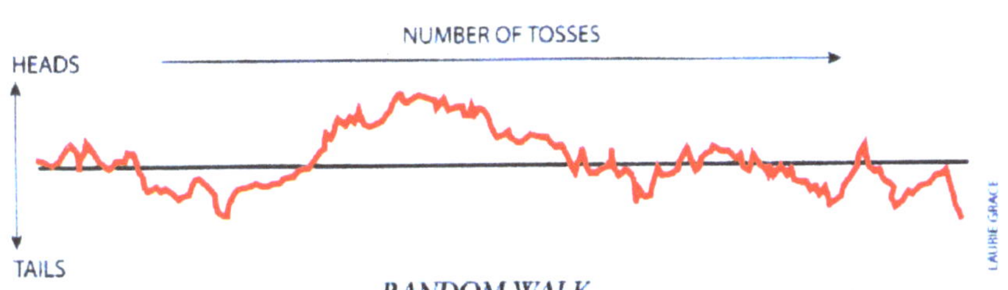

Figure 255 B

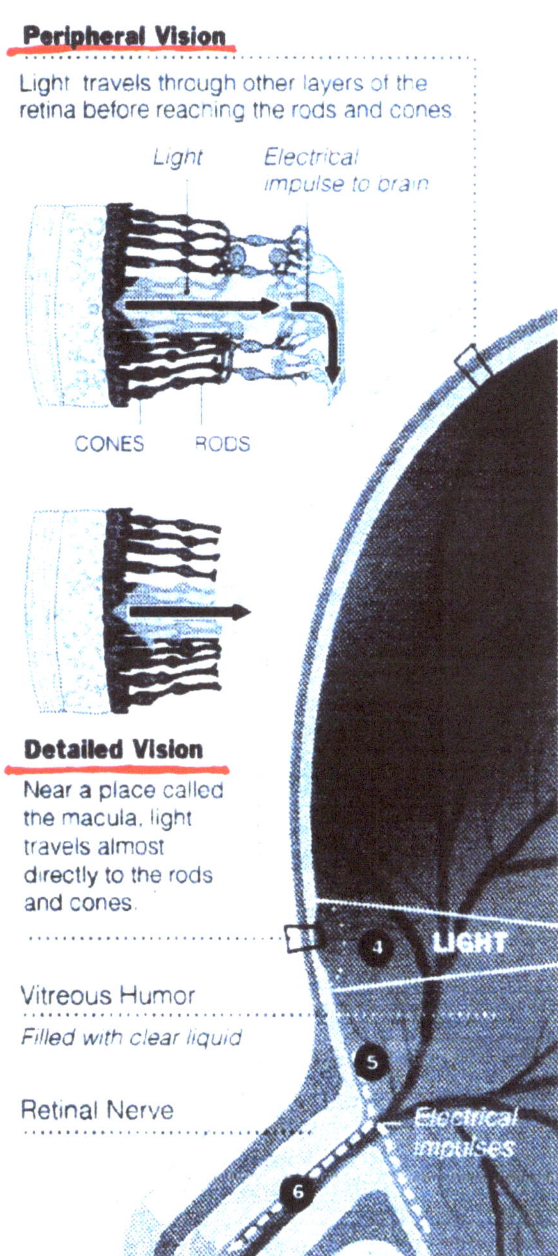

The *New York Times*, October 14, 1997

Figure 259

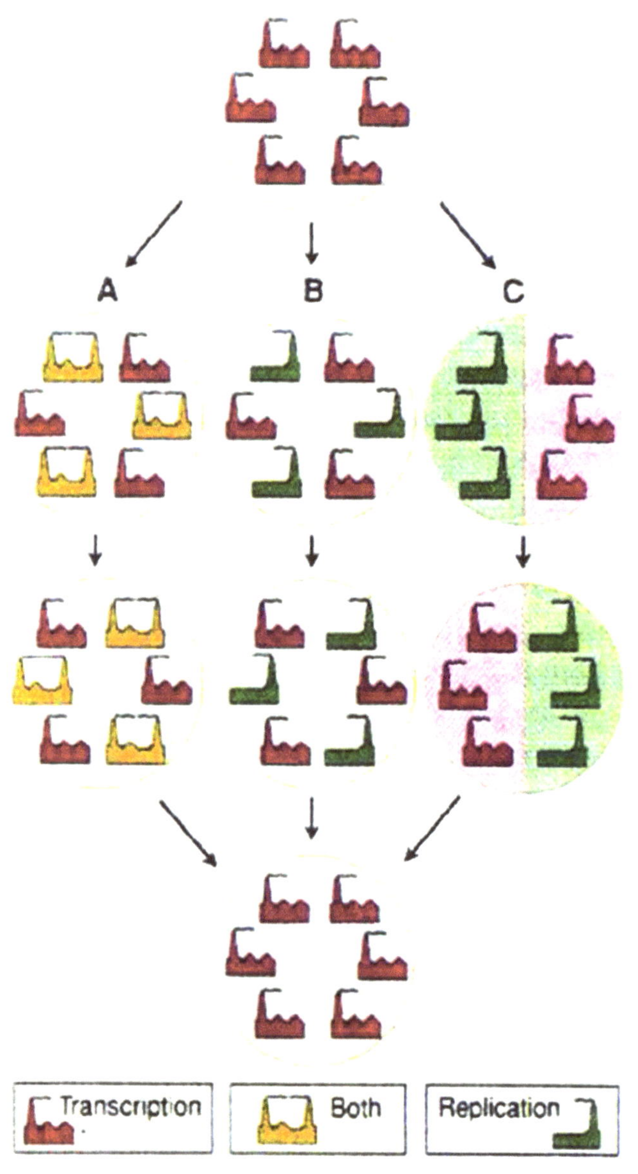

One or the other. Organizing replicating machines in nuclei. (A) Replicating machines are installed in functioning transcription factories. (B) Existing transcription factories are decommissioned and replaced by dedicated replication factories. (C) Zoning regulations ensure dedicated replication factories are grouped together.

"Duplicating a Tangled Genome," by Peter Cook, *Science*, September 4, 1998

Figure 261

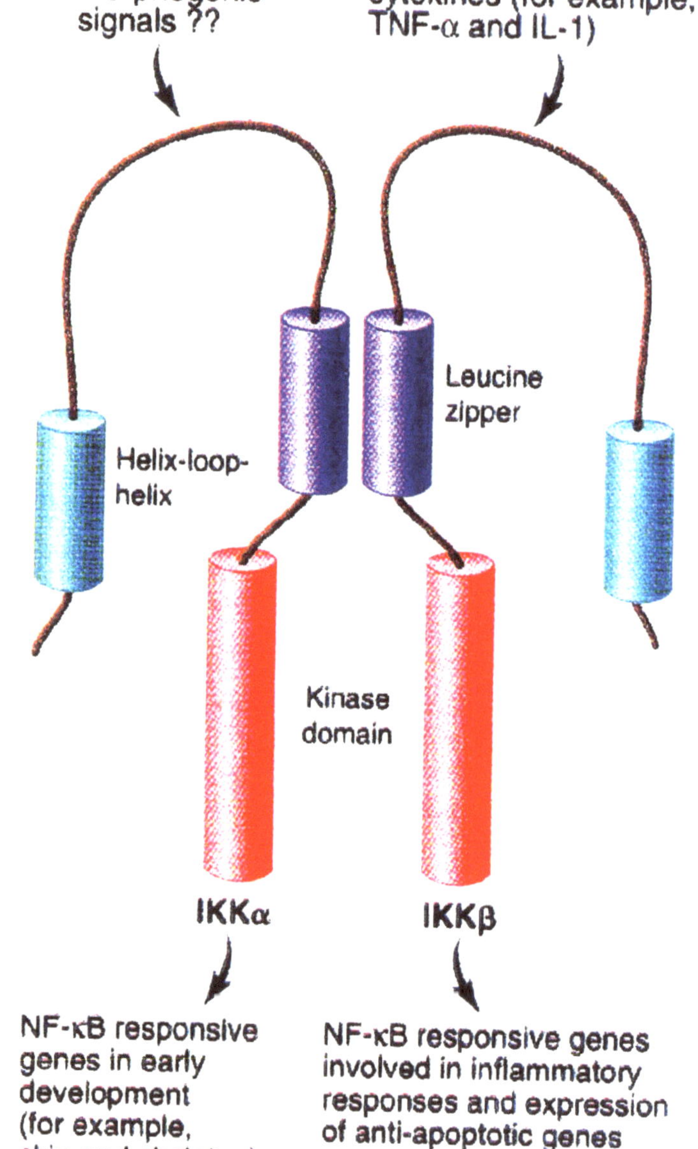

Figure 263

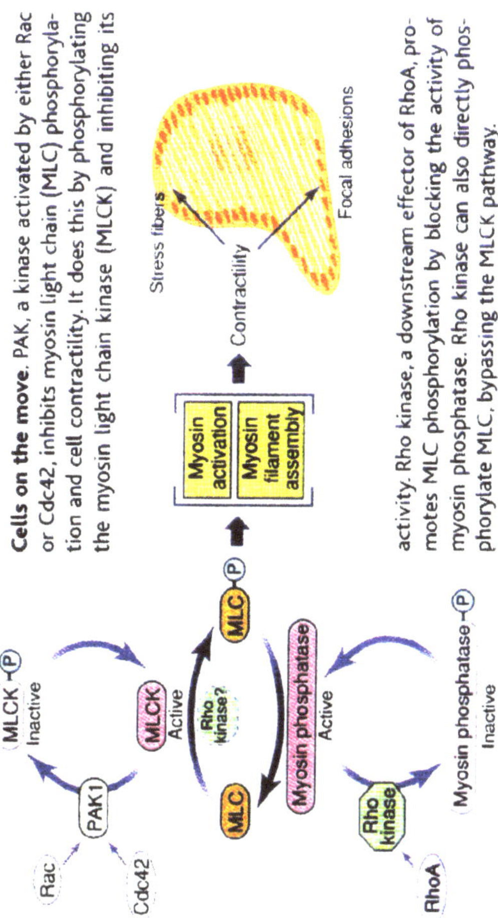

Figure 264

A Greek Letter, Demystified

In principle, screening out statistical noise from a particle physics experiment is a lot like determining whether a coin is biased toward heads or tails. Random events such as coin flips and particle detections tend to follow a bell curve distribution. Sigma—the standard deviation—is related to the fatness of the curve and gives a handy way to quantify how far from the center of the distribution your results are.

With the coin experiment, the center of the curve represents what you'd ideally get with a perfectly fair coin—50 heads and 50 tails, which is zero sigma away from the center. As you rack up a larger and larger surplus of heads (or tails), you move away from the center of the curve. A result of 55 heads and 45 tails is one sigma away from the center; there's about a 16% chance that an unbiased coin will give that result. On the other hand, 60 heads and 40 tails is a two-sigma result, which has only about a 2% chance of happening with an unbiased coin. A run of 65 heads and 35 tails would be pretty damning evidence for bias; it's a three-sigma result, which a fair coin would yield only 0.1% of the time. In a sense, a three-sigma result means that you're 99.9% sure that your coin is biased. At five sigma—the acid test applied to fundamental particles and extrasolar planets—the odds of a fluke dwindle to 1 in 3.5 million. –C.S.

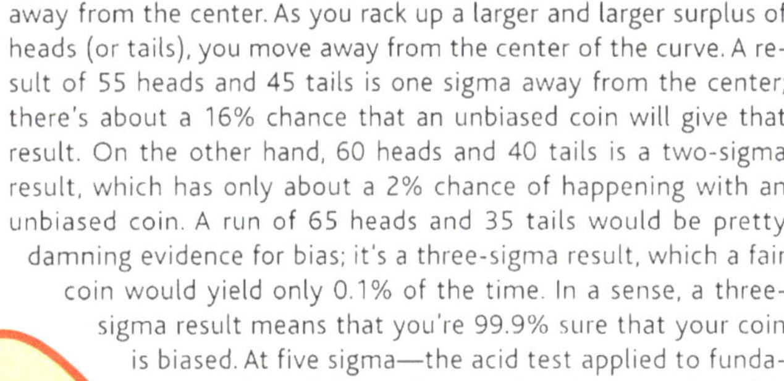

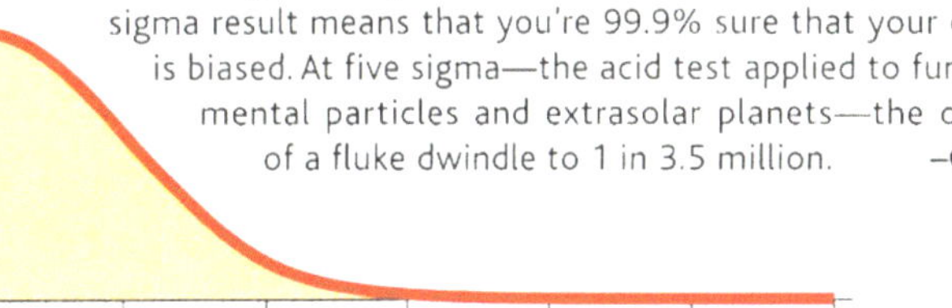

Figure 265

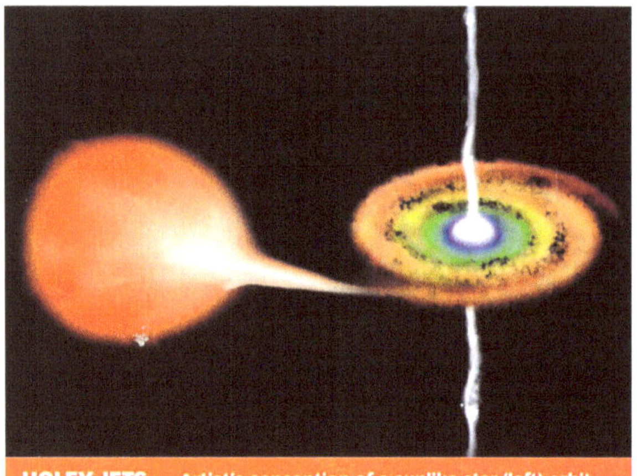

HOLEY JETS — Artist's conception of a sunlike star (left) orbiting a black hole (right). As the black hole pulls gas from the star, the material forms a disk heated to millions of degrees. Some of the energy may be emitted as jets perpendicular to the disk.

Figure 266

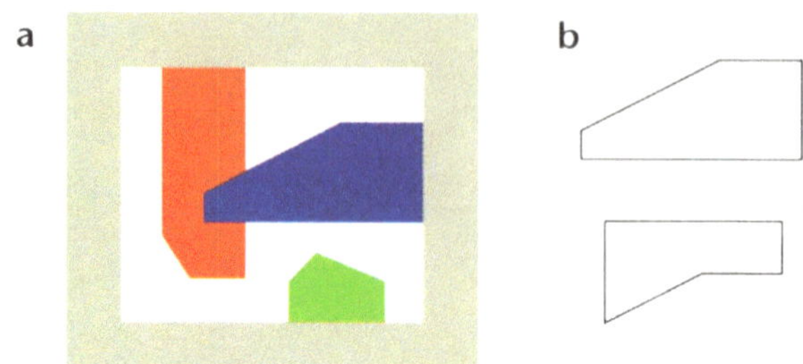

Fig. 1. The assignment of image regions as 'figure' or 'background' has a dramatic effect on their perceived shape. The top shape in (**b**) is easily found in (**a**), but the region that corresponds to the bottom shape in (**b**) is often unnoticed, because it is part of the background (top right corner) in (**a**). Two papers appearing this week report a neural correlate for this perceptual phenomenon.

Figure 269

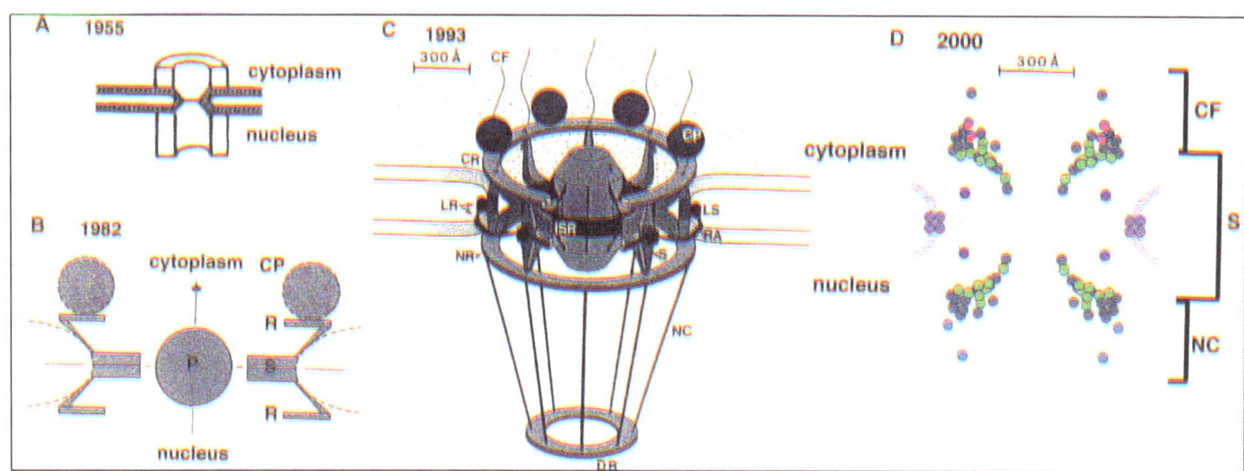

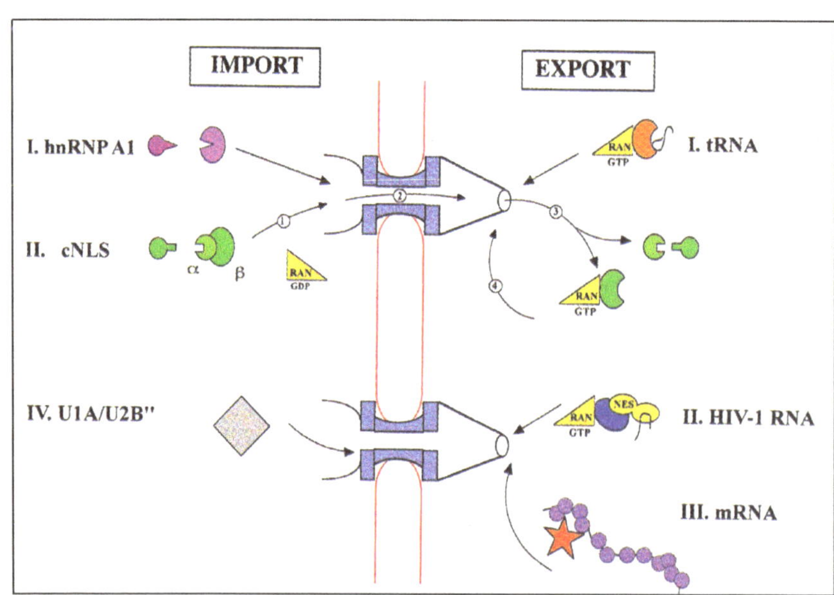

Figure 275

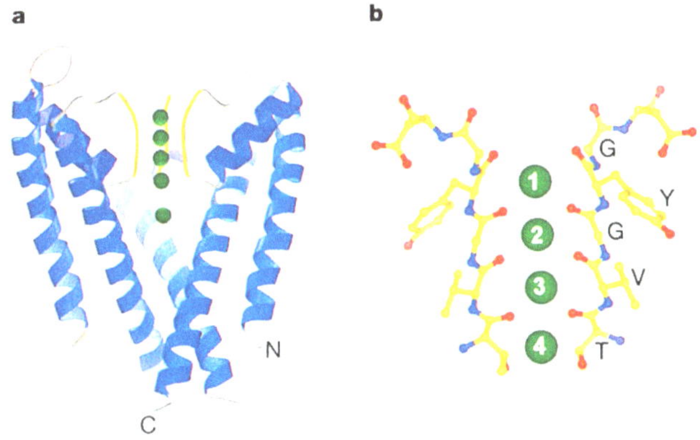

Figure 1 Binding sites for K⁺ ions in the KcsA K⁺ channel. **a**, Ribbon representation of the KcsA channel with the subunit closest to the viewer removed. Potassium ions (green spheres) bind at four locations in the selectivity filter (yellow) and in the water-filled cavity at the membrane centre (bottom ion). **b**, Close-up view of the selectivity filter in ball-and-stick representation, with the front and back subunits removed. The four K⁺ ions are numbered to indicate the location of binding sites in the filter; position 1 is closest to the extracellular solution and position 4 is closest to the cavity. Key amino acids forming the selectivity filter are shown. Pictures were prepared with GL-Render[26,27].

Figure 276

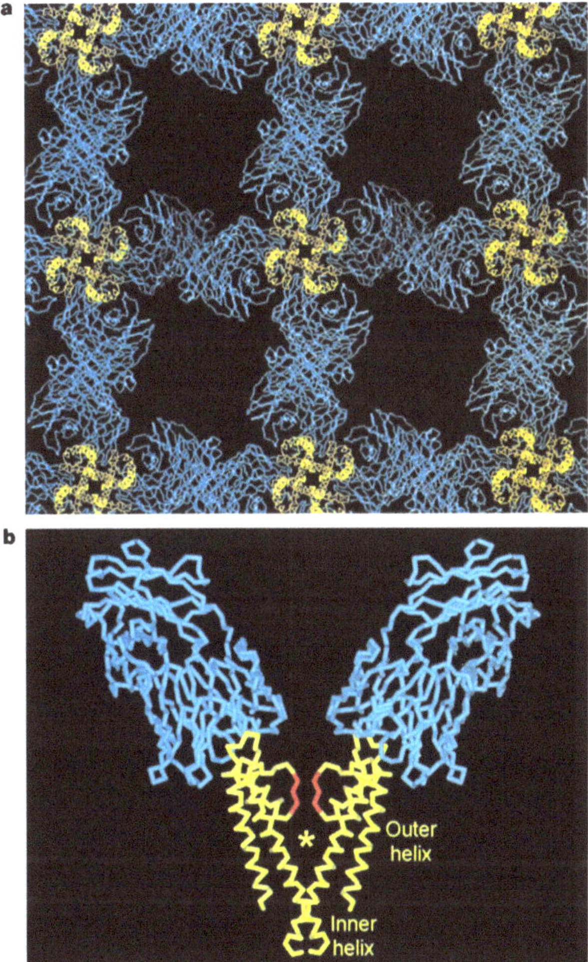

Figure 1 Fab attachment and crystal packing. KcsA (yellow) was crystallized as a complex with an antibody Fab fragment (blue). One Fab fragment is bound to the extracellular-facing turret on each K⁺ channel subunit. **a**, View down the four-fold crystallographic axis of the *I*4 cell, which corresponds to the molecular four-fold axis of the K⁺ channel. **b**, Two subunits and attached Fab fragments viewed perpendicular to the four-fold axis. The transmembrane outer and inner helices are labelled. The asterisk denotes the location of the central cavity, below the selectivity filter (red).

Figure 277

a

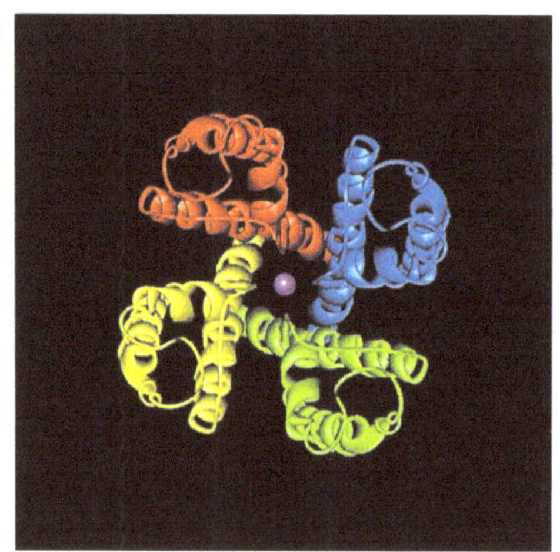

b

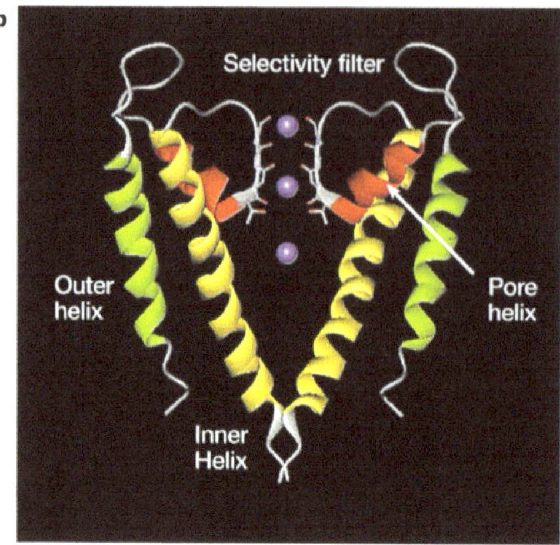

Figure 2 | **Structure of the KcsA channel. a** | Top view of the KcsA channel showing four subunits, each in a different colour. A K^+ ion is shown in purple. **b** | Side view of two diagonally opposed subunits. Inner helices, shown in yellow, line the permeation pathway. Three K^+ ions in the permeation pathway are shown in purple. Oxygens involved in coordinating K^+ ions are shown for the amino acids Thr-Val-Gly-Tyr-Gly in the selectivity filter. Adapted with permission from REF. 7 © 1998 American Association for the Advancement of Science.

"Cyclic Nucleotide-Gated Channels: Shedding light on the opening of a channel pore," by Galen E. Flynn et al., *Nature Reviews Neuroscience*, September 2001
Figure 278

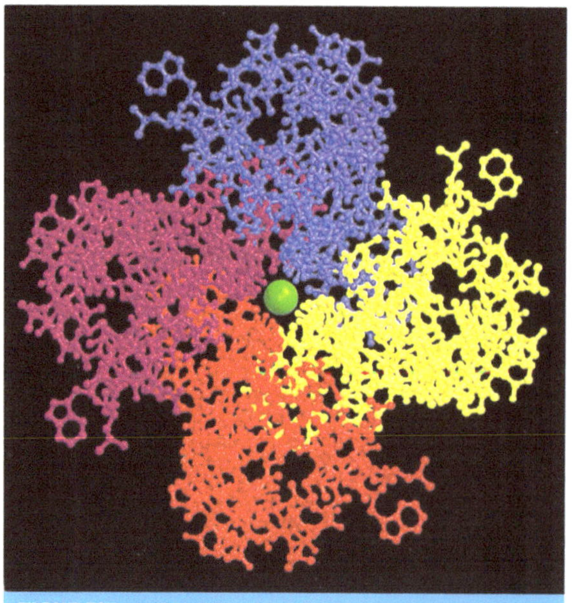

TIGHT FIT — This view of the KcsA potassium channel from the bacterium *Streptomyces lividans* shows the channel's four identical subunits in different colors. The center of the channel holds a potassium ion (green).

"Channel Surfing," by John Travis, *Science News*, March 9, 2002
Figure 279

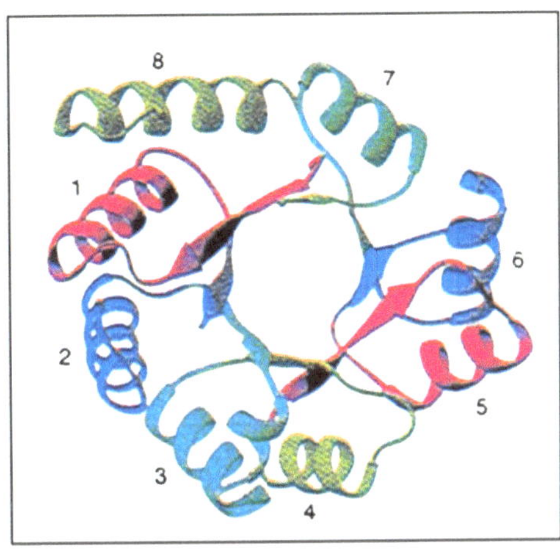

The classic β/α barrel. The structure, also called TIM barrel (named for triose phosphate isomerase), is built from eight repeated β strand/α helix units. Units 1 and 5 are shown in pink, 2 and 6 in blue, 3 and 7 in cyan, and 4 and 8 in green. Lang *et al.* provide evidence that two new β/α barrel structures, HisA and HisF, evolved from a half-barrel ancestor (*3*). This figure is based on the structure of another β/α barrel enzyme, the α subunit of tryptophan synthase from *Salmonella typhimurium* (*12*). The figure was created with the program Ribbons and the coordinates 1bks from the Protein Data Bank. The original structure has three extra α helices, which have been deleted to create a classic β/α barrel.

Figure 280

The Logic of Circuits

Shannon saw the analogy between simple electronic switches and the logic operations that are at the heart of digital computing.

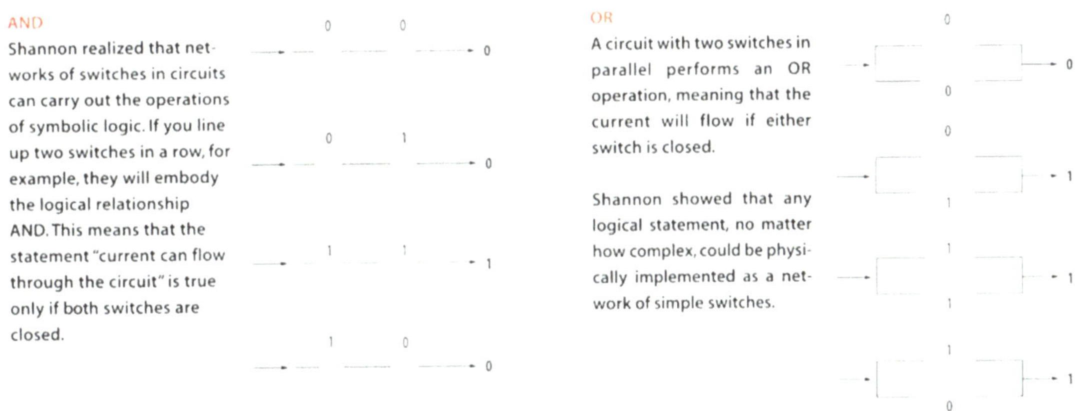

Figure 282

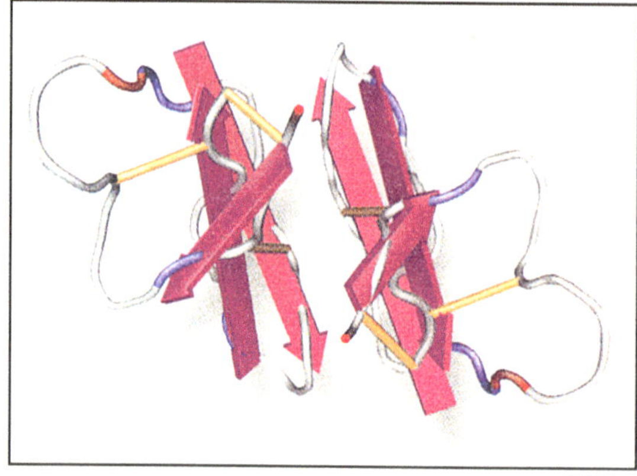

Defensins to the rescue. Human neutrophils secrete the antimicrobial peptides α-defensin-1, -2, and -3 in response to bacterial infection. These peptides form dimeric structures that contain a total of six β-strands (purple arrows). The positively charged side chains of α-defensin-3 are shown in blue, negative charges in red, and disulfide bonds in orange. [Protein structure coordinates from (*12*)]

Figure 290

CONSOLIDATED BALANCE SHEETS

December 31 (in thousands of USD)		2002		2001
ASSETS				
Current assets:				
Cash and cash equivalents (note 1)	$	121,851	$	97,319
Restricted cash		13,854		—
Trade and other receivables		3,559		1,685
Inventories (note 2)		5,906		7,292
Deferred tax asset (note 8)		3,383		14,895
Other current assets		1,619		1,560
Total current assets		150,172		122,751
Property, plant and equipment, net (note 4)		549,336		93,475
Deferred charges		1,912		—
Other assets (note 1)		1,945		1,295
Total assets	$	703,365	$	217,521
LIABILITIES AND SHAREHOLDERS' EQUITY				
Current liabilities:				
Accounts payable, trade and other	$	7,284	$	5,311
Accrued and other liabilities (note 6)		14,856		22,752
Total current liabilities		22,140		28,063
Deferred income taxes (note 8)		159,131		6,454
Other long-term liabilities (note 7)		28,876		35,791
Shareholders' equity:				
Common Shares, no par value, authorized unlimited shares, 98,985,974 and 75,853,183 shares issued and outstanding in 2002 and 2001, respectively (note 11)		380,309		76,829
Additional paid-in capital		3,658		3,658
Minority interest		1,009		—
Retained earnings		108,242		66,726
Total shareholders' equity		493,218		147,213
Total liabilities and shareholders' equity	$	703,365	$	217,521

See notes to consolidated financial statements.

Signed on behalf of the Board of Directors:

Dr. David S. Robertson
DIRECTOR

Brian J. Kennedy
DIRECTOR

CONSOLIDATED STATEMENTS OF CASH FLOWS

Years Ended December 31 (in thousands of USD)	2002	2001	2000
Cash flows from operating activities:			
Net income	$ 41,516	$ 38,725	$ 40,600
Adjustments to reconcile net income to net cash provided by (used in) operating activities:			
Provision for depreciation, depletion and amortization	22,785	22,507	24,028
Gain on sale of assets, net	(161)	(1,863)	(100)
Provision for reclamation, net of costs incurred	(2,506)	1,596	1,316
Stock compensation expense	1,697	1,438	771
Provision for pension cost, net of contributions	510	276	287
Deferred taxes, net	9,781	859	(859)
(Increase) decrease in assets:			
Trade and other receivables	(1,074)	(263)	3,676
Inventories	1,386	(1,519)	2,955
Other current assets	(121)	(305)	174
Other assets	(2,800)	873	1,073
(Decrease) increase in liabilities:			
Accounts payable, trade and other	1,973	2,324	(2,437)
Accrued and other liabilities	478	1,116	1,516
Other long-term liabilities	(2,385)	4,300	(1,468)
Net cash provided by operating activities	71,079	70,064	71,532
Cash flows from investing activities:			
Capital spending	(37,539)	(24,783)	(17,754)
Proceeds from sale of assets, net	—	1,978	200
Net cash used in investing activities	(37,539)	(22,805)	(17,554)
Cash flows from financing activities:			
Proceeds from exercise of stock options	4,795	4,847	426
Minority interest	51	—	—
Repayments of long-term debt	—	(18,000)	(12,000)
Deposit of restricted funds for collateral	(13,854)	—	—
Net cash used in financing activities	(9,008)	(13,153)	(11,574)
Increase in cash and cash equivalents	24,532	34,106	42,404
Cash and cash equivalents, beginning of year	97,319	63,213	20,809
Cash and cash equivalents, end of year	$ 121,851	$ 97,319	$ 63,213
Supplemental disclosure of cash flow information:			
Cash paid and (refunds received) for income taxes	$ 1,700	$ —	$ —
Cash paid for interest	$ —	$ 900	$ 2,100
Non-cash investing and financing activities:			
Common Shares issued in purchase acquisition	$ 296,988	$ —	$ —

See notes to consolidated financial statements.

Figure 291

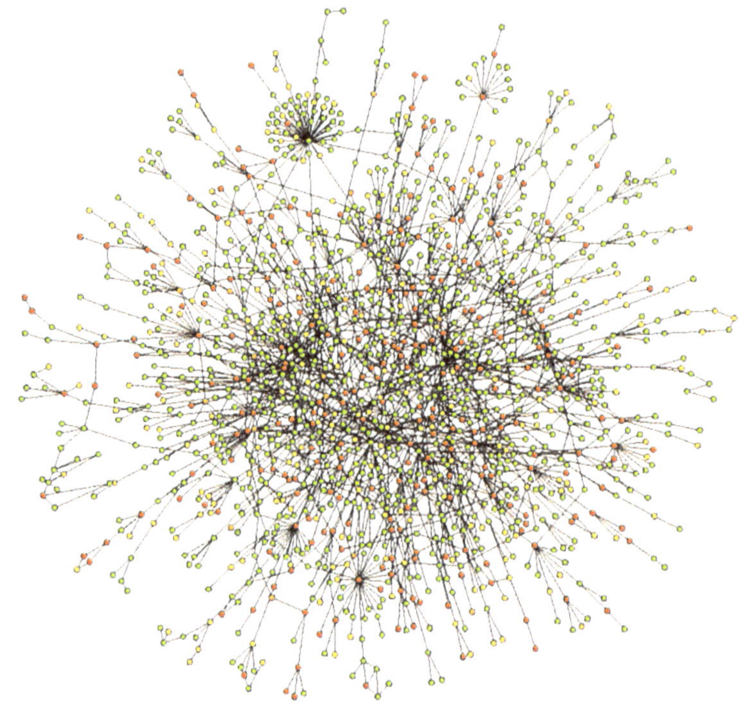

Figure 293

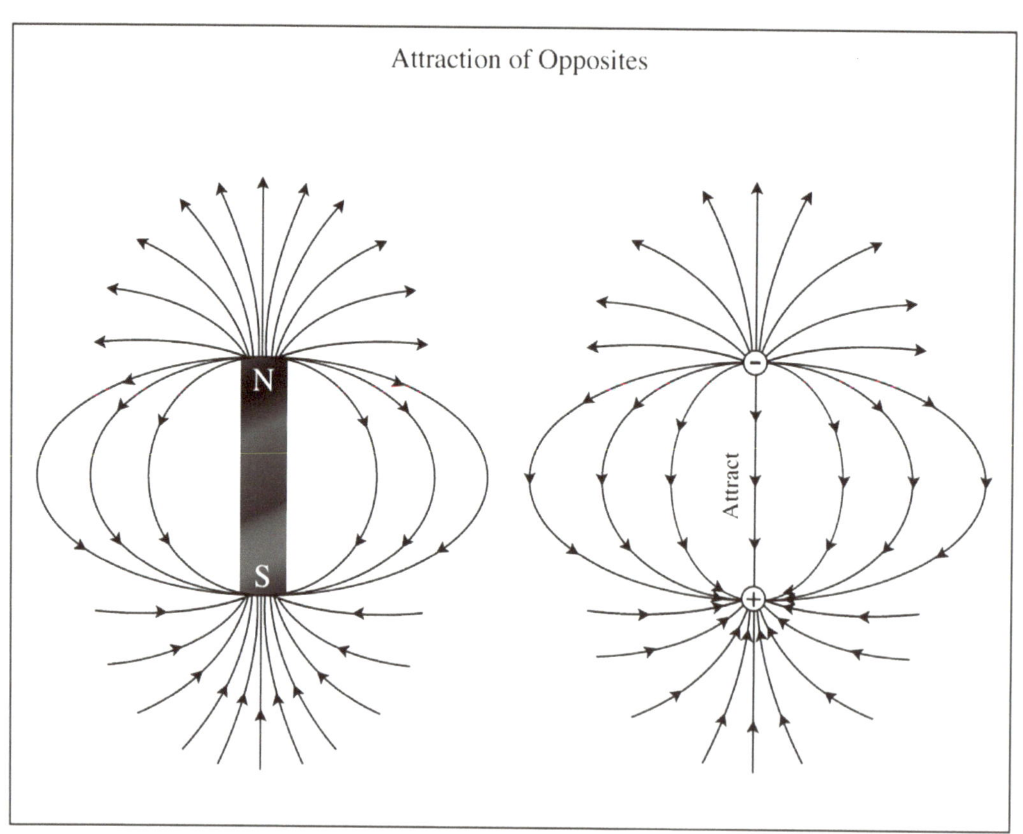

Figure 294a

Repulsion of Sameness

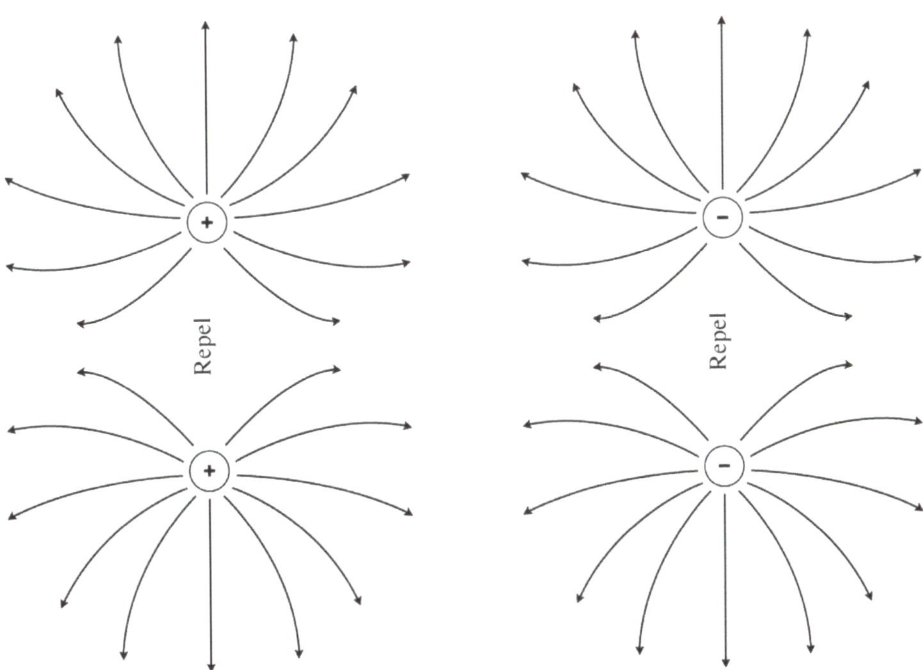

FIGURE 294 b

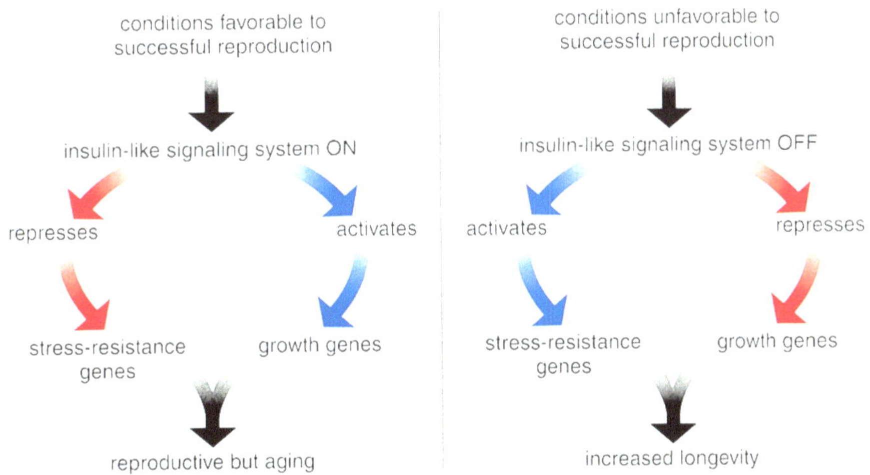

Figure 6. Environmental conditions influence aging. Conditions favorable to successful reproduction, such as a rich food supply and a dearth of potential competitors (in the case of nematodes) sparks the insulin-like signaling system. That system represses stress-resistance genes, which make products that help an organism fight stress, and activates growth genes, which stimulate rapid body growth and a high reproductive rate. As a result, an organism can reproduce well but ages quicker because it is not putting energy into maintenance. Conditions unfavorable to successful reproduction, on the other hand, repress the insulin-like signaling system, which turns off growth genes but turns on stress resistance genes. That pattern supports increased longevity. So, switching metabolism from a pro-growth stance to a pro-repair stance by repressing the insulin-like signaling system might delay the onset of senescence in many organisms.

"Aging: A Biological Perspective," by Roberts A. King, *American Scientist*, November–December 2003

Figure 295

Gyroscopes

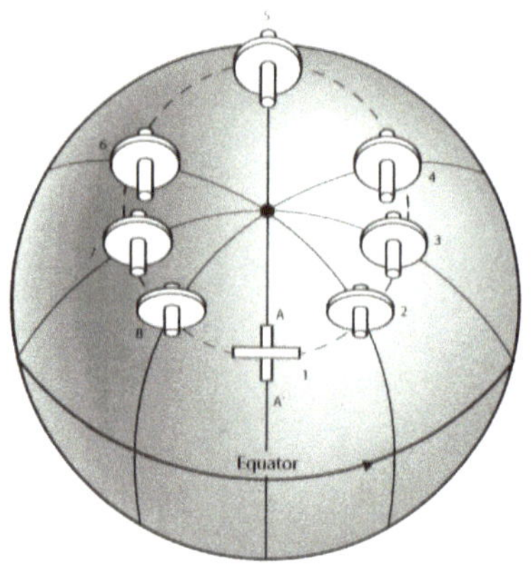

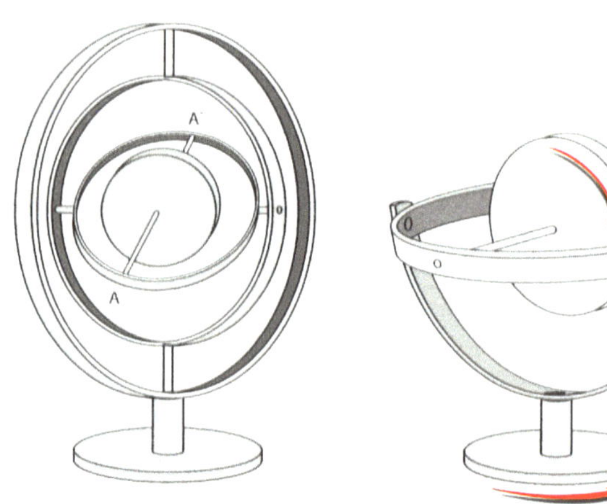

Encyclopedia Britannica
Figure 298

A Tiny, Multi-Tasking Computer

In computers, information is described as a series of 1's and 0's, called bits. In quantum computing, scientists manipulate individual atomic spins, with different directions representing 1 or 0.

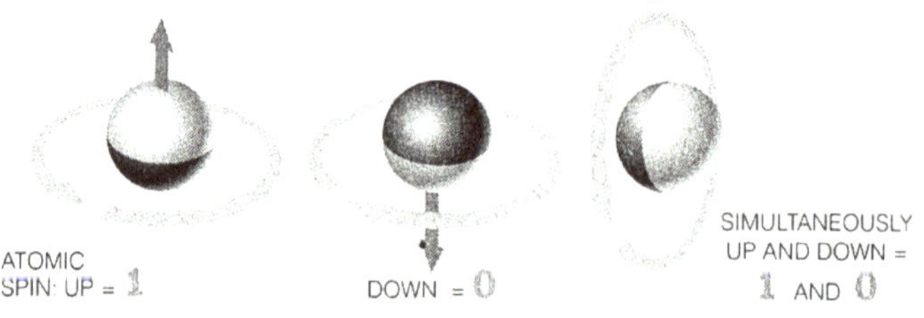

SIMULTANEOUS VALUE The laws of quantum physics allow atoms to be in both states, 1 and 0, simultaneously. This phenomenon, called superposition, lets quantum computers perform multiple operations at the same time. (Ordinary computers can make only one calculation at a time.) Computational power based on such super-positions increases exponentially with the number of atoms involved:

Two atoms, with a total of four digits, yield four quantities

1, 0
1, 0

BITS	DECIMAL EQUIVALENT
00	0
01	1
10	2
11	3

Three atoms yield eight quantities

1, 0
1, 0
1, 0

000 = 0　100 = 4
001 = 1　101 = 5
010 = 2　110 = 6
011 = 3　111 = 7

The New York Times

Figure 299

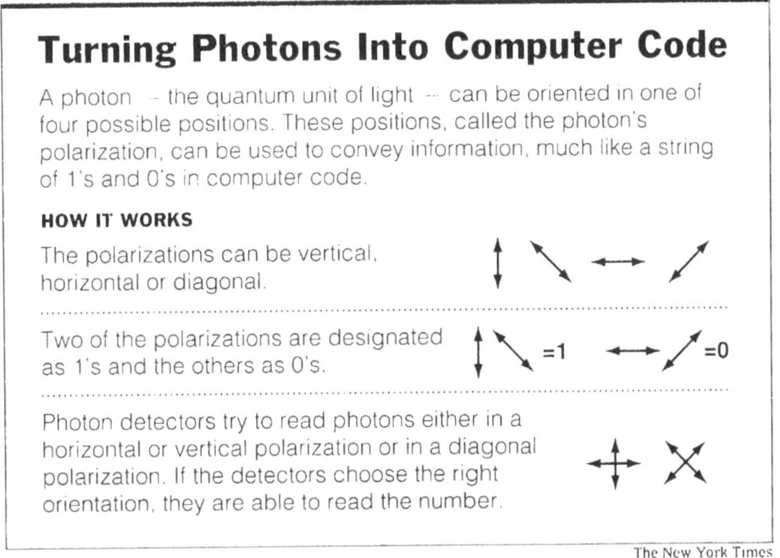

Turning Photons Into Computer Code

A photon -- the quantum unit of light -- can be oriented in one of four possible positions. These positions, called the photon's polarization, can be used to convey information, much like a string of 1's and 0's in computer code.

HOW IT WORKS

The polarizations can be vertical, horizontal or diagonal.

Two of the polarizations are designated as 1's and the others as 0's. =1 =0

Photon detectors try to read photons either in a horizontal or vertical polarization or in a diagonal polarization. If the detectors choose the right orientation, they are able to read the number.

The New York Times

Figure 300

Plane of Complex Numbers

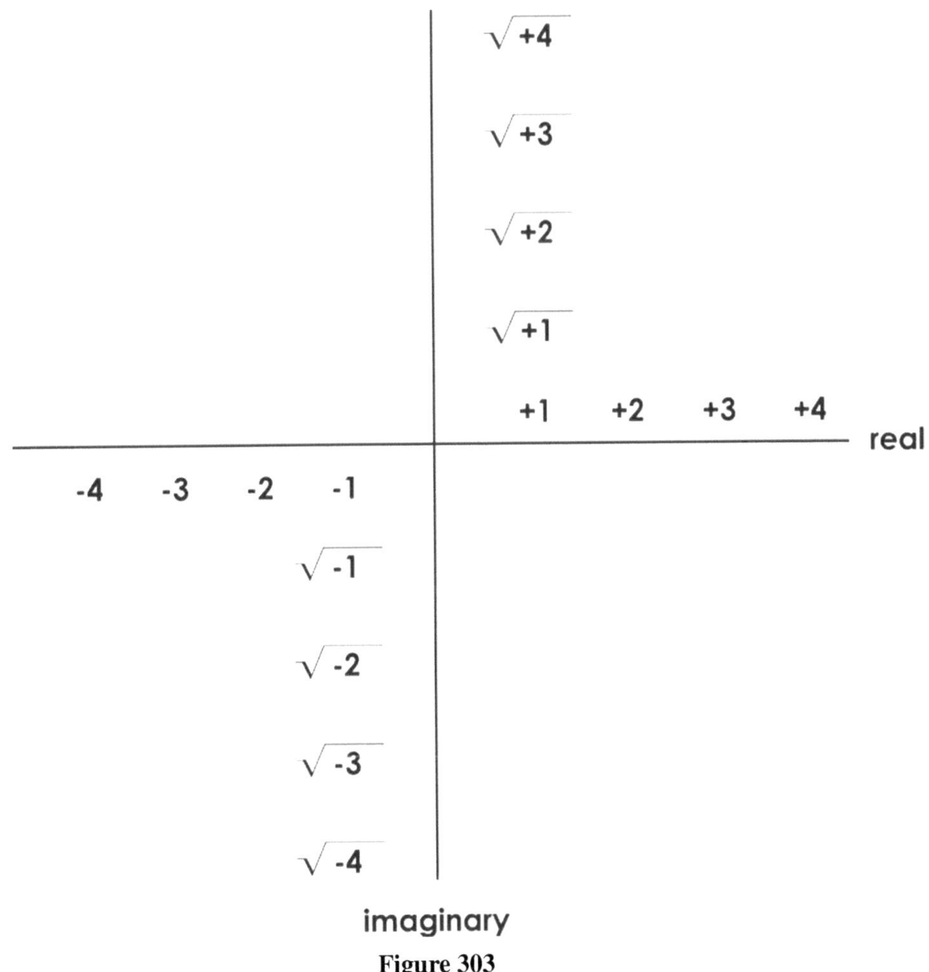

Figure 303

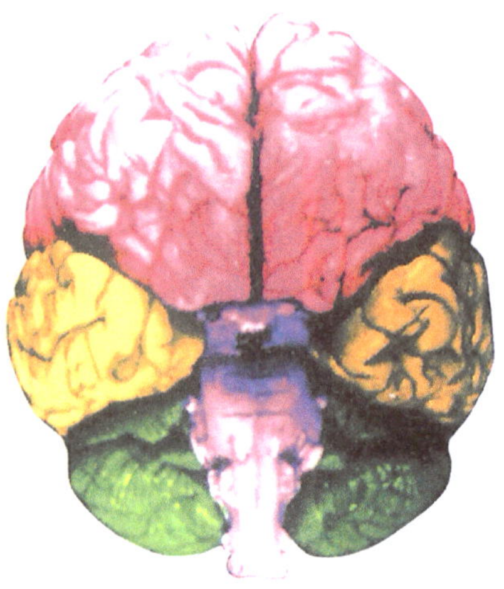

Figure 306 See also Figure 215

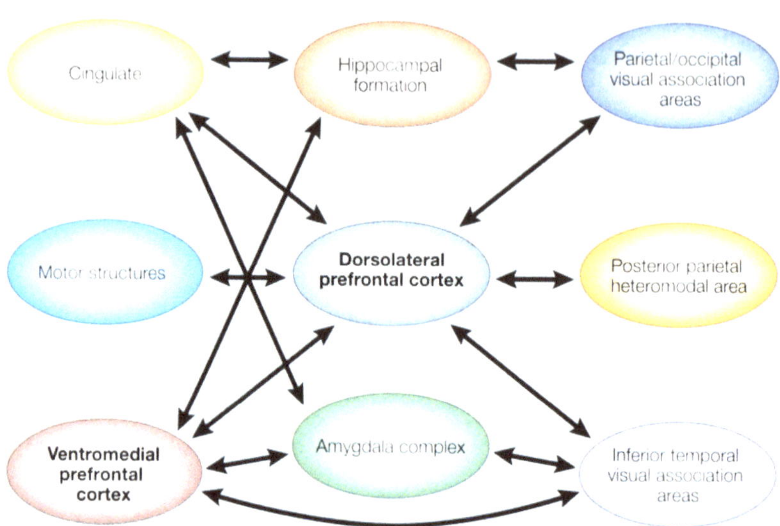

Figure 1 | **A summary of the connectivity between prefrontal cortex and other brain regions.** The ventromedial and dorsolateral prefrontal cortices exhibit reciprocal connectivity with different posterior brain regions, with ventromedial prefrontal regions being associated with emotional processing areas (for example, amygdala) and dorsolateral prefrontal regions with non-emotional sensory and motor areas (for example, basal ganglia and parietal cortex). Modified, with permission, from REF. 107 © 1999 Guilford Press.

Figure 310

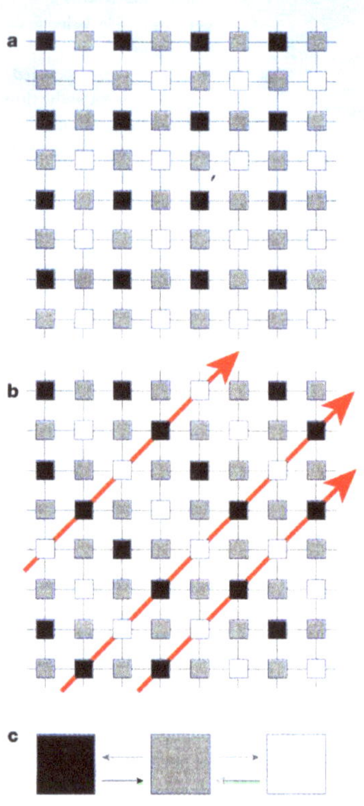

Figure 4 **A square lattice, with nearest-neighbour coupling between nodes.** a, **The balanced equivalence relation is indicated by the three-colour scheme. All couplings are bidirectional, and nodes of the same colour are synchronous.** b, **Arbitrary parity changes along diagonals (red arrows, interchanging black and white squares) also produce a balanced equivalence relation.** c, **The quotient network has three nodes and is the same in all cases. Its left–right symmetry implies that there are robust, 'multirhythm', time-periodic states, in which the black and white nodes are half a period out of phase and the grey node has twice the frequency. Networks** a **and** b **therefore possess corresponding states in which identically coloured nodes are synchronous, and which can all be stable simultaneously**[33].

Figure 313

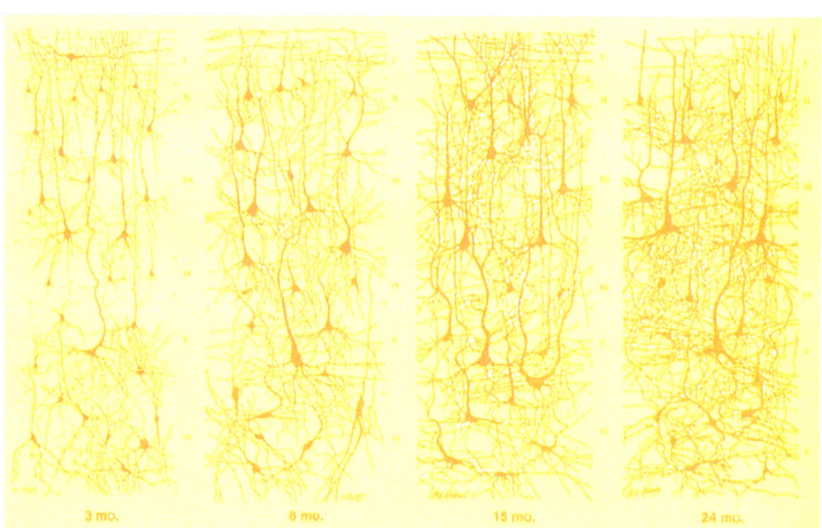

Growth factor: neurons develop rapidly in the human cortex between 3 and 24 months of age.

Figure 315

neuro quest EXPLORE YOUR BRAIN'S INNER WORKINGS

BEAUTY SECRET

What separates homely from comely?

BY ERIC HASELTINE

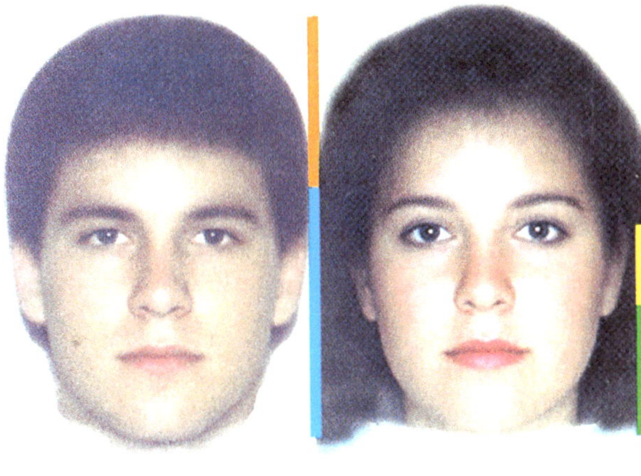

DEVELOPMENTAL PSYCHOLOGISTS RECENTLY MADE THE startling discovery that the faces most people regard as beautiful are actually quite average mathematically speaking, as is the case with these composites created by Judith Langlois of the University of Texas at Austin. She purposely blended the facial features of 32 men and 32 women. So what exactly is appealing?

Perfect symmetry is probably a factor, but a secret known to ancient Greek architects may play a larger role. The following tests will help you discover for yourself why you instinctively consider some people more beautiful than others.

EXPERIMENT 1 All these faces are symmetrical but not equally appealing. Which one is most attractive to you?

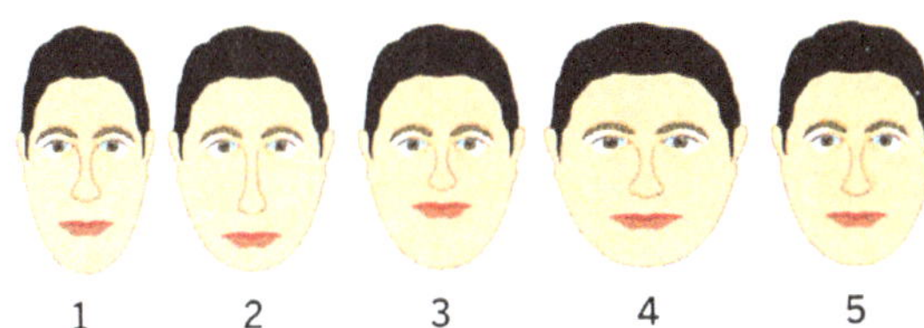

You probably found number 5 to be the handsomest because its proportions exactly represent what the ancient Greeks called the golden number (1.618...), a transcendental number like pi. That quantity is reached by dividing a line into two unequal parts, known as a golden section, so that the ratio of the long segment to the short segment equals the ratio of the entire line to the long segment. Golden ratios were designed into many ancient buildings, including those on the Acropolis in Athens.

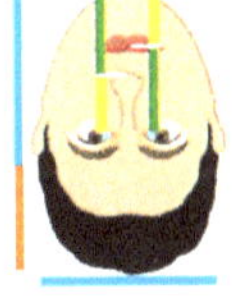

The blue-and-orange line lying between Judith Langlois's composite faces is divided according to the golden ratio. Notice that the width of each head is equal to the distance from the chin to the eyebrows, as measured by the longer segment of the golden section. And if you divide both faces into halves top to bottom, the distance from the chin to the mouth and from the eyes to the tip of the nose can be measured by a smaller golden section (the green-and-yellow line). Turn the page upside down to see how these relationships hold for handsome Mr. 5.

EXPERIMENT 2 Do golden ratios influence our appreciation of the human body as well? Check out these five specimens and judge which one seems most well proportioned.

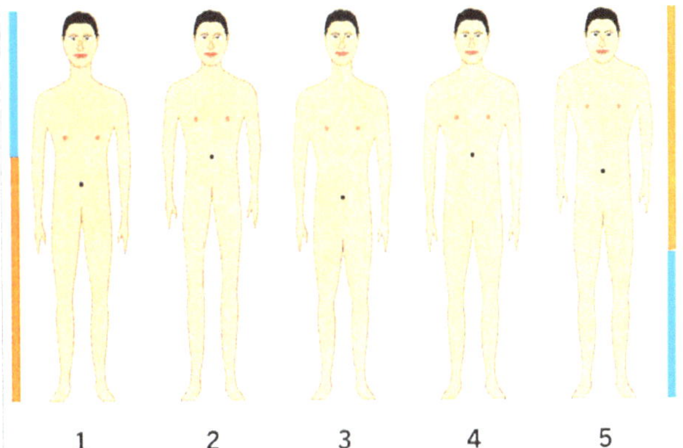

Body number 4 should appear to be the most well sculptured. Examine it closely using the golden section that flanks the group of figures for reference. When a well-proportioned body is measured with a golden section, the larger segment defines both the distance from the top of the head to the tips of the fingers and from the soles of the feet to the navel.

Artists and architects have long suspected that our sense of what is aesthetically pleasing in artificial things, like buildings, grows out of an unconscious sense of what is beautiful in faces and bodies. Whether we are born with this sense or learn it is a subject of debate among scientific researchers. But our genes (at least on average) do carry a program for growing various body parts in "golden" proportions. Some biologists believe the reason for this is that genes don't store enough information to specify exactly what a mature organism will look like but instead cheat with simple rules like "grow this body part to the next stage until it's as big as the sum of its last two growth stages." This rule is equivalent to forming a mathematical series that starts with 0 and 1, then progressively grows by adding new numbers that are the sum of two preceding numbers (i.e., 0, 1, 1, 2, 3, 5, 8, 13, 21...). The Italian merchant and mathematician Fibonacci, who developed this progression in the 13th century, also discovered that the ratio of any two adjacent numbers in the series converges on the golden mean of 1.618.

If Fibonacci's mathematical series does turn out to be the key to understanding how our bodies grow into their final proportions, it will be so simple as to be beautiful! ☒

"The Mating Mind: How Sexual Choice Shaped the Evolution of Human Nature," by Geoffrey Miller (Heinemann/Doubleday, 2000)

Figure 317

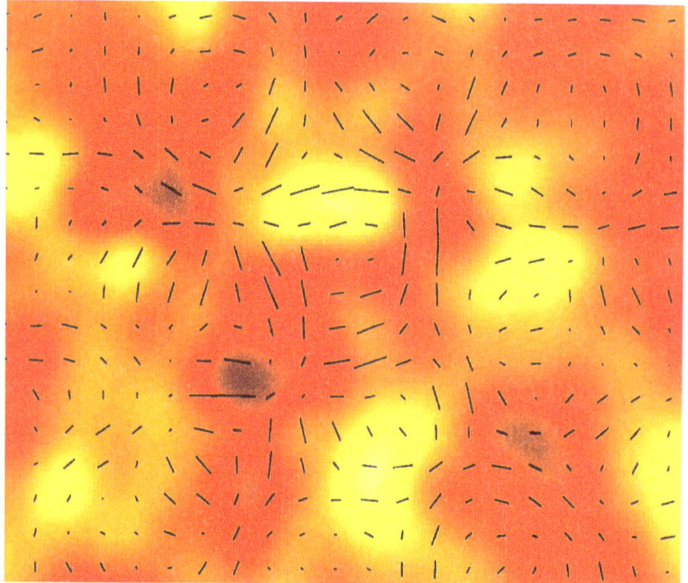

Figure 1 Polarization at the pole. The Degree Angular Scale Interferometer (DASI) in Antarctica (right) has detected[2,3] the polarization of the cosmic microwave background (CMB). The temperature fluctuations in the CMB are shown above — yellow for hot, red for cold. Superimposed is the polarization of the CMB photons measured by DASI: the polarization at each point is represented by a black line, whose orientation and length correspond to the direction and intensity of polarization, respectively. This degree of polarization of the CMB is consistent with the cosmological model of inflation, and further measurements could reveal more about this period of rapid expansion in the early Universe.

Figure 318

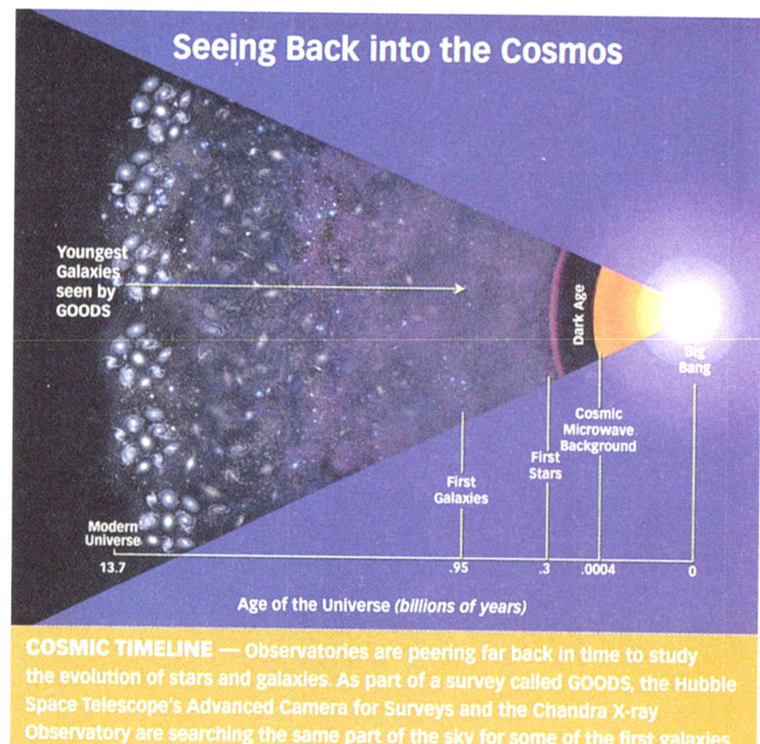

Figure 319

Revising Einstein

Some theorists say Einstein's theory of relativity needs an overhaul to fit extreme conditions that prevailed during the Big Bang, when nature was presumably ruled by a single unified force.

EINSTEIN'S EQUATION One consequence of relativity is that mass can be converted to energy, and vice versa. This kind of conversion powers the atomic bomb.

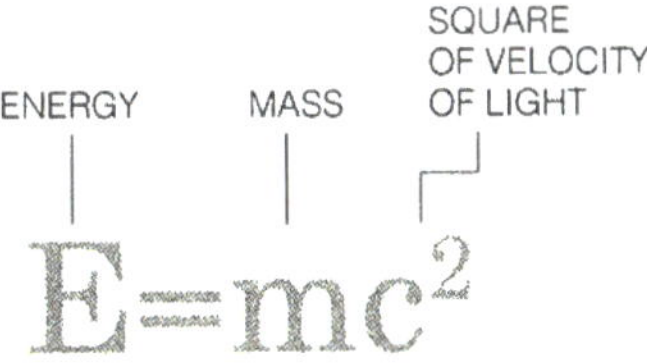

THE NEW EQUATION In a proposed new version called doubly special relativity, Einstein's formula is revised to ensure that no elementary particle can have an energy greater than the so-called Planck energy (E_p, about 10^{19} electron volts). At that energy, experts theorize, all forces are unified.

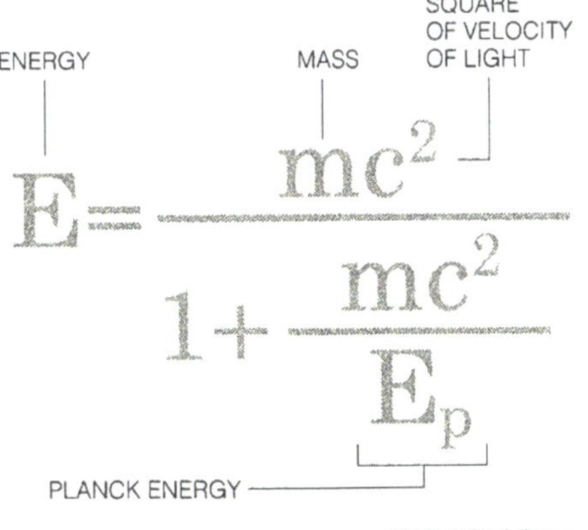

The New York Times

Figure 338

Mirror Reflections

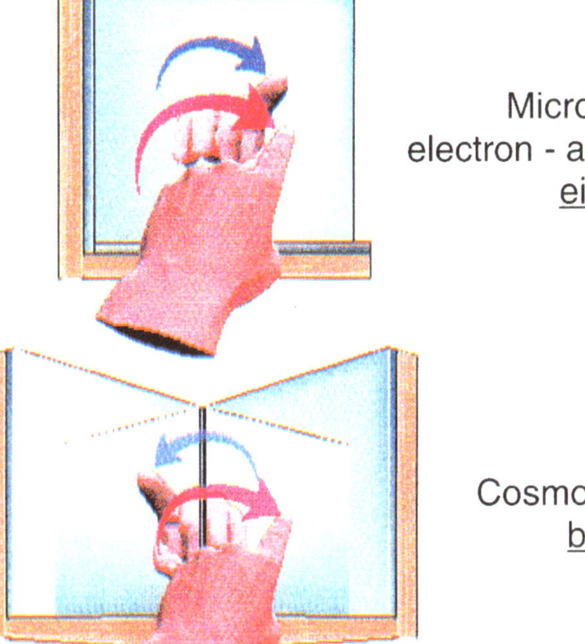

Micro - Particle
electron - antielectron spins
<u>either-or</u>

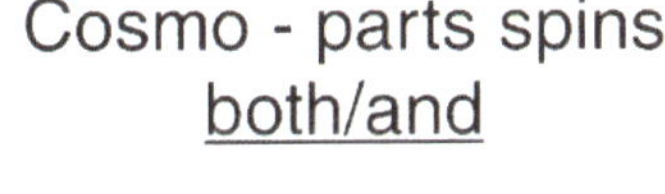

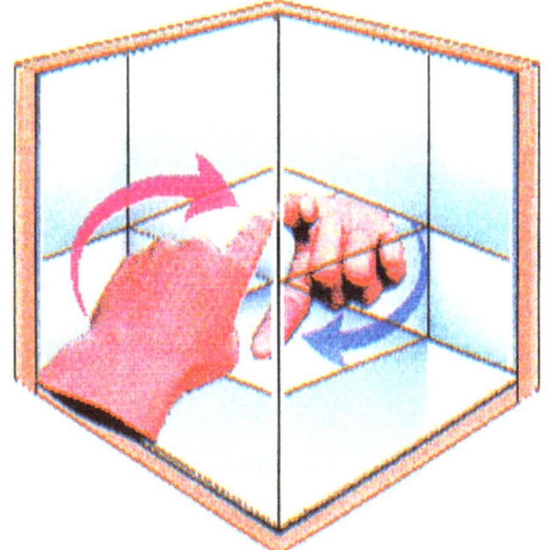

Macro - parts spins
<u>either/both</u>

Figure 321

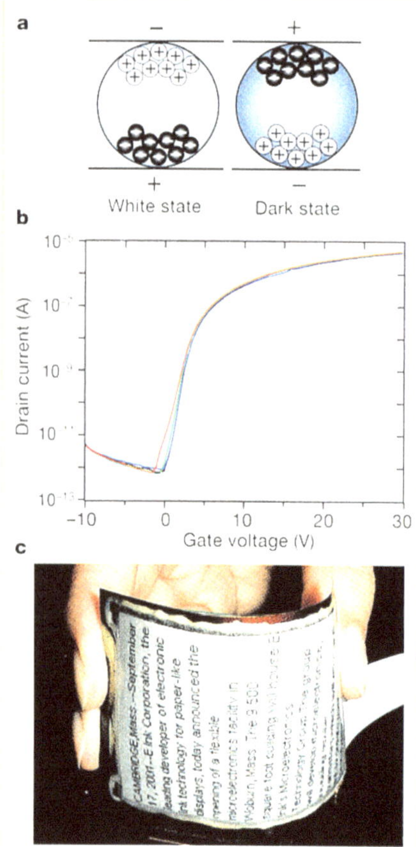

Figure 1 Flexible active-matrix electronic-ink displays. **a,** Operating principle of electronic ink. The relative movement of negatively charged black and positively charged white particles inside their microcapsules is controlled by the direction of the applied voltage. **b,** A backplane thin-film transistor measured *in situ* under compressive stress. The transistor is bent to three different radii of curvature: green, 2.0 cm (0.19% strain); blue, 1.3 cm (0.29% strain); and red, 1.0 cm (0.38% strain). The thin-film transistor has identical characteristics when measured without bending (black curve) and at a radius of curvature of 2.0 cm; degradation is minimal even at 1.0 cm. Results were similar under tensile stress. **c,** Text image shown on a bent display whose resolution is 96 d.p.i. and which has a white-state reflectance of 43% and a contrast ratio of 8.5:1.

Figure 323

Figure 326

2 • BUCS GAMEDAY • SUNDAY, JANUARY 12, 2003 • THE TAMPA TRIBUNE • TBO.com

BUCS VS. 49ERS

49ers Offense

WR 81 Terrell Owens
LT 63 Derrick Deese
FB 40 Fred Beasley
LG 66 Eric Heitmann
QB 5 Jeff Garcia
RB 20 Garrison Hearst
RG 65 Ron Stone
C 62 Jeremy Newberry
RT 78 Scott Gragg
TE 82 Eric Johnson
WR 89 Tai Streets

Buccaneers Defense

RCB 20 Ronde Barber
WLB 55 Derrick Brooks
RE 97 Simeon Rice
DT 99 Warren Sapp
MLB 53 Shelton Quarles
NT 91 Chartric Darby
FS 34 Dexter Jackson
LE 94 Greg Spires
SS 47 John Lynch
SLB 51 Al Singleton
LCB 25 Brian Kelly

Buccaneers Offense

WR 19 Keyshawn Johnson
LT 72 Roman Oben
FB 40 Mike Alstott
LG 71 Kerry Jenkins
QB 14 Brad Johnson
RB 32 Michael Pittman
C 62 Jeff Christy
RG 60 Cosey Coleman
RT 67 Kenyatta Walker
TE 85 Ken Dilger
WR 87 Keenan McCardell

49ers Defense

RCB 36 Jason Webster
RLB 53 Jeff Ulbrich
RE 96 Andre Carter
RT 94 Dana Stubblefield
MLB 50 Derek Smith
LT 97 Bryant Young
FS 31 Zack Bronson
LE 91 Chike Okeafor
OLB 98 Julian Peterson
SS 33 Tony Parrish
LCB 29 Ahmed Plummer

Buccaneers

Punter/Holder	Kicker	Long Snapper	Punt Returner	Kick Returner
9 Tom Tupa	7 Martin Gramatica	66 Ryan Benjamin	86 Karl Williams	27 Aaron Stecker

7 Gramatica, Martin	31 Wansley, Tim	55 Brooks, Derrick	77 Washington, Todd	94 Spires, Greg
9 Tupa, Tom	32 Pittman, Michael	58 Golden, Jack	80 Yoder, Todd	95 Warner, Ron
10 King, Shaun	34 Jackson, Dexter	59 Smith, Justin	83 Jurevicius, Joe	96 Wyms, Ellis
11 Johnson, Rob	35 Ivy, Corey	60 Coleman, Cosey	84 Barlow, Reggie	97 Rice, Simeon
14 Johnson, Brad	38 Howell, John	62 Christy, Jeff	85 Dilger, Ken	99 Sapp, Warren
19 Johnson, Keyshawn	40 Alstott, Mike	64 Goodspeed, Dan	86 Williams, Karl	
20 Barber, Ronde	41 Wilcox, Daniel	66 Benjamin, Ryan	87 McCardell, Keenan	
23 Phillips, Jermaine	43 Cook, Jameel	67 Walker, Kenyatta	88 Dudley, Rickey	
25 Kelly, Brian	47 Lynch, John	71 Jenkins, Kerry	89 Crawford, Casey	
26 Smith, Dwight	51 Singleton, Al	72 Oben, Roman	90 Gurley, Buck	
27 Stecker, Aaron	52 Webster, Nate	74 Green, Cornell	91 Darby, Chartric	
30 Barnes, Darian	53 Quarles, Shelton	75 Brown, Lomas	93 Claybrooks, DeVone	

49ers

Punter/Holder	Kicker	Long Snapper	Punt Returner	Kick Returner
4 Bill LaFleur	3 Jeff Chandler	86 Brian Jennings	85 Vinny Sutherland	85 Vinny Sutherland

3 Chandler, Jeff	32 Barlow, Kevan	56 Moore, Brandon	78 Gragg, Scott	95 Engelberger, John
4 LaFleuer, Bill	33 Parrish, Tony	57 Anthony, Cornelius	81 Owens, Terrell	96 Carter, Andre
5 Garcia, Jeff	36 Webster, Jason	60 Lynch, Ben	82 Johnson, Eric	97 Young, Bryant
11 Doman, Brandon	38 Heard, Ronnie	62 Newberry, Jeremy	83 Stokes, J.J.	98 Peterson, Julian
13 Rattay, Tim	40 Beasley, Fred	63 Deese, Derrick	84 Wilson, Cedrick	99 Moran, Sean
20 Hearst, Garrison	43 Hawthorne, Duane	65 Stone, Ron	85 Sutherland, Vinny	
24 Rumph, Mike	45 Hauck, Tim	66 Heitmann, Eric	86 Jennings, Brian	
26 Holman, Rashad	50 Smith, Derek	71 Osika, Craig	88 Swift, Justin	
27 Smith, Paul	51 Rasheed, Saleem	72 Kosier, Kyle	89 Streets, Tai	
28 Keith, John	52 Killens, Terry	75 Flanigan, Jim	91 Okeafor, Chike	
29 Plummer, Ahmed	53 Ulbrich, Jeff	76 Davis, Jerome	93 Shaw, Josh	
31 Bronson, Zack	54 Stewart, Quincy	77 Willig, Matt	94 Stubblefield, Dana	

Figure 327

Bill Jacobson/Dia Art Foundation

An installation view of a Jo Baer exhibition, "The Minimalist Years, 1960-75," at Dia Center for the Arts.

Figure 331

The classic Mandelbrot set, a fractal that displays self-similarity at various scales. *Courtesy Richard F. Voss*

Figure 334

A Brief History of Fractals

- Fractal geometry developed from Benoit Mandelbrot's studies of complexity in the 1960s and 1970s. Mandelbrot coined the term "fractal" from the Latin *fractus* ("broken") to highlight the fragmented, irregular nature of these forms.
- Fractals display self-similarity—that is, they have a similar appearance at any magnification. A small part of the structure looks very much like the whole.
- Self-similarity comes in two flavors: exact and statistical. The artificial tree (*left series*) displays an exact repetition of patterns at different magnifications. For the real tree (*right*), the patterns don't repeat exactly; instead the statistical qualities of the patterns repeat. Most of nature's patterns obey statistical self-similarity, and so do Pollock's paintings.
- Fractals are characterized in terms of their "dimension," or complexity. The dimension is not an integer, such as the 1, 2 and 3 dimensions familiar from Euclidean geometry. Instead fractal dimensions are fractional; for example, a fractal line has a dimension between 1 and 2.

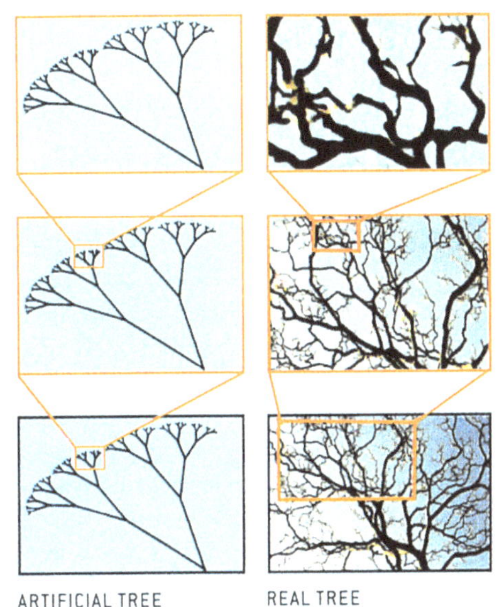

Scientific American, December 2002

Figure 335

Knight Ridder/Tribune

The new symbol of Lady Justice on the courthouse in DeLand, in Volusia County, includes a dress with a strap slipping off the shoulder. The artist modeled the relief after his girlfriend.

Figure 336

THE LOGIC OF ENERGY IN CHEMISTRY
Deduction energy (contraction)
base
general
non-metal
ENDOERGIC
acid
specific
metal
absorbs heat
ion
oppositeness
sameness
Induction energy (expansion)
metal
specific
acid
EXOERGIC
non-metal
general
base
emits heat
bond
©2007 Frederick S. Johnston, Jr.

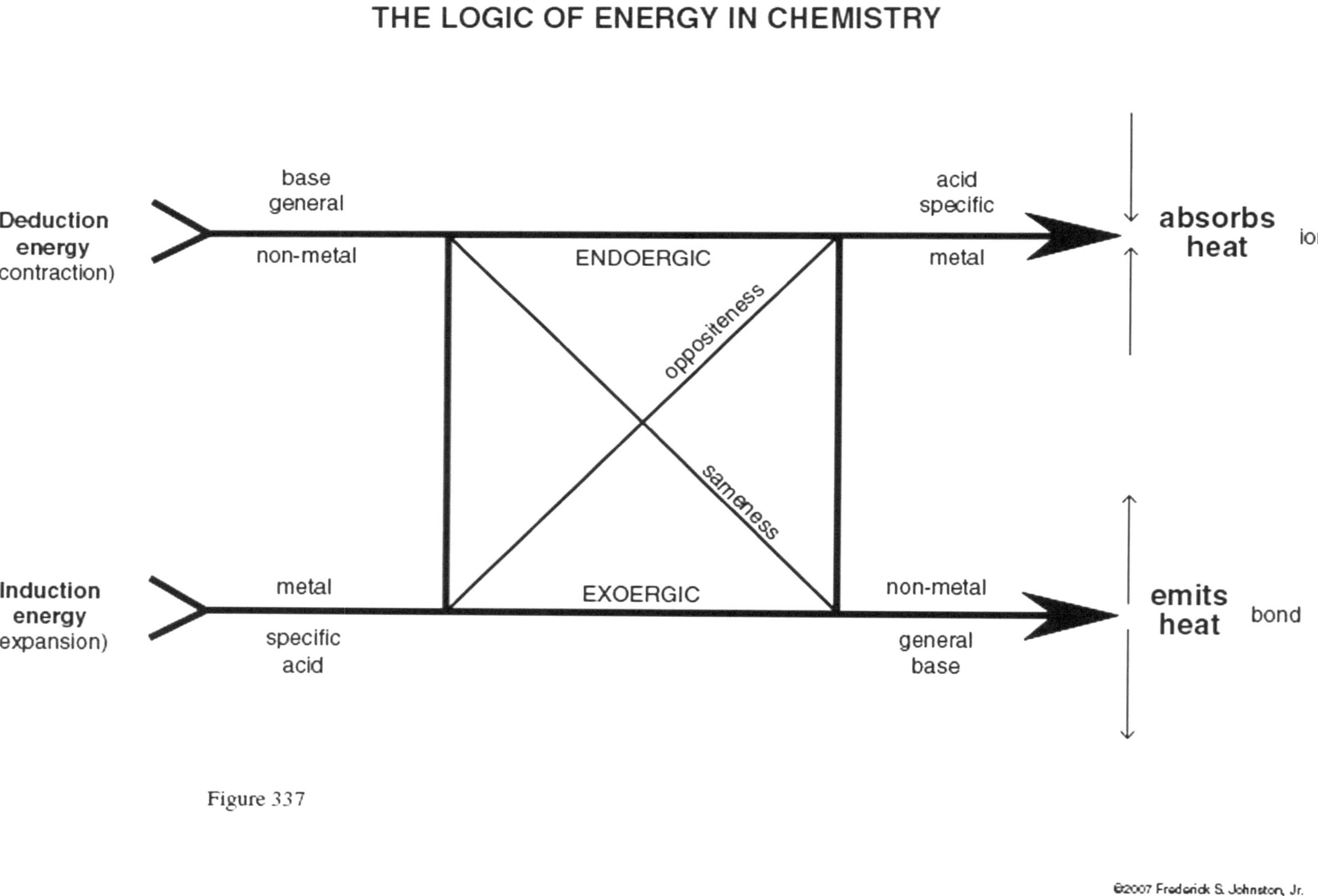

Figure 337

LOGIC OF PHYSICS
electron
quarks
from general
a
both
and
to specific
b
either
or
from specific
c
either
or
to general
d
and
both
dynamics
chromodynamics
inseparable
separable
sameness
oppositeness
(contraction)
singular
conclusive
point
(expansion)
collective
inconclusive
pointless

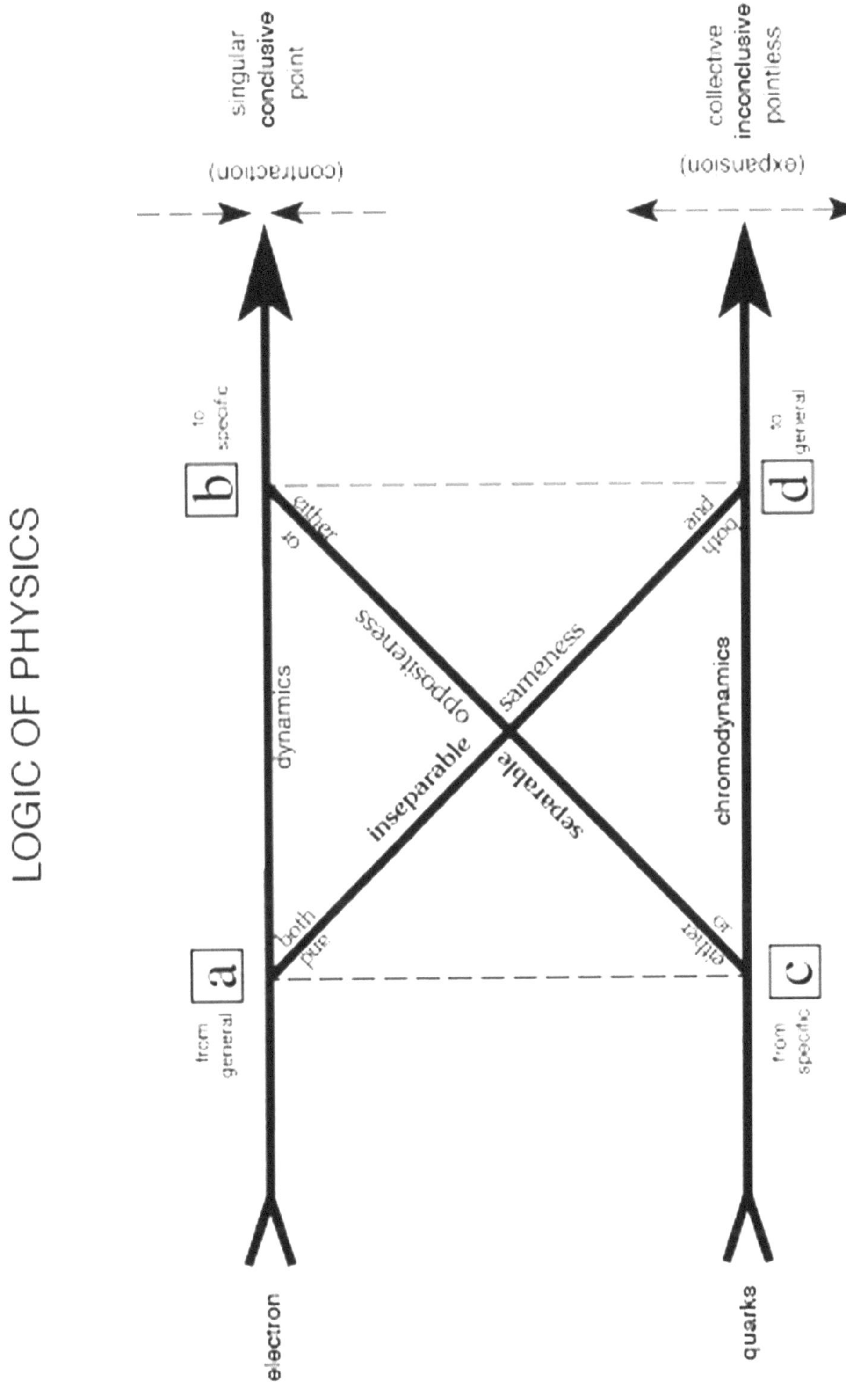

Figure 338

www.ingramcontent.com/pod-product-compliance
Ingram Content Group UK Ltd.
Pitfield, Milton Keynes, MK11 3LW, UK
UKHW060119300726
14090UKWH00002B/261

* 9 7 8 1 4 3 6 3 4 7 1 9 8 *